MATH FOR WASTEWATER TREATMENT OPERATORS GRADES 3 AND 4

MATH FOR WASTEWATER TREATMENT OPERATORS GRADES 3 AND 4

Practice Problems to Prepare for Wastewater
Treatment Operator Certification Exams

John Giorgi

FIRST EDITION

This work is dedicated to my wife, Flora Zhou Giorgi; my children, Sara, Stephanie, and Steve; my mother, Thelma Giorgi; and my father, Albert Peter Giorgi.

Math for Wastewater Treatment Operators Grades 3 and 4
Copyright © 2009 American Water Works Association

AWWA Publications Manager: Gay Porter De Nileon
Technical Editor/Project Manager: Martha Ripley Gray
Production: SquareOne Publishing and Cheryl Armstrong

Disclaimer

The authors, contributors, editors, and publisher do not assume responsibility for the validity of the content or any consequences of their use. In no event will AWWA be liable for direct, indirect, special, incidental, or consequential damages arising out of the use of information presented in this book. In particular, AWWA will not be responsible for any costs, including, but not limited to, those incurred as a result of lost revenue. In no event shall AWWA's liability exceed the amount paid for the purchase of this book.

Although this study guide has been extensively reviewed for accuracy, there may be an occasion to dispute an answer, either factually or in the interpretation of the question. Both AWWA and the author have made every effort to correct or eliminate any questions that may be confusing or ambiguous. If you do find a question that you feel is confusing or incorrect, please contact the AWWA Publishing Group.

Additionally, it is important to understand the purpose of this study guide. It does not guarantee certification. It is intended to provide the operator with an understanding of the types of math questions he or she will be presented with on a certification exam and the areas of knowledge that will be covered. AWWA highly recommends that you make use of the additional resources listed at the end of this study guide and any other resources recommended by your state certification board in preparing for your exam.

Library of Congress Cataloging-in-Publication Data

Giorgi, John.
   Math for wastewater treatment operators grades 3 and 4 : practice problems to prepare for wastewater treatment operator certification exams / John Giorgi and prepared by the editors of American Water Works Association. -- 1st ed.
       p. cm.
   Includes bibliographical references and index.
   ISBN 978-1-58321-586-9
   1. Sewage--Purification--Mathematics. 2. Sewage--Purification--Problems, exercises, etc. 3. Sewage disposal plants--Employees--Certification. 4. Water--Purification--Mathematics. 5. Water--Purification--Problems, exercises, etc. 6. Water treatment plants--Employees--Certification. 7. Engineering mathematics--Formulae. I. Title. II. Title: Math for wastewater treatment operators grades three and four.

TD745.G56192 2008
628.301'51--dc22

                    2008016935

American Water Works Association

6666 West Quincy Avenue
Denver, Colorado 80235-3098
800.926.7337

CONTENTS

FIGURES

TABLES

ACKNOWLEDGMENTS

I would like to thank the staff and editors of the American Water Works Association, publications manager Gay Porter De Nileon, editor Martha Ripley Gray, and reviewers Radenko Odzakovic and Tim McCandless for their help and guidance in making this book possible. Their assistance is greatly appreciated. Thanks also to Dean Bugher for his assistance with the flow charts.

I am grateful to my wife, Flora, and my children, Steve, Stephanie, and Sara. Their patience and support for my long hours working on this book will always be greatly appreciated.

PREFACE

The first edition of *Math for Wastewater Treatment Operators Grades 3 and 4* was written to provide students and operators with examples of a variety of different problems that will be encountered both on certification exams and on the job. This book is divided into four main parts: the introduction, which is a review of significant numbers and rounding; two chapters on math followed by practice tests, one after each math chapter; and appendices. The math problems in chapter two are a little more difficult than the math problems in chapter one and also contain a few new types.

Each problem is presented with easily followed steps and comments to facilitate understanding. One possible way to go through the math problems presented in this study guide is for you to cover the page you are working on with a piece of paper or cardboard. Then, slowly move the cover down until you can read the question. Do the problem on a separate piece of paper. Uncover the worked solution and compare your method and result to the book's method and result. If your answer is the same, but your method is different, that's okay. Remember that there may be more than one way to solve a problem. If there is a certain problem that gives you trouble, try to do the problem again on another day until you completely understand it. Do similar problems that may be found in the other grade in the book. The more math problems you do, the more comfortable you will become with them.

Included after each grade level are tests to help you determine where your strengths or weaknesses are. Each test consists of randomly chosen problems from the associated problems in that chapter. The questions in the test are followed immediately by each individual question, procedure, and result. You can complete all the problems in the test on a separate piece of paper. After completing the test, the procedures and answers can be checked against the provided procedures and solutions.

Common conversion factors are included in appendix A for reference purposes and for doing problems in this book. Appendix B is a summary of the wastewater treatment equations. Appendix C consists of chemistry tables that are needed for some of the problems. Appendix D is the depth-to-diameter table for calculating flow in a pipeline that is not full. Appendix E contains flow charts of wastewater treatment processes. and Appendix F lists abbreviations used in this book.

Any suggestions for improving this math book, including additional types of problems, would be appreciated by the author. Please send your suggestions or questions to John Giorgi at aujourney@hotmail.com, or in care of Publications Manager, AWWA, 6666 West Quincy Avenue, Denver, CO 80235-3098.

SIGNIFICANT FIGURES

When you see an answer to a mathematical problem, laboratory test result, or other measured values, do you ever wonder how accurate they are? The accuracy of any answer is based on how accurate the values are in determining the answer, or the accuracy of the laboratory result depends on the precision of the measuring instruments, and even the laboratory analyst.

The following discussion will show how to determine the number of significant figures or digits an answer to any particular problem should have, i.e., how many decimal places, if any, should the answer have.

The number 30.03 has four significant figures, while the number 33,000 has only two. Why is this so? The number 30.03 has been measured to the hundredth place, so the zeroes that are straddled by the threes are significant. In fact, all figures to the left of a decimal point are significant (for example, 2.000 has four significant figures). The second number, 33,000, has only two because the zeroes are only placeholders and are thus not significant. See the exercise below on significant figures.

"Rounding Off"

"Rounding off" numbers is simply the dropping of figures starting on the right until the appropriate numbers of significant figures remain. Let's look at the three rules and an example for each that governs the process of rounding numbers.

1. When a figure less than five is dropped, the next figure to the left remains unchanged. Thus, the number 11.24 becomes 11.2 when it is required that the four be dropped.

2. When the figure is greater than five that number is dropped and the number to the left is increased by one. Thus 11.26 will become 11.3.

3. When the figure that needs to be dropped is a five, round to the nearest even number. This prevents rounding bias. Thus 11.35 becomes 11.4 and 46.25 becomes 46.2.

The Significance or Insignificance of Zero

A zero may be a significant figure, if it is a measured value, or be insignificant and serve only as a spacer for locating the decimal point. If a zero or zeroes are used to give position value to the significant figures in the number, then the zero or zeroes are not significant. An example of this would be the following expression: 1.23 mm = 0.123 cm = 0.000123 m = 0.00000123 km. The zeroes are insignificant and only give the significant figures, 123, a position that dictates their value.

Addition and Subtraction

In addition and subtraction, only similar units and written to the same number of decimal places may be added or subtracted. Also, the number with the least decimal places, and not necessarily the fewest number of significant figures, places a limit on the number that the sum can justifiably carry, for example, when adding the following numbers: 446 mm + 185.22 cm + 18.9 m.

First, convert the quantities to similar units, which in this case will be the meter (second row below). Next, choose the least accurate number, which is 18.9. It has only one number to the right of the decimal, so the other two values will have to be rounded off (third row below).

$$
\begin{array}{rcrcr}
446 \text{ mm} & = & 0.446 \text{ m} & = & 0.4 \text{ m} \\
185.22 \text{ cm} & = & 1.8522 \text{ m} & = & 1.8 \text{ m} \\
18.9 \text{ m} & = & 18.9 \text{ m} & = & \underline{18.9 \text{ m}} \\
& & & & 21.1 \text{ m}
\end{array}
$$

When adding numbers (including negative numbers), the rule is that the least accurate number will determine the number reported as the sum. In other words, the number of significant figures reported in the sum cannot be greater than the least significant figure in the group being added. Another example is given below in which the least accurate number, 170, dictates how the other three numbers will have to be changed before addition is done.

$$
\begin{array}{rcr}
1.023 \text{ grams (g)} & = & 1 \text{ g} \\
23.22 \text{ g} & = & 23 \text{ g} \\
170 \text{ g} & = & 170 \text{ g} \\
1.008 \text{ g} & = & \underline{1 \text{ g}} \\
195.251 \text{ g} & = & \overline{195 \text{ g}}
\end{array}
$$

However, you cannot report either of these values. The third value, 170 g, has two significant numbers, while all the others have four. The limiting factor is this third value, 170 g. The number 195 has three significant figures, and thus, cannot be used either. The answer must be reported as 200 g even though this looks wrong because it only has one significant figure!

Multiplication and Division

In multiplication or division, the number that has the fewest significant figures will dictate how the answer will be written. Suppose we had a problem where we had to multiply two numbers: (23.88) (7.2) = 171.936. The first number has four significant figures, while the second has only two. The answer should only be written with two significant figures, as 170, because one of the numbers, 7.2, has only two significant figures. In both multiplication and division, "rounding off" never should be done before the mathematical exercise. Only the result should be appropriately "rounded off."

However, the above explanation of rounding multiplication and division problems is really an approximation of the exact rule. The exact rule states that the fractional or percentage error of a multiplication or division problem cannot be any less than the fractional or percentage error of any one factor. The exact rule always has to consider the amount of error that would result when rounding and applies to problems that have only one or two significant figures. These problems may require an additional significant number. An example follows: (0.93)(1.23) = 1.14. If this were rounded to 1.1 (two significant figures as is 0.93), as per the generalized or approximate rule, it would result in an error of approximately 3.5%, i.e., from the 1.14 answer. If the least significant number in the problem, 0.93, were written as 0.93 ± 0.01, it would result in an error of just over 1%, i.e., if it were really 0.94 or 0.92.

$$\frac{(0.01)(100\%)}{0.93} = 1.075\%, \text{ round to } 1.1\%$$

Thus, the best answer would be 1.14 because it introduces less error than the least significant number, if again it were measured wrong and were a little more or less (0.94 or 0.92).

Because this is much more difficult than what we need in this book, we will use the approximate rounding rule throughout this book, except where indicated. I stated it here so that you would be aware of this rule in case you are not already.

In the following exercise, give the number of significant figures from each of the values below:

Value	Answer
a. 8.34 lb/gal	3
b. 0.03 ntu	1
c. 19.08 mgd	4
d. 3 1-ton sulfur dioxide containers	1 or infinite[*]
e. 2.30 mg/L	3
f. 0.00000254	3
g. 80,000 pennies	1 to 5[**]
h. 0.006700	4
i. 43,560 ft³/acre-ft	4 or 5 or infinite[#]
j. 220 m	2 or 3
k. 9.02 mg	3
l. 10,200,050 gal	7 or 8
m. 1,000,000/mil	7 or infinite[#]
n. 1,440 min/day	4 or infinite[+]
o. 7.481 gal/ft³	4
p. 86,400 sec/day	5 or infinite[+]

[*]   Because one would not divide them up in little pieces.

[**]   Because one could argue that 1 or all of the zeroes could be significant, but not likely.

[#]   By definition, so one could extend the zeros past the decimal point to infinity.

[+]   By definition, these are the minutes and seconds, respectively, that are in a day; until the earth's rotational speed slows down enough to warrant a change in these times, I will call these as indicated.

See chapter 1, Significant Figures, for more practice.

ROUNDING IN THIS BOOK

The problems in this book are done in steps so the student can see each operation visually, which facilitates understanding. In so doing, a certain amount of "pre-rounding" has to occur, otherwise the numbers continue in most cases to absurdity. This "pre-rounding" was done as much as possible such that the final answer would not be affected.

SET UP OF PROBLEMS IN THIS BOOK

Note that many of the same problem types have the equation written each time and that the units are written throughout each problem. The reason this is done is to help the student form a good habit, because carrying units throughout a calculation will assist in identifying which units require conversion and as a check on the final result. This habit should be carried over during a certification test or on the job. In some cases, partial credit will be given for a correct written equation. When the units are written down, it will not only help the student in setting up the problem correctly by seeing how the units will cancel, but will also make it easier for the person correcting the test to see your intent and also in correcting the problems.

Pi (π)

The number used for pi (π) will be 3.14 throughout this book.

ppm and mg/L

Since in most cases mg/L = ppm, please note that mg/L will cancel out units that are in millions. For example, mg/L will cancel with the mil (million) in mil gal.

DERIVATION OF THE NUMBER 0.785

The number 0.785 is used extensively in this book in conjunction with the diameter squared. Examples include the determination of the area of a circular reservoir or the volume of the tank. The same answer can be achieved using πr^2, where "r" equals the radius. But how is this number, 0.785, derived? The following applies:

$$(0.785)(\text{Diameter})^2 = \pi r^2$$
$$(\text{Diameter})^2 = 4r^2$$

Proof: Assume x is the number 0.785 but is not yet known. We know some number, x, times the diameter squared equals π times the radius squared. The equation is:

$$x(\text{Diameter})^2 = \pi r^2$$

From 1 above and substituting x for 0.785, substitute $4r^2$ for (Diameter)^2 from 2 above.

$$x(4r^2) = \pi r^2$$

Rearrange equation to solve for x.

$$x = \frac{\pi r^2}{4r^2} = \frac{\pi}{4} = \frac{3.14}{4} = \mathbf{0.785}$$

Thus, x is equal to 0.785, which is what we wanted to prove. Most water treatment operators like to use $(0.785)(\text{Diameter})^2$, while engineers and scientists like to use πr^2. Because both will be encountered, it is advisable to know both methods.

WASTEWATER TREATMENT
Grade 3

*Students preparing for Grade 4 wastewater treatment
certification tests should also understand these problems.*

PERCENT CALCULATIONS

Percent calculations are used throughout this book and are thus essential to understand. They are also a good refresher for the student or operator.

1. **A settleability test is performed on a mixed liquor suspended solids sample. If 376 mL of solids settle out of a 2,000 mL sample in a graduated cylinder, what are the percent settleable solids by volume? Give answer to three significant figures.**

Equation: **Percent settleable solids** $= \dfrac{(\text{Settleable solids, mL})(100\%)}{\text{Total sample volume, mL}}$

Percent settleable solids $= \dfrac{(376 \text{ mL})(100\%)}{2,000 \text{ mL}} =$ **18.8% Settleable solids by volume**

2. **What is the percent removal across a primary clarifier, if the influent biochemical oxygen demand (BOD$_5$) is 365 and the effluent BOD$_5$ is 199?**

Equation: **Percent BOD$_5$ removal** $= \dfrac{(\text{Influent BOD}_5 - \text{Effluent BOD}_5)(100\%)}{\text{Influent BOD}_5}$

Percent BOD$_5$ removal $= \dfrac{(365 \text{ BOD}_5 - 199 \text{ BOD}_5)(100\%)}{365 \text{ BOD}_5} = \dfrac{(166 \text{ BOD}_5)(100\%)}{365 \text{ BOD}_5} =$ **45.5%**

PERCENT STRENGTH BY WEIGHT SOLUTION PROBLEMS

The strength of solution calculations are important to determine so that operators can properly mix chemicals in the percentages they need for dosing a particular wastewater process or other application.

3. **If 250 lb of magnesium hydroxide are dissolved in 125 gallons of water, what is the percent strength by weight of the magnesium hydroxide solution?**

Equation: **Percent strength** $= \dfrac{(\text{Number of lb, chemical})\,(100\%)}{(\text{Number of gal})\,(8.34\ \text{lb/gal}) + \text{Number of lb, chemical}}$

Substitute values and solve.

Percent strength $= \dfrac{(250\ \text{lb})\,(100\%)}{(125\ \text{gal})\,(8.34\ \text{lb/gal}) + 250\ \text{lb}} = \dfrac{25{,}000\ \text{lb}\ \%}{1{,}042.5\ \text{lb} + 250\ \text{lb}}$

Percent strength $= \dfrac{25{,}000\ \text{lb}\ \%}{1{,}292.5\ \text{lb}} = 19.34\%$, round to **19% Mg(OH)$_2$ solution by weight**

4. **If 304 lb of dry alum (Al$_2$SO$_4$) are dissolved in 75.0 gallons of water, what is the percent strength by weight of the alum solution?**

Another way to solve this problem that differs from the problem above is to first convert the number of gallons of water to pounds.

Number of lb $=$ (75.0 gallons)(8.34 lb/gal) $=$ 625.5 lb of water

Next, find the percent strength of the solution.

Equation: **Percent strength** $= \dfrac{(\text{Number of lb of chemical})\,(100\%)}{\text{Number of lb, Water} + \text{lb, Chemical}}$

Substitute values and solve.

Percent strength $= \dfrac{(304\ \text{lb Al}_2\text{SO}_4)\,(100\%)}{625.5\ \text{lb, Water} + 304\ \text{lb Alum}}$

Percent strength $= \dfrac{(304\ \text{lb Al}_2\text{SO}_4)\,(100\%)}{929.5\ \text{lb}} =$ **32.7% Alum solution by weight**

5. If 25.0 grams (g) of magnesium hydroxide are dissolved in 1.000 liters (L) of water, what is the percent strength by weight of the magnesium hydroxide solution?

Equation: **Percent strength** $= \dfrac{(\text{Number of grams, chemical})\,(100\%)}{(\text{Number of L})\,(1{,}000\ \text{grams/L}) + \text{grams, Chemical}}$

Percent strength $= \dfrac{(25.0\ \text{g Mg(OH)}_2)\,(100\%)}{(1.000\ \text{L})\,(1{,}000\ \text{g/L}) + 25.0\ \text{g Mg(OH)}_2}$

Percent strength $= \dfrac{(25.0\ \text{g Mg(OH)}_2)\,(100\%)}{1{,}025\ \text{g}} =$ **2.44% Mg(OH)$_2$ solution by weight**

PERCENT SOLIDS BY WEIGHT CALCULATIONS

Operators use percent solids calculations to determine efficiency of different unit processes, as well as to determine how much waste will require disposal.

6. A wastewater treatment plant pumps an average of 27,480 gpd of sludge. If the percent of solids in the sludge averages 4.59% by weight and the sludge weighs 8.56 lb/gal, what is the lb/day of solids pumped?

First, calculate the number of lb/day of sludge pumped.

Sludge, lb/day $= (27{,}480\ \text{gpd})(8.56\ \text{lb/gal}) = 235{,}228.8\ \text{lb/day}$

Next, calculate the lb/day of solids pumped.

Equation: **Solids, lb/day** $= \dfrac{(\text{Sludge, lb/day})\,(\text{Percentage of solids})}{100\%}$

Solids, lb/day $= \dfrac{(235{,}228.8\ \text{lb/day})\,(4.59\%)}{100\%} = 10{,}797\ \text{lb/day, round to}$ **10,800 lb/day**

7. What is the percent of total inorganic solids by weight in a sludge sample, given the following data?

Sludge sample wet weight = 513 grams
Total solids dry weight = 32.7 grams
Inorganic dry weight = 6.81 grams

Equation: **Percent inorganic solids** $= \dfrac{(\text{Dry sample in grams})\,(100\%)}{\text{Sludge sample in grams}}$

Percent inorganic solids $= \dfrac{(6.81\ \text{grams})\,(100\%)}{513\ \text{grams}} =$

1.327%, round to **1.33% total inorganic solids by weight**

8. A total of 3,570 gpd of sludge (primary sludge) with 5.82% solids by weight content and weighing 8.41 lb/gal is pumped to a wastewater thickener tank. Determine the amount of sludge that should flow from the thickener tank in gpd, if the sludge (secondary sludge) is further concentrated to 7.09% solids by weight and weighs 8.49 lb/gal. *Note:* For convenience, primary sludge is abbreviated to 1° sludge and secondary sludge to 2° sludge.

Equation: **(x gpd)(2° sludge lb/gal)(% 2° sludge) =**

(1° sludge, gpd)(1° sludge lb/gal)(% 1° sludge)

Substitute values and solve.

(x gpd)(8.49 lb/gal)(7.09%/100%) = (3,570 gpd)(8.41 lb/gal)(5.82%/100%)

x gpd $= \dfrac{(3{,}570\ \text{gpd})\,(8.41\ \text{lb/gal})\,(5.82\%)\,(100\%)}{(8.49\ \text{lb/gal})\,(100\%)\,(7.09\%)} = 2{,}902.9$ gpd, round to **2,900 gpd**

PERCENT VOLATILE SOLIDS REDUCTION AND VOLATILE SOLIDS DESTROYED

The percent volatile solids reduction calculations indicate the effectiveness of the digested sludge process when compared to the volatile solids in the influent. The higher the percent volatile solids reduced or destroyed, the more stable the organic matter in the digester becomes and the more gas that is produced. Volatile solids destroyed are a measure of the effectiveness of the digester process. It tells the operator the number of pounds of volatile solids destroyed per cubic foot of digester volume. See figures E-2, E-4, E-5, and E-6 in Appendix E for four types of wastewater plants using a digester.

9. **Calculate the percent volatile solids (VS) reduction, if the digester influent sludge has a VS content of 58.4% by weight and the digester effluent sludge has a VS content of 38.0% by weight.**

First, convert percentage to decimal form by dividing by 100%.

58.4%/100% = 0.584 and 38.0%/100% = 0.380

Equation: **Percent VS reduction** $= \dfrac{(\text{Influent} - \text{Effluent})(100\%)}{\text{Influent} - (\text{Influent})(\text{Effluent})}$

Percent VS reduction $= \dfrac{(0.584 - 0.380)(100\%)}{0.584 - (0.584)(0.380)} = \dfrac{0.204(100\%)}{0.584 - 0.22192}$

Percent VS reduction $= \dfrac{20.4\%}{0.36208} = 56.34\%$, round to **56.3% VS reduction by weight**

10. **If the sludge entering a digester has a volatile solids (VS) content of 62.9% by weight and the digester effluent sludge has a VS content of 42.8% by weight, calculate the percent VS reduction.**

First, convert percentage to decimal form by dividing by 100%.

62.9%/100% = 0.629 and 42.8%/100% = 0.428

Equation: **Percent VS reduction** $= \dfrac{(\text{Influent} - \text{Effluent})(100\%)}{\text{Influent} - (\text{Influent})(\text{Effluent})}$

Percent VS reduction $= \dfrac{(0.629 - 0.428)(100\%)}{0.629 - (0.629)(0.428)} = \dfrac{0.201(100\%)}{0.629 - 0.269212}$

Percent VS reduction $= \dfrac{20.1\%}{0.359788} = 55.866\%$, round to **55.9% VS reduction by weight**

11. Given the following data, calculate the amount of volatile solids (VS) destroyed in lb/day/ft³ of digester capacity:

Digester radius = 25.4 ft
Average sludge height = 17.0 ft
Sludge flow (Flow) = 3,485 gpd
Sludge solids concentration (SSC) = 6.33% by weight
Volatile solids content (VSC) = 68.4% by weight
Volatile solids reduction (VSR) = 56.1% by weight
Specific gravity of sludge = 1.02

First, determine the number of lb/gal for the sludge.

Sludge, lb/gal = (8.34 lb/gal)(1.02 sp gr) = 8.507 lb/gal

Next, calculate the digester capacity in ft³.

Digester capacity, ft³ = π(radius)²(Height, ft)

Digester capacity, ft³ = 3.14(25.4 ft)(25.4 ft)(17.0 ft) = 34,438.64 ft³

Next, write the equation.

$$\textbf{VS destroyed, lb/day/ft}^3 = \frac{\text{(Flow, gpd) (Sludge, lb/gal) (SSC, \%) (VSC, \%) (VSR, \%)}}{\text{Digester capacity, ft}^3}$$

Substitute values and solve.

$$\text{VS destroyed, lb/day/ft}^3 = \frac{(3,485 \text{ gpd}) (8.507 \text{ lb/gal}) (6.33\%) (68.4\%) (56.1\%)}{(34,438.64 \text{ ft}^3) (100\%) (100\%) (100\%)}$$

VS destroyed, lb/day/ft³ = **0.021 lb/day/ft³ VS destroyed by weight**

PERCENT SEED SLUDGE PROBLEMS

These calculations are used by operators when starting up a new digester or one that has been cleaned, and are based on volume. The other method presented later in this chapter is based on volatile solids added per pound of volatile solids contained in the digester.

12. **How many gallons of seed sludge are needed, if the digester is 48.5 ft in diameter, has a maximum side depth of 24.9 ft, has a wastewater depth of 17.2 ft, and requires the seed sludge to be 24.5% of the digester volume?**

First, determine the volume of the digester in gallons.

Number of gallons = $(0.785)(\text{Diameter})^2(\text{Depth, ft})(7.48 \text{ gal/ft}^3)$

Number of gallons = $(0.785)(48.5 \text{ ft})(48.5 \text{ ft})(24.9 \text{ ft})(7.48 \text{ gal/ft}^3) = 343{,}917 \text{ gal}$

Next, calculate the seed sludge required in gallons.

Equation: **Percent seed sludge** $= \dfrac{(\text{Seed sludge, gal})(100\%)}{\text{Total digester volume, gal}}$

Rearrange the equation to solve for the number of gallons.

Seed sludge, gal $= \dfrac{(\text{Total digester volume, gal})(\text{Percent seed sludge})}{100\%}$

Seed sludge, gal $= \dfrac{(343{,}917 \text{ gal})(24.5\%)}{100\%} = 84{,}260 \text{ gal, round to } \mathbf{84{,}300 \text{ gal}}$

13. **What must have been the percent seed sludge by volume added to a digester, given the following data?**

Diameter of digester = 44.7 ft
Maximum side height = 21.5 ft
Seed sludge added to digester = 63,450 gal

First, determine the volume of the digester in gallons.

Number of gallons = $(0.785)(\text{Diameter})^2(\text{Depth, ft})(7.48 \text{ gal/ft}^3)$

Number of gallons = $(0.785)(44.7 \text{ ft})(44.7 \text{ ft})(21.5 \text{ ft})(7.48 \text{ gal/ft}^3) = 252{,}246 \text{ gal}$

Next, calculate the seed sludge required in gallons.

Equation: **Percent seed sludge** $= \dfrac{(\text{Seed sludge, gal})(100\%)}{\text{Total digester volume, gal}}$

Percent seed sludge $= \dfrac{(63{,}450 \text{ gal})(100\%)}{252{,}246 \text{ gal}} =$ **25.2% Seed sludge by volume**

VOLATILE SOLIDS PUMPING CALCULATIONS

These calculations are used as a planning tool by the operator. By knowing the pumping rate of volatile solids into a digester, an operator can make sure it is not overloaded, which would adversely affect the digester's operation and performance.

14. **Given the following data, how many lb/day of volatile solids (VS) are pumped to a digester?**

Influent pumping rate = 4,274 gpd
Solids content = 5.62%
Volatile solids (VS) = 60.5%
Specific gravity of sludge = 1.06

First, determine the lb/gal for the sludge.

Sludge, lb/gal = (8.34 lb/gal)(1.06 sp gr) = 8.84 lb/gal

Equation: **VS, lb/day =**

$$\textbf{(Number of gpd to digester)}\frac{\textbf{(Percent solids)}}{\textbf{100\%}}\frac{\textbf{(Percent VS)}}{\textbf{100\%}}\textbf{(Sludge, lb/gal)}$$

$$\text{VS, lb/day} = (4{,}274 \text{ gpd Solids})\frac{(5.62\%)}{100\%}\frac{(60.5\% \text{ VS})}{100\%}(8.84 \text{ lb/gal})$$

VS, lb/day = 1,284.63 lb/day, round to **1,280 lb/day VS**

15. **Given the following data, how many lb/day of volatile solids (VS) are pumped to a digester?**

Pumping rate = 4.1 gpm
Solids content = 4.74%
Volatile solids (VS) = 59.3%
Specific gravity of sludge = 1.03

First, determine the lb/gal for the sludge.

Sludge, lb/gal = (8.34 lb/gal)(1.03 sp gr) = 8.59 lb/gal

Next, convert gpm to gpd.

Number of gpd = (4.1 gpm)(1,440 min/day) = 5,904 gpd

Equation: **VS, lb/day =**

$$\textbf{(Number of gpd to digester)}\frac{\textbf{(Percent solids)}}{\textbf{100\%}}\frac{\textbf{(Percent VS)}}{\textbf{100\%}}\textbf{(Sludge, lb/gal)}$$

$$\text{VS, lb/day} = (5{,}904 \text{ gpd Solids})\frac{(4.74\%)}{100\%}\frac{(59.3\% \text{ VS})}{100\%}(8.59 \text{ lb/gal})$$

VS, lb/day = 1,425.52 lb/day, round to **1,400 lb/day VS**

SOLUTION MIXTURE CALCULATIONS

These calculations are used when mixing two of the same solutions that have different strengths given a volume target. They are important for the operator to understand because there most probably will be times when solutions will require mixing. There are three ways to solve dilution problems. The dilution triangle is perhaps the easiest, and is shown below for the next two problems. The method scientists and chemists use is $C_1V_1 + C_2V_2 = C_3V_3$.

16. **How many gallons of a 64.5% solution must be mixed with a 30.0% solution to make exactly 250 gallons of a 48.5% solution?**

How to solve the problem using the dilution triangle: The two numbers on the left are the existing concentrations of 64.5% and 30.0%. The number in the center, 48.5%, is the desired concentration. The numbers on the right are determined by subtracting diagonally the existing concentrations from the desired concentration.

64.5%	18.5*[1]	18.5 parts of the 64.5% solution are required for every 34.5 parts.
48.5%		
30.0%	16.0*[2]	16.0 parts of the 30.0% solution are required for every 34.5 parts.
	34.5 total parts	

*[1] **18.5 is determined by subtracting diagonally 30.0% from 48.5%.**
*[2] **16.0 is determined by subtracting diagonally 48.5% from 64.5%.**

$$\frac{18.5 \text{ parts} (250 \text{ gal})}{34.5 \text{ parts}} = 134.06 \text{ gallons, round to } \textbf{134 gallons of the 64.5\% solution}$$

$$\frac{16.0 \text{ parts} (250 \text{ gal})}{34.5 \text{ parts}} = 115.94 \text{ gallons, round to } \underline{116} \textbf{ gallons of the 30.0\% solution}$$
$$\overline{250} \text{ gallons—added here to cross check math.}$$

To make the 250 gallons of the 48.5% solution, mix 134 gallons of the 64.5% solution with 116 gallons of the 30.0% solution.

17. **A 7.5% hypochlorite solution is required. If exactly 475 gallons are needed, how many gallons of a 12% solution must be mixed with a 2.5% solution to make the required solution? Give answer to nearest gallon.**

Solve the problem using the dilution triangle.

12%		5.0	5.0 parts of the 12% solution are required for every 9.5 parts.
	7.5%		
2.5%		4.5	4.5 parts of the 2.5% solution are required for every 9.5 parts.
		9.5 total parts	

$$\frac{5.0 \text{ parts} (475 \text{ gal})}{9.5 \text{ parts}} = \textbf{250 gallons of the 12\% solution}$$

$$\frac{4.5 \text{ parts} (475 \text{ gal})}{9.5 \text{ parts}} = \frac{225 \text{ gallons of the 2.5\% solution}}{475 \text{ gallons}}$$

Mix 250 gal of the 12% solution with 225 gal of the 2.5% solution to get the final solution of 7.5% hypochlorite.

18. **What percent hypochlorite solution would result, if 325 gallons of a 14% solution were mixed with exactly 170 gallons of a 3.5% solution? Assume both solutions have the same density.**

First, find the total volume that would result from mixing these two solutions.

Total Volume $= 325$ gal $+ 170$ gal $= 495$ gal

Then, write the equation.

$$(\text{Concentration}_1)(\text{Volume}_1) + (\text{Concentration}_2)(\text{Volume}_2) = (\text{Concentration}_3)(\text{Volume}_3)$$

Condensed as $C_1V_1 + C_2V_2 = C_3V_3$, where C_1 and $C_2 = \%$ Concentration of the two solutions before being mixed, V_1 and $V_2 =$ Volume of the two solutions before being mixed, and C_3 and $V_3 =$ the resulting % Concentration and Volume, respectively.

Substitute values and solve.

$$\frac{(14\%)\,(325\text{ gal})}{100\%} + \frac{(3.5\%)\,(170\text{ gal})}{100\%} = \frac{C_3\,(495\text{ gal})}{100\%}$$

$$45.5\text{ gal} + 5.95\text{ gal} = \frac{C_3\,(495\text{ gal})}{100\%}$$

Solve for C_3.

$$C_3 = \frac{(45.5\text{ gal} + 5.95\text{ gal})\,100\%}{495\text{ gal}} = \frac{(51.45\text{ gal})\,(100\%)}{495\text{ gal}}$$

$C_3 = 10.39$, round to **10% Final solution**

19. **What percent polymer solution would result, if 355 gallons of a 15.5% solution were mixed with 935 gallons of a 2.50% solution? Assume both solutions have the same density.**

First, find the total volume that would result from mixing these two solutions.

Total Volume $= 355$ gal $+ 935$ gal $= 1{,}290$ gal

Write the equation.

$C_1V_1 + C_2V_2 = C_3V_3$, where C_1 and C_2 = % Concentration of the two solutions before being mixed, V_1 and V_2 = Volume of the two solutions before being mixed, and C_3 and V_3 = the resulting % Concentration and Volume, respectively.

Substitute values and solve.

$$\frac{(15.5\%)\,(355\text{ gal})}{100\%} + \frac{(2.50\%)\,(935\text{ gal})}{100\%} = \frac{C_3\,(1{,}290\text{ gal})}{100\%}$$

Solving for C_3

$$C_3 = \frac{(55.025\text{ gal} + 23.375\text{ gal})\,(100\%)}{1{,}290\text{ gal}} = \frac{(78.4\text{ gal})\,(100\%)}{1{,}290\text{ gal}}$$

C_3 = 6.078, round to **6.08% Final solution**

20. **What percent strength of a solution mixture results, when 45 lb of a 12% solution are mixed with 75 lb of a 5.5% solution?**

Equation: **% Mixture strength =**

$$\frac{(\text{Solution}_1\text{ lb})\,(\text{Available }\%/100\%) + (\text{Solution}_2,\text{ lb})\,(\text{Available }\%/100\%)\,(100\%)}{\text{Solution}_1,\text{ lb} + \text{Solution}_2,\text{ lb}}$$

Substitute values and solve.

$$\%\text{ Mixture strength} = \frac{[(45\text{ lb})\,(12\%/100\%) + (75\text{ lb})\,(5.5\,\%/100\%)]\,(100\%)}{45\text{ lb} + 75\text{ lb}}$$

$$\%\text{ Mixture strength} = \frac{[5.4\text{ lb} + 4.125\text{ lb}]\,(100\%)}{120\text{ lb}}$$

$$\%\text{ Mixture strength} = \frac{[9.525\text{ lb}]\,(100\%)}{120\text{ lb}} = 7.9375\%,\text{ round to }\textbf{7.9\% Mixture strength}$$

21. Given the following data, calculate the percent solids content that results when a primary sludge is mixed with a secondary sludge.

Average of primary sludge = 2,560 gpd
Average of secondary sludge = 2,975 gpd
Primary sludge = 6.38% solids
Secondary sludge = 3.77% solids
Primary sludge = 8.36 lb/gal
Secondary sludge = 8.58 lb/gal

Equation: **Percent sludge mixture =**

$$\frac{[(\text{Sludge}_1 \text{ gal})(\text{lb/gal})(\text{Avail }\%/100\%) + (\text{Sludge}_2 \text{ gal})(\text{lb/gal})(\text{Avail }\%/100\%)]\,100\%}{(\text{Sludge}_1, \text{gal})(\text{lb/gal}) + (\text{Sludge}_2, \text{gal})(\text{lb/gal})}$$

Where Avail = Available

% Mixture strength =

$$\frac{[(2,560 \text{ gal})(8.36 \text{ lb/gal})(6.38\%/100\%) + (2,975 \text{ gal})(8.58 \text{ lb/gal})(3.77\%/100\%)]\,100\%}{(2,560 \text{ gal})(8.36 \text{ lb/gal}) + (2,975 \text{ gal})(8.58 \text{ lb/gal})}$$

$$\% \text{ Mixture strength} = \frac{(1,365.42 \text{ lb} + 962.31 \text{ lb})(100\%)}{21,401.6 \text{ lb} + 25,525.5 \text{ lb}} = \frac{(2,327.73 \text{ lb})(100\%)}{46,927.1 \text{ lb}}$$

% Mixture strength = **4.96% Mixture strength**

22. **What percent strength of a solution mixture results, when 125 gallons of a 19.5% solution that weighs 8.94 lb/gal are mixed with 275 gallons of an 8.25% solution that weighs 9.01 lb/gal?**

Equation: **Percent mixture strength** $=$

$$\frac{[(\text{Solution}_1\ \text{gal})(\text{lb/gal})(\text{Avail }\%/100\%) - (\text{Solution}_2\ \text{gal})(\text{lb/gal})(\text{Avail }\%/100\%)]100\%}{(\text{Solution}_1,\ \text{gal})(\text{lb/gal}) + (\text{Solution}_2,\ \text{gal})(\text{lb/gal})}$$

Where Avail $=$ Available

% Mixture strength $=$

$$\frac{[(125\ \text{gal})(8.94\ \text{lb/gal})(19.5\%/100\%) + (275\ \text{gal})(9.01\ \text{lb/gal})(8.25\%/100\%)]100\%}{(125\ \text{gal})(8.94\ \text{lb/gal}) + (275\ \text{gal})(9.01\ \text{lb/gal})}$$

% Mixture strength $= \dfrac{(217.91\ \text{lb} + 204.41\ \text{lb})(100\%)}{1,117.5\ \text{lb} + 2,477.75\ \text{lb}} = \dfrac{(422.32\ \text{lb})(100\%)}{3,595.25\ \text{lb}}$

% Mixture strength $= 11.75\%$, round to **11.8% Mixture strength**

ARITHMETIC MEAN, MEDIAN, RANGE, MODE, GEOMETRIC MEAN, AND STANDARD DEVIATION

These calculations are good tools for planning and evaluating plant processes.

23.   Given the following data, calculate the unknowns:

Note: A scientific calculator is required for determining the geometric mean and standard deviation.

Day	Chlorine dose, mg/L	Unknown
Monday	11.5	a. Arithmetic mean, mg/L
Tuesday	11.8	b. Median, mg/L
Wednesday	12.4	c. Range, mg/L
Thursday	11.8	d. Mode, mg/L
Friday	11.4	e. Geometric mean, mg/L
Saturday	10.9	f. Standard deviation, mg/L
Sunday	10.7	

a.   Calculate the arithmetic mean, mg/L.

Equation: $\textbf{Arithmetic Mean} = \dfrac{\textbf{Sum of all measurements}}{\textbf{Number of measurements}}$

$\text{Arithmetic Mean} = \dfrac{11.5 + 11.8 + 12.4 + 11.8 + 11.4 + 10.9 + 10.7}{7} = \textbf{11.5 mg/L Cl}_2$

b.   Determine the median, mg/L.

To determine the median, put the chlorine dosages in ascending order and choose the middle value.

1	2	3	4	5	6	7
10.7	10.9	11.4	**11.5**	11.8	11.8	12.4

In this case the middle value is **11.5 mg/L Cl$_2$**

c.   Determine the range, mg/L.

Equation: $\textbf{Range} = \textbf{Largest value} - \textbf{Smallest value}$

$\text{Range, mg/L} = 12.4 \text{ mg/L} - 10.7 \text{ mg/L} = \textbf{1.7 mg/L Cl}_2$

d.   Determine the mode, mg/L.

Mode is the measurement that occurs most frequently.

In this case it is **11.8 mg/L Cl$_2$**

e.   Calculate the geometric mean.

Equation: **Geometric mean** $= [(x_1)(x_2)(x_3)(x_4).....(x_n)]^{1/n}$

Geometric mean, mg/L $= [(10.7)(10.9)(11.4)(11.5)(11.8)(11.8)(12.4)]^{1/7}$

Geometric mean, mg/L $= [26399680.269168]^{1/7}$

Geometric mean, mg/L $=$ **11.5 mg/L Cl$_2$**

f.   Calculate the standard deviation.

Dose, mg/L	Frequency (f)	$(X - \overline{X})$	$(X - \overline{X})^2$	$f(X - \overline{X})^2$
10.7	1	$10.7 - 11.5$	0.64	0.64
10.9	1	$10.9 - 11.5$	0.36	0.36
11.4	1	$11.4 - 11.5$	0.01	0.01
11.8	2	$11.8 - 11.5$	0.09	0.18
12.4	1	$12.4 - 11.5$	0.81	0.81

$$\sum f(X - \overline{X})^2 = 2.00$$

Equation: **Standard deviation** $= [\Sigma f(X - \overline{X})^2/n - 1]^{1/2}$

Substitute values and solve.

Standard deviation $= [\Sigma f(2.00)^2/7 - 1]^{1/2}$

Standard deviation $= (4/6)^{1/2} =$ **0.82 mg/L Cl$_2$**

24. Given the following data, calculate the unknowns. *Note:* **A scientific calculator is required for determining the geometric mean and standard deviation.**

Day	Flow mgd	Unknown
Monday	1.25	a. Arithmetic mean, mg/L
Tuesday	1.19	b. Median, mg/L
Wednesday	1.24	c. Range, mg/L
Thursday	1.35	d. Mode, mg/L
Friday	1.48	e. Geometric mean, mg/L
Saturday	1.46	f. Standard deviation, mg/L
Sunday	1.57	

a. **Calculate the arithmetic mean, mg/L.**

Equation: **Arithmetic Mean** $= \dfrac{\textbf{Sum of all measurements}}{\textbf{Number of measurements}}$

$$\text{Arithmetic Mean} = \frac{1.25 + 1.19 + 1.24 + 1.35 + 1.48 + 1.46 + 1.57}{7} = \textbf{1.36 mgd}$$

b. **Determine the median, mg/L.**

To determine the median, put the chlorine dosages in ascending order and choose the middle value.

1	2	3	4	5	6	7
1.19	1.24	1.25	**1.35**	1.46	1.48	1.57

In this case the middle value is **1.35 mgd**

c.   **Determine the range, mg/L.**

Equation: **Range = Largest value − Smallest value**

Range, mg/L = 1.57 mg/L − 1.19 mg/L = **0.38 mgd**

d.   **Determine the mode, mg/L.**

Mode is the measurement that occurs most frequently.

In this case it is all of them because they all occur once. You can think of it as a tie, so they all win first place.

e.   **Calculate the geometric mean.**

Equation: **Geometric mean = $[(x_1)(x_2)(x_3)(x_4).....(x_n)]^{1/n}$**

Geometric mean, mg/L = $[(1.19)(1.24)(1.25)(1.35)(1.46)(1.48)(1.57)]^{1/7}$

Geometric mean, mg/L = $[8.447469874]^{1/7}$

Geometric mean, mg/L = **1.36 mgd**

e. Calculate the standard deviation.

mgd	frequency(f)	$(X - \bar{X})$	$(X - \bar{X})^2$	$f(X - \bar{X})^2$
1.19	1	$1.19 - 1.36$	0.0289	0.0289
1.24	1	$1.24 - 1.36$	0.0144	0.0144
1.25	1	$1.25 - 1.36$	0.0121	0.0121
1.35	1	$1.35 - 1.36$	0.0001	0.0001
1.46	1	$1.46 - 1.36$	0.01	0.01
1.48	1	$1.48 - 1.36$	0.0144	0.0144
1.57	1	$1.57 - 1.36$	0.0441	0.0441

$$\Sigma f(X - \bar{X})^2 = 0.124$$

Equation: **Standard deviation $= [\Sigma f(X - \bar{X})^2/n - 1]^{1/2}$**

Substitute values and solve.

Standard deviation $= [\Sigma f(0.124)^2/7 - 1]^{1/2}$

Standard deviation $= (0.002563)^{1/2} = $ **0.0506 mgd**

AREA PROBLEMS

Areas are important to determine for a number of reasons including knowing the "footprint" of a tank or pond or the area of a particular process to make further calculations in other wastewater problems.

25. What is the area of a circular pond, if it has a diameter of 194 ft?

Equation: **Area $= \pi r^2$ where $\pi = 3.14$**

First, find the radius. Radius, ft $=$ Diameter/2 $= 194/2 = 97$ ft

Area of tank $= (3.14)(97 \text{ ft})(97 \text{ ft}) = 29{,}544.26 \text{ ft}^2$, round to **29,500 ft^2**

26. **Find the area in cm² for a parallelogram, if the base is 48 cm and the height is 16.5 cm.**

Know: **Area = (Base)(Height)**

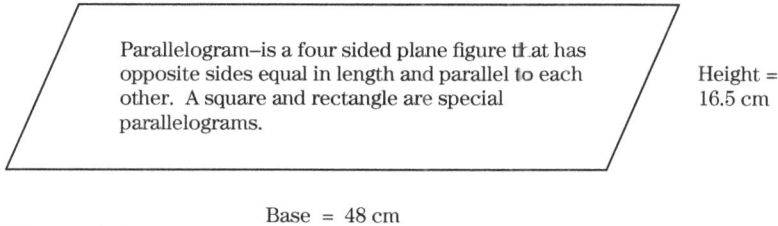

Parallelogram–is a four sided plane figure that has opposite sides equal in length and parallel to each other. A square and rectangle are special parallelograms.

Height = 16.5 cm

Base = 48 cm

Figure 1-1.

Substitute values and solve.

Area = (48 cm)(16.5 cm) = 792 cm², round to **790 cm²**

27. **Find the area of a water channel that has a trapezoidal shape in square feet, if the altitude is 4.55 ft, the base on the bottom (Base₁) is 8.22 ft, and the base on the top (Base₂) is 12.03 ft.**

Know: **Area = (Altitude)$\dfrac{(\text{Base}_1 + \text{Base}_2)}{2}$**

Substitute values and solve.

Area = $(4.55 \text{ ft})\dfrac{(8.22 \text{ ft} + 12.03 \text{ ft})}{2} = (4.55 \text{ ft})\dfrac{(20.25 \text{ ft})}{2}$

Area = (4.55 ft)(10.125 ft) = 46.07 ft², round to **46.1 ft²**

Base 2 = 12.03 ft

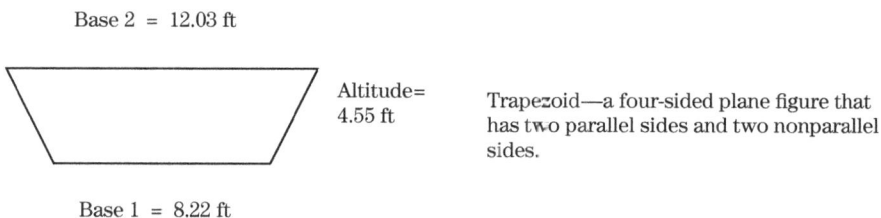

Altitude=
4.55 ft

Trapezoid—a four-sided plane figure that has two parallel sides and two nonparallel sides.

Base 1 = 8.22 ft

Figure 1-2.

28. What is the diameter of a tank, if the surface area is 1,962 ft^2?

Know: **Area = (0.785)(Diameter)^2**

Solve for the diameter by rearranging the equation as shown.

$$D^2 = \frac{Area}{0.785} = \frac{1,962 \text{ ft}^2}{0.785} = 2,499.36 \text{ ft}^2$$

Then, take the square root of 2,499.36 ft^2.

D = 49.99 ft, round to **50.0 ft in diameter**

VOLUME PROBLEMS

Volumes are very important to determine because many problems in the wastewater field require the volume to be known first before the rest of the calculations can be made. Knowing the volume of a particular process can also help the operator plan and make proper decisions in the treatment of wastewater.

29. **Find the capacity of a cylindrical tank in cubic feet and in liters, if it has a diameter of 10.3 ft and has a height of 18.5 ft.**

Equation: **Volume = (0.785)(Diameter)^2(Height)**

Volume = (0.785)(10.3 ft)(10.3 ft)(18.5 ft) = 1540.69 ft^3, round to **1,540 ft^3**

Next, find the number of liters.

(1540.69 ft^3)(7.48 gal/ ft^3)(3.785 liters/gal) = 43,619.71 liters, round to **43,600 liters**

30. **Calculate the volume in cubic feet for a pipeline that is 18.0 in. in diameter and 1,704 ft long, using both formulas in bold, below.**

First, convert the diameter to feet.

$$(18.0 \text{ in.})\frac{(1 \text{ ft})}{12 \text{ in.}} = 1.50 \text{ ft (Diameter)}$$

Then, convert the diameter to the radius.

Radius = Diameter /2 = 1.50 ft/2 = 0.75 ft in radius

Formula for the volume of a pipe in cubic feet is πr^2(**Length**) or (**0.785**)(**Diameter**)^2(**Length**).

Using the first equation, the Volume, ft^3 = (3.14)(0.750 ft)(0.750 ft)(1,704 ft)

Volume, ft^3 = 3,009.69 ft^3, round to **3,010 ft³**

Using the second equation, the Volume, ft^3 = (0.785)(1.50 ft)(1.50 ft)(1,704 ft)

Volume, ft^3 = 3,009.69 ft^3, round to **3,010 ft³**

31. **How many gallons would be in the pipe for the previous problem?**

(3,009.69 ft^3)(7.48 gal/ft^3) = 22,512.48 gal, round to **22,500 gal**

32. **What is the volume of a conical tank in cubic feet that has a radius of 10.25 ft and a height of 18.1 ft?**

Equation: **Volume, ft³ = 1/3πr^2(Height or Depth)**

Volume, ft^3 = 1/3(3.14)(10.25 ft)(10.25 ft)(18.1 ft) = 1,990.37 ft^3, round to **1,990 ft³**

33. **Determine the volume in cubic feet for a pipe with a diameter of 3.0 ft and a length of 7.22 miles.**

First, determine the number of feet in 7.22 miles.

$(5,280 \text{ ft/mile})(7.22 \text{ miles}) = 38,121.6 \text{ ft}$

Equation: **Volume, ft³ = (0.785)(Diameter)²(Length)**

Volume, ft³ = (0.785)(3.0 ft)(3.0 ft)(38,121.6 ft) = 269,329 ft³, round to **270,000 ft³**

34. **A chemical storage tank is conical at the bottom and cylindrical at the top. If the diameter of the cylinder is 9.51 ft, with a depth of 22.8 ft, and the cone depth is 12.5 ft, what is the volume of the tank in cubic feet?**

First, find the volume of the cone in cubic feet.

Equation: **Volume, ft³ = 1/3(0.785)(Diameter)²(Depth)**

Volume, ft³ = 1/3(0.785)(9.51 ft)(9.51 ft)(12.5 ft) = 295.81 ft³

Next, find the volume of the cylindrical part of the tank.

Volume = (0.785)(Diameter)²(Depth) = (0.785)(9.51 ft)(9.51 ft)(22.8 ft) = 1,618.7 ft³

Lastly, add the two volumes for the answer.

Total volume, ft³ = 295.81 ft³ + 1,618.7 ft³ = 1,914.51 ft³, round to **1,910 ft³**

35. Calculate the volume of a rectangular pond, given the following data:

Average pond depth = 5.74 ft
Surface water dimensions = 285 ft by 522 ft
Bottom pond dimensions = 265 ft by 502 ft
Slope of sides average = 2 ft horizontal to 1 ft vertical (2:1)

Equation: **Volume, gal** $= \dfrac{(\text{Length}_1 + \text{Length}_2)}{2} \dfrac{(\text{Width}_1 + \text{Width}_2)}{2}$ **(Depth, ft)(7.48 gal/ft³)**

Where Length$_1$ and Width$_1$ are the bottom dimensions of the pond and Length$_2$ and Width$_2$ are the surface water dimensions of the pond.

Substitute values and solve.

Volume, gal $= \dfrac{(502 \text{ ft} + 522 \text{ ft})}{2} \dfrac{(265 \text{ ft} + 285 \text{ ft})}{2}(5.74 \text{ ft})(7.48 \text{ gal/ft}^3)$

Volume, gal $= \dfrac{(1,024 \text{ ft})}{2} \dfrac{(550 \text{ ft})}{2}(5.74 \text{ ft})(7.48 \text{ gal/ft}^3) = 6,045,276 \text{ gal},$ round to **6,050,000 gal**

36. **Digester gas is stored in a spherical tank that is 21.3 ft in diameter. What is the capacity of the sphere in cubic feet?**

First, convert the diameter in feet to the radius in feet.

Radius, ft = Diameter/2 = 21.3 ft/2 = 10.65 ft

Next, calculate the sphere's capacity.

Equation: **Sphere volume ft³** $= \dfrac{4\pi r^3}{3}$

Sphere volume ft³ $= \dfrac{4(3.14)(10.65 \text{ ft})(10.65 \text{ ft})(10.65 \text{ ft})}{3} = 5,057.28 \text{ ft}^3 = \mathbf{5,060 \text{ ft}^3}$

37. **Calculate the volume in gallons for an oxidation ditch, given the following data:**

Average top width of ditch at water surface = 10.8 ft
Average depth = 4.2 ft
Average bottom width = 6.2 ft
Length of ditch = 110 ft
Diameter of half circles = 85.5 ft

Equation: $\dfrac{[(b_1 + b_2)\,\text{Depth}]}{2}$ **(Length of 2 sides + Length of 2 half circles)(7.48 gal/ft³)**

Where b_1 = bottom width of ditch and b_2 = top width of ditch at water surface, and the length of the two half circles is π(Diameter)

Substitute values and solve.

$$\text{Volume, gal} = \frac{[(6.2\ \text{ft} + 10.8\ \text{ft})\,(4.2\ \text{ft})]}{2}[(2)(110\ \text{ft}) + (3.14)(85.5\ \text{ft})](7.48\ \text{gal/ft}^3)$$

$$\text{Volume, gal} = (35.7\ \text{ft}^2)(220\ \text{ft} + 268.47\ \text{ft})(7.48\ \text{gal/ft}^3)$$

$$\text{Volume, gal} = (35.7\ \text{ft}^2)(488.47\ \text{ft})(7.48\ \text{gal/ft}^3)$$

Volume, gal = 130,439 gal, round to **130,000 gal**

DENSITY AND SPECIFIC GRAVITY OF LIQUIDS AND SOLIDS

The density of a substance is the amount of mass for a given volume. It is usually expressed as lb/gal or lb/ft³ in the English system or as g/cm³, kg/L, or kg/m³ in the metric system. Mass is defined as the quantity of matter as determined from Newton's second law of motion or by its weight. Weight is defined as the force that gravitation exerts upon a body and is equal to the mass of the body times the local acceleration of gravity. Specific gravity compares the density of one substance to another. Water is the standard for liquids and solids and is equal to 1.

38. **The density of an unknown substance is 2.54 grams/cm³. How much space would this substance occupy in cm³, if it weighed 7.41 lb?**

First, convert the number of lb to grams (g).

Number of g = (Number of lb)(454 g/1 lb)

Substitution: Number of g = (7.41 lb)(454 g/1 lb) = 3,364.14 g

We know that 2.54 grams of the substance occupies 1 cm³ by knowing its density. To get the space 3,364.14 grams occupies, we only need to divide by the density.

Space occupied by substance $= \dfrac{3,364.14 \text{ g}}{2.54 \text{ g/cm}^3} = 1{,}324 \text{ cm}^3$, round to **1,320 cm³**

39. What is the specific gravity for a solution that weighs 895 grams/liter?

Know: Density of water equals 1 gram per milliliter

$$\text{Sp gr} = \frac{(895 \text{ g/L})(1 \text{ L})}{(1 \text{ g/mL})(1{,}000 \text{ mL})} = \textbf{0.895 sp gr}$$

40. What is the specific gravity of a rock, if it weighs 991 grams in air and weighs 712 grams in water?

First, subtract the weight in air from the weight in water to determine the loss of weight in water.

Number of kilograms = 991 g − 642 g = 349 g is weight loss in water

Next, find the specific gravity by dividing the weight of the rock in air by the weight loss in water.

Sp gr = 991 g/349 g = 2.8395, round to **2.84 sp gr**

HYDRAULIC PRESS CALCULATIONS

Hydraulic press calculations have two fundamental principles: total force equals pressure applied times area the pressure is applied to, and the force applied to a liquid will be equally distributed within that liquid.

41. A small cylinder on a hydraulic jack is 10.0 in. in diameter. A force of 285 pounds is applied to the small cylinder. If the diameter of the large cylinder is 3.0 ft, what is the total lifting force?

Equation: **Pressure** $= \dfrac{\text{Total force, lb}}{\text{Area, ft}^2}$ for pressure on the small cylinder.

First, convert 10.0 inches to feet.

(10.0 in)(1 ft/12 in.) $= 0.833$ ft

$$\text{Pressure} = \frac{285 \text{ lb}}{(0.785)(0.833 \text{ ft})(0.833 \text{ ft})} = 523.22 \text{ lb/ft}^2$$

Next, calculate the total force on the large cylinder.

Equation: **Total Force = (Pressure)(Area)**

Total Force $= (523.22 \text{ lb/ft}^2)(0.785)(3.0 \text{ ft})(3.0 \text{ ft}) = 3{,}696.5$ lb, round to **3,700 lb**

42. A small cylinder on a hydraulic jack is 8.0 in. in diameter. A force of 342 pounds is applied to the small cylinder. If the diameter of the large cylinder is 2.75 ft, what is the total lifting force?

Equation: **Pressure** $= \dfrac{\text{Total force, lb}}{\text{Area, ft}^2}$ for pressure on the small cylinder.

First, convert 8.0 inches to feet.

(8.0 in.)(1 ft/12 in.) $= 0.667$ ft

$$\text{Pressure} = \frac{342 \text{ lb}}{(0.785)(0.667 \text{ ft})(0.667 \text{ ft})} = 979 \text{ lb/ft}^2$$

Next, calculate the total force on the large cylinder.

Equation: **Total Force = (Pressure)(Area)**

Total Force $= (979 \text{ lb/ft}^2)(0.785)(2.75 \text{ ft})(2.75 \text{ ft}) = 5{,}812$ lb, round to **5,800 lb**

SCREENING MATERIAL REMOVAL CALCULATIONS

The amount of screening debris should be calculated by operators so that they can plan and properly dispose of the material. A record should be kept each time for the amount of material removed from the screening pits. Screenings are usually disposed of by landfill, incinerated, or ground and returned to the wastewater process. They are very odorous and will attract flies. See the figures in Appendix E for placement of wastewater screens.

43. **Given the following data, calculate the number of ft³/mil gal of screening material removed at a wastewater treatment plant:**

Total screenings = 112 gallons
Corresponding wastewater processed = 4,275,000 gallons

First, determine the amount of cubic feet in 112 gallons.

Number of ft³ = 112 gal/7.48 gal/ft³ = 14.97 ft³

Next, convert gallons to mil gal.

$$\text{Number of mil gal} = \frac{4,275,000 \text{ gal}}{1,000,000/\text{mil}} = 4.275 \text{ mil gal}$$

Next, determine the screenings removed in ft³/mil gal.

Equation: **Screenings, ft³/mil gal** $= \dfrac{\text{Number of ft}^3}{\text{Number of mil gal}}$

Screenings, ft³/mil gal $= \dfrac{14.97 \text{ ft}^3}{4.275 \text{ mgd}} =$ **3.50 ft³/mil gal**

44. **How many gallons of screenings were removed from a wastewater plant, if the plant processed 3,465,000 gallons and the screenings removed per million gallons were 3.68 ft³/mil gal?**

First, convert gallons to mil gal.

$$\text{Number of mil gal} = \frac{3,465,000 \text{ gal}}{1,000,000/\text{mil}} = 3.465 \text{ mil gal}$$

Next, calculate the number of ft³ of screenings removed by rearranging the following equation.

Equation: **Screenings, ft³/mil gal** $= \dfrac{\text{Number of ft}^3}{\text{Number of mil gal}}$

Rearrange the equation.

Number of ft³ = (Screenings, ft³/mil gal)(Number of mil gal)

Substitute values and solve.

Number of ft³ = (3.68 ft³/mil gal)(3.465 mil gal) = 12.7512 ft³

Lastly, convert the ft³ of screenings removed to gallons of screenings removed.

Screenings removed, gal = (12.7512 ft³)(7.48 gal/ft³) = 95.38 ft³, round to **95.4 ft³**

SCREENING PIT CAPACITY CALCULATIONS

The operator needs to know the capacity of a screening pit so he or she knows when it should be cleaned based on past records of material removed (above calculations).

45. **How many cubic feet is a screening pit, if it would fill in 68.4 days and the average screenings each day are 2.08 ft³?**

Equation: **Number of days to fill** $= \dfrac{\text{Pit volume, ft}^3}{\text{Screenings removed, ft}^3/\text{day}}$

Rearrange the equation to solve for pit volume.

Pit volume, ft^3 = (Number of days to fill)(Screenings removed, ft^3/day)

Substitute values and solve.

Pit volume, ft^3 = (68.4 days)(2.08 ft^3/day) = 142.272 ft^3, round to **142 ft^3**

46. **How many days will it take to fill a screening pit, if the pit is 4.8 ft by 14.2 ft, 4.25 ft deep, and the average screenings are 3.05 ft^3/day?**

First, determine the volume of the pit in ft^3.

Pit volume, ft^3 = (4.8 ft)(14.2 ft)(4.25 ft) = 289.68 ft^3

Equation: **Number of days to fill** $= \dfrac{\textbf{Pit volume, ft}^3}{\textbf{Screenings removed, ft}^3/\textbf{day}}$

Substitute values and solve.

Number of days to fill $= \dfrac{289.68 \text{ ft}^3}{3.05 \text{ ft}^3/\text{days}} = 94.98$ days, round to **95 days**

GRIT REMOVAL CALCULATIONS

Grit removal is important for the same reason as screening removal—planning for proper disposal. Grit channels are important in wastewater treatment because removing the grit from the waste prevents wear on pumps and deposition in pipelines or channels. It also prevents grit from accumulating in other processes such as digesters or biological contactors. Not all wastewater treatment plants have grit channels, and they are not always placed after screens or comminutors. See the figures in Appendix E for where grit channels are commonly placed in different treatment plants.

47. A wastewater plant removes 268 liters of grit during the processing of 1,733,000 gallons. What is the ft³/mil gal removal rate during this interval?

First, convert liters to gallons.

Number of gallons = (268 liters)(1 gal/3.785 liters) = 70.8 gal

Next, convert gallons to mil gal.

$$\text{Number of mil gal} = \frac{1,733,000 \text{ gal}}{1,000,000/\text{mil}} = 1.733 \text{ mil gal}$$

Equation: **Grit removal, ft³/mil gal** $= \dfrac{\text{Number of gallons removed}}{(7.48 \text{ gal/ft}^3)(\text{mil gal treated})}$

$$\text{Grit removal, ft}^3/\text{mil gal} = \frac{70.8 \text{ gal}}{(7.48 \text{ gal/ft}^3)(1.733 \text{ mil gal})} = 5.462 \text{ ft}^3/\text{mil gal, round to } \mathbf{5.46 \text{ ft}^3/\text{mil gal}}$$

48. How many mil gal of waste was treated by a plant, if the number of gallons of grit removed was 314 liters and the grit removal rate was 4.64 ft³/mil gal?

First, convert liters to gallons.

Number of gallons = (314 liters)(1 gal/3.785 liters) = 82.959 gal

Equation: **Grit removal, ft³/mil gal** $= \dfrac{\text{Number of gallons removed}}{(7.48 \text{ gal/ft}^3)(\text{mil gal treated})}$

Rearrange to solve for mil gal treated.

$$\text{Mil gal treated} = \frac{\text{Number of gallons removed}}{(\text{Grit removal, ft}^3/\text{mil gal})(7.48 \text{ gal/ft}^3)}$$

Substitute values and solve.

$$\text{Mil gal treated} = \frac{82.959 \text{ gal}}{(4.64 \text{ ft}^3/\text{mil gal})(7.48 \text{ gal/ft}^3)} = \mathbf{2.39 \text{ mil gal}}$$

DETENTION TIME CALCULATIONS

Detention time is simply the time period that starts when wastewater flows into a basin or tank and that ends when it flows out of the basin or tank. Detention time is usually calculated for wastewater ponds, oxidation (aerobic) ditches, and clarifiers. Detention times are theoretical, because basins begin to fill with settled sludge and other debris. This causes the true detention time to constantly change (decrease). While it is true that sludge removals will cause the detention time to increase, the true detention time will always be less than theoretical. Also, flows through a basin are never perfectly laminar and thus cause a further decrease in the true detention time. See Figures E-5 and E-6 in Appendix E for two types of wastewater plants using ponds.

49. **What is the detention time in days for a wastewater treatment pond, given the following data?**

Pond = averages 308 ft by 128 ft with a depth of 5.85 ft
Flow = 37,100 gpd

First, calculate the volume of the waste pond in gallons.

Equation: **Volume, gal = (Length, ft)(Width, ft)(Depth, ft)(7.48 gal/ft³)**

Volume, gal = (308 ft)(128 ft)(5.85 ft)(7.48 gal/ft³) = 1,725,115 gal

Equation: **Detention time, days $= \dfrac{\text{Volume, gal}}{\text{Flow, gpd}}$**

Detention time, days $= \dfrac{1,725,115 \text{ gal}}{37,100 \text{ gpd}} =$ **46.5 days**

50. **What is the detention time in days for a wastewater treatment pond, given the following data?**

Pond = averages 285 ft by 117 ft with a depth of 6.75 ft
Flow = 3.95 ft³/min

First, convert ft³/s to gpd

Number of gpd = (3.95 ft³/min)(1,440 min/day)(7.48 gal/ft³) = 42,546 gpd

Next, calculate the volume of the waste pond in gallons.

Equation: **Volume, gal = (Length, ft)(Width, ft)(Depth, ft)(7.48 gal/ft³)**

Volume, gal = (285 ft)(117 ft)(6.75 ft)(7.48 gal/ft³) = 1,683,589 gal

Equation: **Detention time, days** $= \dfrac{\text{Volume, gal}}{\text{Flow, gpd}}$

Detention time, days $= \dfrac{1,683,589 \text{ gal}}{42,546 \text{ gpd}}$ = 39.57 days, round to **39.6 days**

51. **What is the detention time in hours, if an oxidation ditch has an influent flow of 0.186 mgd and the volume of the oxidation ditch is 82,500 ft³?**

First, convert mgd to gpd.

Number of gpd = (0.186 mgd)(1,000,000 gal/mil) = 186,000 gpd

Next, convert the volume of the oxidation ditch from ft³ to gallons.

Oxidation ditch volume, gal = (82,500 ft³)(7.48 gal/ft³) = 617,100 gal

Lastly, calculate the detention time in hours.

Equation: **Detention time, hr** $= \dfrac{(\text{Volume, gal})(24 \text{ hr/day})}{\text{Flow, gpd}}$

Detention time, hr $= \dfrac{(617,100 \text{ gal})(24 \text{ hr/day})}{186,000 \text{ gpd}} = \textbf{79.6 hr}$

WEIR AND SURFACE OVERFLOW RATE PROBLEMS

A weir is like a small dam, gate, notch, or other barrier placed across a basin to help regulate water out of the basin. The weir overflow rate is used to determine the velocity of wastewater over the weir. The velocity informs the operator about the efficiency of the sedimentation process. At constant wastewater flow, the shorter the length of the weir, the faster the water velocity will be out of the basin. Conversely, the longer the weir length, the slower the velocity will be out of the basin. See Figures E-1, E-2, E-3, E-7, and E-8 in Appendix E for five types of wastewater plants using a clarifier.

52. What is the weir overflow rate in gpd/ft, if the flow is 0.357 mgd and the radius of the clarifier is 34.85 ft?

First, calculate the length of the weir.

Weir length, ft $= 2\pi(\text{radius, ft})$

Weir length, ft $= 2(3.14)(34.85 \text{ ft}) = 218.858 \text{ ft}$

Next, convert mgd to gpd.

Number of gpd $= (0.357 \text{ mgd})(1,000,000 \text{ gal/mil}) = 357,000 \text{ gpd}$

Next, solve for the weir overflow rate.

Equation: **Weir overflow rate, gpd/ft** $= \dfrac{\text{Flow, gpd}}{\text{Weir length, ft}}$

Weir overflow rate, gpd/ft $= \dfrac{357,000 \text{ gpd}}{218.858 \text{ ft}} = 1,631.19 \text{ gpd/ft, round to } \textbf{1,630 gpd/ft}$

53. What is the surface overflow rate in gpd/ft², if the clarifier is 48.75 ft in radius and the flow into the basin is 641 gpm?

First, determine the area of the clarifier.

Area $= \pi r^2$ where $\pi = 3.14$

Area $= (3.14)(48.75 \text{ ft})(48.75 \text{ ft}) = 7{,}462.41 \text{ ft}^2$

Next, convert gpm to gpd.

Number of gpd $= (641 \text{ gpm})(1{,}440 \text{ min/day}) = 923{,}040 \text{ gpd}$

Lastly, calculate the surface overflow rate.

Equation: **Surface overflow rate** $= \dfrac{\text{Flow, gpd}}{\text{Area, ft}^2}$

Surface overflow rate $= \dfrac{923{,}040 \text{ gpd}}{7{,}462.41 \text{ ft}^2} = 123.69 \text{ gpd/ft}^2$, round to **124 gpd/ft²**

FLOW AND VELOCITY CALCULATIONS

Operators need to know the flow and velocity of the wastewater throughout the different plant processes, for example to feed proper dosages of chemicals to treat wastewaters, to know how many clarifiers or ponds to use or how much supernatant to recirculate, and for settling purposes, among other uses.

54. **If a 3.35-ft diameter chemical tank drops 7.72 in. in exactly 1.5 hours, what is the pumping rate for the chemical in gpm?**

First, convert 1.5 hours to min.

Number of minutes = (1.5 hr)(60 min/hr) = 90 min

Next, determine the amount in feet the tank level dropped.

Drop, ft = (7.72 in.)(1 ft/12 in.) = 0.643 ft

Then, determine the volume in gallons for the drop in level of the tank.

Equation: **Volume, gal = (0.785)(Diameter)^2(Drop, ft)(7.48 gal/ft^3)**

Substitute values and solve.

Volume, gal = (0.785)(3.35 ft)(3.35 ft)(0.643 ft)(7.48 gal/ft^3) = 42.37 gal

Now, calculate the pumping rate in gpm.

Equation: **Pumping rate = Flow, gal/Time, min**

Pumping rate = 42.37 gal/90 min = **0.47 gpm**

55. If a 39.75-ft diameter clarifier drops 29.7 in. in 4.5 hours, what is the pumping rate out of the tank in gpm?

First, convert 4.5 hours to min.

Number of minutes = (4.5 hr)(60 min/hr) = 270 min

Next, determine the amount in feet the tank level dropped.

Drop, ft = (29.7 in.)(1 ft/12 in.) = 2.475 ft

Then, determine the volume in gallons for the drop in level of the tank.

Equation: **Volume, gal = (0.785)(Diameter)²(Drop, ft)(7.48 gal/ft³)**

Substitute values and solve.

Volume, gal = (0.785)(39.75 ft)(39.75 ft)(2.475 ft)(7.48 gal/ft³) = 22,962.58 gal

Now, calculate the pumping rate in gpm.

Equation: **Pumping rate, gpm = Flow, gal/Time, min**

Pumping rate = 22,962.58 gal/270 min = 85.05 gpm, round to **85 gpm**

56. What is the flow velocity in feet per second (ft/s) for a trapezoidal channel, given the following data?

Bottom width, w₁ = 4.45 ft
Water surface width, w₂ = 7.68 ft
Depth 3.66 ft
Flow = 31.2 ft³/s

Equation: **Flow (Q), ft³/s = $\frac{(w_1 + w_2)}{2}$(Depth, ft)(Velocity, ft/s)**

Rearrange the formula to solve for velocity in ft/s.

$$\text{Velocity, ft/s} = \frac{2\,(Q,\,\text{ft}^3/\text{s})}{(w_1 + w_2)(\text{Depth, ft})}$$

Substitute values and solve.

$$\text{Velocity, ft/s} = \frac{2\,(31.2\,\text{ft}^3/\text{s})}{(4.45\,\text{ft} + 7.68\,\text{ft})(3.66\,\text{ft})} = \frac{62.4\,\text{ft}^3/\text{s}}{(12.13\,\text{ft})(3.66\,\text{ft})} = \mathbf{1.41\,ft/s}$$

57. **Water is flowing at a velocity of 1.22 ft/s in a 16.0-inch diameter pipe. If the pipe changes from the 16.0-inch to a 10.0-inch pipe, what will the velocity be in the 10.0-inch pipe?**

Flow in 16.0-inch pipe equals the flow in the 10.0-inch pipe as the flow must remain constant: $\mathbf{Q_1 = Q_2}$

Since Q, flow = (Area)(Velocity), it follows that: $\mathbf{(Area_1)(Velocity_1) = (Area_2)(Velocity_2)}$

First, find the diameters in feet for the 16.0-inch and 10.0-inch pipes.

Diameter for 16.0-in. = 16.0-in.(1 ft/12-in.) = 1.333 ft

Diameter for 10.0-in. = 10.0-in.(1 ft/12-in.) = 0.833 ft

Then, determine the areas of each size pipe.

Area = (0.785)(Diameter)^2

Area 1 (16.0-in.) = (0.785)(1.333 ft)(1.333 ft) = 1.395 ft^2

Area 2 (10.0-in.) = (0.785)(0.833 ft)(0.833 ft) = 0.545 ft^2

Lastly, substitute areas calculated and known velocity in 16.0-inch pipe.

(1.395 ft^2)(1.22 ft/s) = (0.545 ft^2)(x, ft/s)

Solve for x, ft/s.

$$x, \text{ft/s} = \frac{(1.395 \text{ ft}^2)(1.22 \text{ ft/s})}{(0.545 \text{ ft}^2)} = \textbf{3.12 ft/s in 10.0-in. pipe}$$

****

The following two problems involve flow through a pipeline that is **not** flowing full. The calculations are almost the same as determining flow in a full pipeline, except the multiplication factor of 0.785 is not used. A new factor is used and is based on the liquid level in the pipe divided by the pipe's diameter. These factors are presented in the depth/Diameter table in Appendix D. **Please note that answers are only approximate.**

58. **A 12-in. sewage pipeline is flowing at a velocity of 1.55 ft/s, and the depth of the sewage averages 4.5 in. Determine the flow in the pipeline in ft³/s and gpm.**

First, divide the depth of sewage flow by the diameter of the pipe. Converting inches to feet is not necessary.

Ratio = depth/Diameter = 4.5 in./12 in. = 0.375, round to 0.38

Note: Extrapolation can also be used if more accuracy is required.

Next, determine the factor that needs to be used.

In Appendix D, look up 0.38 under the column d/D. The number immediately to the right will be the factor that needs to be used. In this case it is 0.2739. This will be the number used rather than 0.785.

Next, convert the pipe's diameter from inches to feet.

$$\text{Number of feet} = \frac{12 \text{ inches}}{12 \text{ inches/ft}} = 1.0 \text{ ft}$$

Equation: **Flow, ft³/s = (Area, ft²)(Velocity, ft/s)**

Where: **Area = (Factor)(Diameter)²**

Substitute values and solve.

Flow, ft³/s = (0.2739)(1.0 ft)(1.0 ft)(1.55 ft/s) = 0.4245 ft³/s, round to **0.42 ft³/s**

Now, convert ft³/s to gpm.

Flow, gpm = (0.4245 ft³/s)(60 s/min)(7.48 gal/ft³) = 190.52 gpm, round to **190 gpm**

59. **A 24-inch sewage pipeline is flowing at a velocity of 1.29 ft/s and the depth of the sewage averages 11.8 in. Determine the flow in the pipeline in gpm.**

First, divide the depth of sewage flow by the diameter of the pipe. Converting inches to feet is not necessary.

Ratio = depth/Diameter = 11.8 in./24 in. = 0.492, round to 0.49

Note: Again, extrapolation can also be used if more accuracy is required.

Next, determine the factor that needs to be used.

In Appendix D, look up 0.49 under the column d/D. The number immediately to the right will be the factor that needs to be used. In this case it is 0.3827. This will be the number used rather than 0.785.

Next, convert the pipe's diameter from inches to feet.

$$\text{Number of feet} = \frac{24 \text{ in.}}{12 \text{ in./ft}} = 2.0 \text{ ft}$$

Equation: **Flow, ft³/s = (Area, ft²)(Velocity, ft/s)**

Where the area = (Factor)(Diameter)²

Substitute values and solve.

Flow, ft³/s = (0.3827)(2.0 ft)(2.0 ft)(1.29 ft/s) = 1.975 ft³/s

Now, convert ft³/s to gpm.

Flow, gpm = (1.975 ft³/s)(60 s/min)(7.48 gal/ft³) = 886.38 gpm, round to **890 gpm**

PUMP DISCHARGE PROBLEMS

Operators need to understand pump discharge calculations, which helps in planning treatment processes and time it will take to complete the process; in determining how long a pump will take to discharge a certain amount of wastewater or chemical to treat the wastewater; and maybe in changing the size of a pump to fit the need better.

60. **A reclaim water tank has a radius of 17.8 ft. How long will it take in hours and minutes for a pump to fill the tank to the 19.5-ft level, if it already has a water level of 4.25 ft and the pumping rate is 145 gpm?**

First, determine the number of gallons in the tank if it were filled to the 19.5-ft level.

Equation: Volume, gal $= \pi(\text{radius})^2(\text{Depth, ft})(7.48 \text{ gal/ft}^3)$

Volume, gal $= 3.14(17.8 \text{ ft})(17.8 \text{ ft})(19.5 \text{ ft})(7.48 \text{ gal/ft}^3) = 145{,}112.85$ gal

Next, calculate the number of gallons already in the tank.

Volume, gal $= 3.14(17.8 \text{ ft})(17.8 \text{ ft})(4.25 \text{ ft})(7.48 \text{ gal/ft}^3) = 31{,}627.16$ gal

Next, determine the number of gallons that are required to fill the tank by subtracting the volume of water at the 4.25-ft level from the volume of water at the 19.5-ft level.

Number of gallons $= 145{,}112.85$ gal $- 31{,}627.16$ gal $= 113{,}485.69$ gal

Next, calculate the pump's discharge rate in gpm.

Equation: **Pumping time, min** $= \dfrac{\textbf{Discharge, gal}}{\textbf{Pump rate, gpm}}$

Substitute values and solve.

Pump's discharge rate, gal $= \dfrac{113{,}485.69 \text{ gal}}{145 \text{ gpm}} = 782.66$ min

Now, divide by 60 (60 min/hr) to determine the number of hours.

Number of hours = (782.66)/60 min/hr) = 13.0443 hr

Now, determine the number of minutes in 0.0443 hours by multiplying by 60 (60 min/hr).

Number of minutes = (0.0443 hr)(60 min/hr) = 2.658 min, round to 3 min

Thus, the tank will be filled to the 19.5-ft level in **13 hours and 3 minutes.**

61. **Given the following data, determine the rate a pump discharges from the tank in gpm:**

Duration pump operates = 21 hr and 14 minutes
Tank diameter = 60.1 ft
Wastewater level at beginning of pumping = 14.72 ft
Wastewater level at end of pumping = 10.89 ft

First, find the number of minutes the pump worked.

Number of min = (21 hr)(60 min/hr) + 14 min = 1,274 min

Next, calculate the change in level during pumping.

Level change, ft = 14.72 ft − 10.89 = 3.83 ft

Next, calculate the volume in gallons added to the tank by the pump.

Equation: Volume, gal = (0.785)(Diameter)^2(Level change, ft)(7.48 gal/ft^3)

Volume, gal = (0.785)(60.1 ft)(60.1 ft)(3.83 ft)(7.48 gal/ft^3) = 81,230.47 gal

Now, calculate the pump's discharge rate in gpm.

Equation: **Pump's discharge rate, gpm** $= \dfrac{\text{Discharge, gal}}{\text{Time, min}}$

Substitute values and solve.

Pump's discharge rate, gpm $= \dfrac{81,230.47 \text{ gal}}{1,274 \text{ min}} = 63.76$ gpm, round to **64 gpm**

62. **How long will it take in hours and minutes to empty a tanker truck with aluminum sulfate (alum), if the truck's pump unloads the alum at 71.2 gpm and a total of 18,029 liters needs to be unloaded? The tank's capacity is 10,000 gallons and it already has about 4,000 gallons of alum. Will the tank be able to take the entire load?**

First, determine the number of gallons in 18,029 liters.

$$\text{Number of gal} = \frac{18,029 \text{ liters}}{3.785 \text{ liters/gal}} = 4,763.28 \text{ gal}$$

Then, divide the number of gallons by the pumping rate.

Time to pump $= 4,763.28$ gal/71.2 gpm $= 66.9$ min

Divide by 60 min/hr.

66.9 min/60 min/hr $= 1.115$ hr

Next, find how many minutes are in 0.115 hr by multiplying by 60 min/hr.

$(0.155 \text{ hr})(60 \text{ min/hr}) = 6.9$ min, round to 7 min.

The unloading time $=$ **1 hr and 7 min.**

Yes, the chemical tank can take the entire load.

63. **A pump increases the level in a 12-ft diameter chemical storage tank by 49.5 in. in 1 hr and 37 minutes. What is the pumping rate of the pump in gpm?**

First, determine the number of minutes the pump worked.

Number of min = (1 hr)(60 min/hr) + 37 min = 97 min

Next, convert the level increase in inches to feet.

Number of ft = (49.5 in.)(1 ft/12 in.) = 4.125 ft

Then, calculate the volume using the following equation.

Volume, gal = (0.785)(Diameter)^2(Drop, ft)7.48 gal/ft^3)

Substitute values and solve.

Volume, gal = (0.785)(12 ft)(12 ft)(4.125 ft)(7.48 gal/ft^3) = 3,487.85 gal

Lastly, determine the pumping rate in gpm.

Equation: **Pumping rate or Flow, gpm** $= \dfrac{\text{Number of gal}}{\text{Time, min}}$

Substitute values and solve.

Flow, gpm $= \dfrac{3,487.85 \text{ gal}}{97 \text{ min}} = 35.957$ gpm, round to **36 gpm**

SOLIDS AND HYDRAULIC LOADING RATE CALCULATIONS

Solids and hydraulic loading rate calculations are used to determine the solids or hydraulic loading on clarifiers, trickling filters, and other processes. These calculations are important to know so operators, for example, can determine when to discharge sludge from a clarifier or know the contact time between organisms in a trickling filter and the food entering that trickling filter.

64. **What is the solids loading rate on a secondary clarifier with a diameter of 59.8 ft, if the flow rate is 946,000 gpd, with a mixed liquor suspended solids (MLSS) of 3,178 mg/L?**

First, determine the area of the clarifier.

Area $= \pi r^2$ where r $=$ Diameter/2 $=$ 59.8 ft/2 $=$ 29.9 ft

Area $=$ (3.14)(29.9 ft)^2 $=$ 2,807.19 ft^2

Next, convert gpd to mgd.

Number of mgd $= \dfrac{946,000 \text{ gpd}}{1,000,000 \text{ gal/mil}} = 0.946$ mgd

Finally, calculate the solids loading rate.

Equation: **Solids loading rate** $= \dfrac{(\text{MLSS, mg/L})\,(\text{mgd})\,(8.34 \text{ lb/gal})}{\text{Area, ft}^2}$

Solids loading rate $= \dfrac{(3,178 \text{ mg/L})\,(0.946 \text{ mgd})\,(8.34 \text{ lb/gal})}{2,807.19 \text{ ft}^2}$

Solids loading rate $=$ **8.93 lb of solids/d/ft^2**

65. **What is the solids loading rate in lb/d/ft^2, if a 74.9-ft diameter gravity thickener receives 42,800 gpd of biosolids and the biosolids contain 1.52% solids?**

First, determine the surface area of the gravity thickener.

Surface area of gravity thickener, ft^2 $=$ (0.785)(74.9 ft)(74.9 ft) $=$ 4,403.858 ft^2

Now, calculate the solids loading rate.

Equation: **Solids loading rate, lb/d/ft^2** $= \dfrac{(\text{Percent solids})\,(\text{Biosolids added, gpd})\,(8.34 \text{ lb/gal})}{(\text{Surface area, ft}^2)}$

Solids loading rate, lb/d/ft^2 = $\dfrac{(1.52\%/100\%)\,(42{,}800\text{ gpd})\,(8.34\text{ lb/gal})}{(4{,}403.858\text{ ft}^2)}$

Substitute values and solve.

Solids loading rate, lb/d/ft^2 = **1.23 lb/d/ft^2**

66. **A trickling filter has a diameter of 124.75 ft. If the flow through the filter is 1.49 mgd and the recirculation rate is 24% of the flow rate, what is the hydraulic loading rate on a trickling filter in gallons per day per square foot (gpd/ft^2)?**

First, determine the total flow in gallons per day (gpd) through the trickling filter.

Total flow, gal = [1.49 mgd + 1.49 mgd(24%/100%)](1,000,000/mil)

Total flow, gal = [1.49 mgd + 0.3576 mgd](1,000,000/mil) = 1,847,600 gpd

Next, determine the surface area in ft^2 for the clarifier.

Area = πr^2 where r = Diameter/2 = 124.75 ft/2 = 62.375 ft

Clarifier surface area, ft^2 = (3.14)(62.375 ft)(62.375 ft) = 12,216.61 ft^2

Lastly, calculate the hydraulic loading rate.

Hydraulic loading rate = $\dfrac{\textbf{Total flow, gpd}}{\textbf{Surface area, ft}^2}$

Hydraulic loading rate = $\dfrac{1{,}847{,}000,\text{ gpd}}{12{,}216.61\text{ ft}^2}$ = 151.188 gpd/ft^2, round to **150 gpd/ft^2**

67. **What is the hydraulic loading rate for a pond that is 11.8 acre-ft in gallons per day per ft² (gpd/ft²), if the flow into the pond is 1.36 mgd?**

Since the problem asks for gpd, first convert the volume of the pond in acre-ft to gallons.

Know: 1 acre-ft = 43,560 ft²

Area of pond, ft² = (11.8 acre-ft)(43,560 ft²/acre-ft) = 514,008 ft²

Next, convert mgd to gallons.

Flow into pond, gal = (1.36 mgd)(1,000,000/mil) = 1,360,000 gpd

Lastly, divide the flow.

Equation: **Hydraulic loading rate** $= \dfrac{\textbf{Total flow, gpd}}{\textbf{Surface area, ft}^2}$

Hydraulic loading rate $= \dfrac{1,360,000 \text{ gpd}}{514,008 \text{ ft}^2} = 2.646$ gpd/ft², round to **2.65 gpd/ft²**

68. **Calculate the solids loading rate in lb of solids/d/ft², given the following data:**

Secondary clarifier radius = 40.14 ft
Primary effluent flow = 1,470,000 gpd
Return of activated sludge is 0.598 mgd
Mixed liquor suspended solids (MLSS) = 2,825 mg/L
Specific gravity of the solids is 1.05

First, convert the primary effluent flow in gallons per day to mgd.

$\text{mgd} = \dfrac{1,470,000 \text{ gpd}}{1,000,000/\text{mil}} = 1.47 \text{ mgd}$

Next, determine the total flow.

Total flow = Primary flow + Return of activated sludge

Total flow = 1.47 mgd + 0.598 mgd = 2.068 mgd

Next, calculate the area of the clarifier.

Area $= \pi r^2$

Area = (3.14)(40.14 ft)(40.14 ft) = 5,059.23 ft^2

Next, determine the lb/gal of solids.

Solids, lb/gal = (8.34 lb/gal)(1.05 sp gr) = 8.757 lb/gal

Next, calculate the solids loading rate in lb of solids/d/ft^2.

Equation: **Solids loading rate** $= \dfrac{(\text{MLSS, mg/L})\,(\text{mgd})\,(8.34\text{ lb/gal})}{\text{Area, ft}^2}$

Solids loading rate $= \dfrac{(2,825\text{ mg/L})\,(2.068\text{ mgd})\,(8.757\text{ lb/gal})}{5,059.23\text{ ft}^2} =$

Solids loading rate = 10.11 lb of solids/d/ft^2, round to **10.1 lb of solids/d/ft^2**

69. Given the following data, calculate the hydraulic loading rate on a high-rate trickling filter in gpd/ft^2.

Influent flow = 2,330 gpm
Trickling filter diameter = 99.9 ft

First, determine the area in ft^2 for the trickling filter.

Equation: **Trickling filter area, ft^2 = (0.785)(Diameter)^2**

Trickling filter area, ft^2 = (0.785)(99.9 ft)(99.9 ft) = 7,834.31 ft^2

Next, convert gpm to gpd.

Number of gpd = (2,330 gpm)(1,440 min/day) = 3,355,200 gpd

Next, determine the hydraulic loading rate.

Hydraulic loading rate, gpd/ft2 $= \dfrac{3,355,200 \text{ gpd}}{7,834.31 \text{ ft}^2} = 428.27$ gpd/ft^2, round to **428 gpd/ft^2**

SLUDGE PUMPING PROBLEMS

Sludge pumping calculations are important for operators to determine so they know how much sludge and solids are being loaded into a digester so that underloading or overloading does not occur. Also, operators need to know how much sludge is being pumped to other sludge processing applications such as sludge thickening, filter presses, or for land application.

70. **Given the following data, calculate the lb/day of solids pumped to a sludge thickener:**

Sludge sample = 1,985.183 grams (g)
Solids content after drying = 81.759 g
Pump operates exactly 8 minutes every 1.0 hour
Pump rate = 14.9 gpm
Specific gravity (sp gr) = 1.03
Clarifier effluent flow = 1.82 mgd

First, determine the percent solids in the sludge.

Equation: **Solids, percent = (Dry solids, g)(100%) / Sludge sample, g**

$$\text{Solids, percent} = \frac{(81.759 \text{ g})(100\%)}{1,985.183 \text{ g}} = 4.118\% \text{ solids}$$

Now, calculate the solids pumped in lb/day.

Equation: **Solids, lb/day =**

(Pumping, min/day)(24 hr/day)(Pump rate, gpm)(8.34 lb/gal)(sp gr of sludge)(% solids)

Solids, lb/day = (8 min/hr)(24 hr/day)(14.9 gpm)(8.34 lb/gal)(1.03 sp gr)(4.118%/100 %)

Solids, lb/day = 1,011.99 lb/day, round to **1,010 lb/day of Solids**

71. Given the following parameters, calculate how long a primary sludge pump should operate in minutes per hour:

Plant flow = 1,250 gpm
Sludge pump = 48 gpm
Influent suspended solids (SS) = 295 mg/L
Effluent SS = 115 mg/L
Sludge = 4.6% solids

First, convert gpm to mgd.

$$\text{Number of mgd} = \frac{(1{,}250\text{ gpm})(1{,}440\text{ min/day})}{1{,}000{,}000/\text{mil}} = 1.8\text{ mgd}$$

Equation: **Operating time, min/hr =**

$$\frac{(\text{Flow, mgd})(\text{Influent SS, mg/L} - \text{Effluent SS, mg/L})(100\%)}{(\text{Sludge pump, gpm})(\text{Percent Solids})(24\text{ hr/day})}$$

Substitute values and solve.

$$\text{Operating time, min/hr} = \frac{(1.8\text{ mgd})(295\text{ mg/L, SS} - 115\text{ mg/L, SS})(100\%)}{(48\text{ gpm})(4.6\%)(24\text{ hr/day})}$$

$$\text{Operating time, min/hr} = \frac{(1.8\text{ mgd})(180\text{ mg/L, SS})(100\%)}{(48\text{ gpm})(4.6\%)(24\text{ hr/day})}$$

Operating time, min/hr = **6.1 min/hr**

72. How many lb/day of solids were pumped to a digester, if a sludge pump operates exactly 10 minutes every hour at a rate of 18.7 gpm, the percent solids in the sludge was 5.09%, and the specific gravity of the sludge was 1.03?

Equation: **Solids, lb/day =**

(Pumping, min/day)(24 hr/day)(Pump rate, gpm)(8.34 lb/gal)(sp gr of sludge)(% solids)

Solids, lb/day = (10 min/hr)(24 hr/day)(13.7 gpm)(8.34 lb/gal)(1.03 sp gr)(5.09%/100 %)

Solids, lb/day = 1,962.34 lb/day, round to **1,960 lb/day of Solids**

BIOSOLIDS PUMPING, PRODUCTION, AND RETENTION TIME PROBLEMS

Biosolids pumping calculations provide operators accurate process control data for the sedimentation process. Biosolids are mostly composed of water with the biosolids ranging from only 3 to 7 percent by volume.

73.   What is the estimated biosolids pumping rate for the following system?

Assume sludge is 8.34 lb/gal
Plant flow = 1.56 mgd
Removed biosolids = 1.37%
Influent total suspended solids (TSS) = 333 mg/L
Effluent TSS = 124 mg/L

Equation: **Estimated pumping rate =**

$$\frac{(\text{Influent TSS, mg/L} - \text{Effluent TSS, mg/L})(\text{Flow, mgd})(8.34 \text{ lb/gal})}{(\text{Percent solids in sludge})(\text{Sludge, lb/gal})(1,440 \text{ min/day})}$$

Substitute values and solve.

$$\text{Estimated pumping rate} = \frac{(333 \text{ TSS mg/L} - 124 \text{ TSS, mg/L})(1.56 \text{ mgd})(8.34 \text{ lb/gal})}{(1.37\%/100\%)(8.34 \text{ lb/gal})(1,440 \text{ min/day})}$$

$$\text{Estimated pumping rate} = \frac{(209 \text{ TSS mg/L})(1.56 \text{ mgd})(8.34 \text{ lb/gal})}{(1.37\%/100\%)(8.34 \text{ lb/gal})(1,440 \text{ min/day})}$$

Estimated pumping rate = **16.5 gpm**

74.   **What is the estimated biosolids pumping rate for the following system?**

Plant flow = **1.93 mgd**
Removed biosolids = **1.25%**
Effluent TSS = **132 mg/L**
Weight of biosolids = **8.44 lb/gal**
Influent total suspended solids (TSS) = **301 mg/L**

Equation: **Estimated pumping rate** =

$$\frac{(\text{Influent TSS, mg/L} - \text{Effluent TSS, mg/L})(\text{Flow, mgd})(8.34\,\text{lb/gal})}{(\text{Percent solids in sludge})(\text{Sludge, lb/gal})(1,440\,\text{min/day})}$$

Substitute values and solve.

$$\text{Estimated pumping rate} = \frac{(301\,\text{TSS mg/L} - 132\,\text{TSS, mg/L})(1.93\,\text{mgd})(8.34\,\text{lb/gal})}{(1.25\%/100\%)(8.44\,\text{lb/gal})(1,440\,\text{min/day})}$$

$$\text{Estimated pumping rate} = \frac{(169\,\text{TSS mg/L})(1.93\,\text{mgd})(8.34\,\text{lb/gal})}{(1.25\%/100\%)(8.44\,\text{lb/gal})(1,440\,\text{min/day})}$$

Estimated pumping rate = **17.9 gpm**

75. A wastewater treatment plant produces 126,000 gallons of biosolids in a 31-day month. If the plant treated 1.83 mgd on average, what is the biosolids production in lb/mil gal for this time period?

Equation: **Biosolids, lb/mil gal** $= \dfrac{(\text{Biosolids, gal})(8.34\ \text{lb/gal})}{(\text{Flow, mgd})(\text{Number of days})}$

Biosolids, lb/mil gal $= \dfrac{(126,000\ \text{gal})(8.34\ \text{lb/gal})}{(1.83\ \text{mgd})(31\ \text{days})}$

Biosolids, lb/mil gal = 18,524 lb/mil gal, round to **18,500 lb/mil gal Biosolids**

76. What is the biosolids production in wet tons per year, if the plant flow averages 1.84 mgd and production of biosolids averages 14,175 lb/day?

Equation: **Biosolids, wet tons/yr** $= \dfrac{(\text{Biosolids, lb/mil gal})(\text{mgd})(365\ \text{days/yr})}{2,000\ \text{lb/ton}}$

Substitute values and solve.

Biosolids, wet tons/yr $= \dfrac{(14,175\ \text{lb/mil gal})(1.84\ \text{mgd})(365\ \text{days/yr})}{2,000\ \text{lb/ton}}$

Biosolids, wet tons/yr = 4,759.965 wet tons/yr, round to **4,760 wet tons/yr Biosolids**

77. Given the following data, calculate the amount of solids and volatile solids removed in lb/day:

Pumping rate = **224 gpm**
Pump frequency = **24 times/day**
Pumping cycle = **15 minutes exactly per cycle**
Solids = **3.45%**
Volatile solids (VS) = **62.3%**

First, determine the solids removal in lb/day.

Equation: **Solids, lb/day** =

(Time, min/cycle)(cycles/day)(Pump rate, gpm)(8.34 lb/gal)(Percent solids)

Substitute values and solve.

Solids, lb/day = (15 min/cycle)(24 cycles/day)(224 gpm)(8.34 lb/gal)(3.45%/100%)

Solids, lb/day = 23,203 lb/day, round to **23,200 lb/day Solids**

Next, calculate the amount of volatile solids removed in lb/day.

Equation: **VS, lb/day** =

(Time, min/cycle)(cycles/day)(Pump rate, gpm)(8.34 lb/gal)(Percent, solids)(Percent VS)

Substitute values and solve.

VS, lb/day = (15 min/cycle)(24 cycles/day)(224 gpm)(8.34 lb/gal)(3.45%/100%)(62.3%/100%)

VS, lb/day = 14,455.187 lb/day, round to **14,500 lb/day VS**

78. Determine the amount of solids and volatile solids removed in lb/day, given the following data:

Pumping rate = 266,400 gal/day
Pump frequency = 36 times/day
Pumping cycle = 6 minutes exactly per cycle
Solids = 3.86%
Volatile solids (VS) = 57.8%

First, convert gal/day to gpm.

Pumping rate, gpm $= \dfrac{266,400 \text{ gal/day}}{1,440 \text{ min/day}} = 185 \text{ gpm}$

Next, determine the solids removal in lb/day.

Equation: **Solids, lb/day** =

(Time, min/cycle)(cycles/day)(Pump rate, gpm)(8.34 lb/gal)(Percent solids)

Substitute values and solve.

Solids, lb/day = (6 min/cycle)(36 cycles/day)(185 gpm)(8.34 lb/gal)(3.86/100%)

Solids, lb/day = 12,864 lb/day, round to **12,900 lb/day Solids**

Next, calculate the amount of volatile solids removed in lb/day.

Equation: **VS, lb/day** =

(Time, min/cycle)(cycles/day)(Pump rate, gpm)(8.34 lb/gal)(Percent, solids)(Percent VS)

Substitute values and solve.

VS, lb/day = (6 min/cycle)(36 cycles/day)(185 gpm)(8.34 lb/gal)(3.86%/100%)(57.8%/100%)

VS, lb/day = 7,435 lb/day, round to **7,440 lb/day VS**

79. **What is the biosolids retention time (BRT) in days for a digester that is 78.5 ft in diameter, has a working level of 19.25 ft, and receives an average flow of 11.3 gpm?**

First, convert gpm to gpd.

Digester influent, gpm = (11.3 gpm)(1,440 min/day) = 16,272 gpd

Next, determine the working volume in gallons for the digester.

Equation: **Digester volume, gal = (0.785)(Diameter, ft)^2(Height, ft)(7.48 gal/ft^3)**

Digester volume, gal = (0.785)(78.5 ft)(78.5 ft)(19.25 ft)(7.48 gal/ft^3) = 696,532 gal

Lastly, calculate the BRT.

Equation: **BRT, days** $= \dfrac{\text{Digester working volume, gal}}{\text{Influent flow, gpd}}$

Substitute values and solve.

BRT, days $= \dfrac{696,532 \text{ gal}}{16,272 \text{ gpd}} =$ **42.8 days**

WASTE ACTIVATED SLUDGE PUMPING RATE CALCULATIONS

These calculations are used as a planning tool by the operator. The waste activated sludge (WAS) suspended solids (SS) are pumped out of the secondary clarifier and wasted or returned to the aeration tank. It is better to pump continuously rather than intermittently and not to change the amount by more than 15 percent from one day to the next.

80. **Determine the waste activated sludge (WAS) pumping rate in gal/hr, given the following data:**

Amount of WAS to be wasted = 3,375 lb/day
Suspended solids concentrations = 3,220 mg/L

First, use the "pounds" equation to solve for the number of mgd.

Equation: **Number of lb/day WAS = (SS, mg/L)(Number of mgd)(8.34 lb/gal)**

Rearrange the equation to solve for mg/L.

$$\text{Number of mgd} = \frac{\text{Number of lb/day WAS}}{(\text{Number of mg/L WAS})(8.34\ \text{lb/gal})}$$

Substitute values and solve.

$$\text{Number of mgd} = \frac{(3,375\ \text{lb/day})}{(3,220\ \text{mg/L WAS})(8.34\ \text{lb/gal})} = 0.12568\ \text{mgd}$$

Now, convert mgd to gal/hr.

$$\text{Number of gal/hr} = \frac{(0.12568\ \text{mgd})(1,000,000/\text{mil})}{24\ \text{hr/day}} = 5,236.49\ \text{gal/hr, round to } \textbf{5,240 gal/hr}$$

81. **Calculate the waste activated sludge (WAS) pumping rate in gal/hr, if 3,425 lb/day are to be wasted and the WAS suspended solids concentrations are 3,770 mg/L.**

First, use the "pounds" equation to solve for the number of mgd.

Equation: **Number of lb/day = (WAS, mg/L)(Number of mgd)(8.34 lb/gal)**

Rearrange the equation to solve for mg/L.

$$\text{Number of mgd} = \frac{\text{Number of lb/day WAS}}{(\text{Number of mg/L WAS})(8.34\ \text{lb/gal})}$$

Substitute values and solve.

$$\text{Number of mgd} = \frac{3,425\ \text{lb/day}}{(3,770\ \text{mg/L WAS})(8.34\ \text{lb/gal})} = 0.10893\ \text{mgd}$$

Now, convert mgd to gal/hr.

$$\text{Number of gal/hr} = \frac{(0.10893\ \text{mgd})(1,000,000/\text{mil})}{24\ \text{hr/day}} = 4,538.75\ \text{gal/hr, round to } \textbf{4,540 gal/hr}$$

82. **What was the waste activated sludge (WAS) pumping rate in mgd, if the return-activated sludge (RAS) suspended solids (SS) concentration averaged that day was 6,190 mg/L and the solids removed from the system was 4,950 lb/day?**

Equation: **Solids, lb/day = (RAS, mg/L)(mgd)(8.34 lb/gal)**

Rearrange the equation to solve for mgd.

$$\text{Number of mgd} = \frac{\text{Solids, lb/day}}{(\text{RAS, mg/L})(8.34\ \text{lb/gal})}$$

$$\text{Number of mgd} = \frac{6,190\ \text{lb/day}}{(4,950\ \text{mg/L})(8.34\ \text{lb/gal})} = \textbf{0.150 mgd}$$

83. **Determine the waste activated sludge (WAS) in mgd, given the following data.**

Influent flow = 3.78 mgd
Clarifier radius = 39.9 ft
Clarifier depth = 15.8 ft
Aerator = 1.15 mil gal
Mixed liquor suspended solids (MLSS) = 2,722 mg/L
Return activated sludge (RAS) SS = 6,580 mg/L
Secondary effluent SS = 16.5 mg/L
Target solids retention time (SRT) = 10 days exactly

First, determine the volume in mil gal for the clarifier and add to the aerator volume.

Equation: $\text{Clarifier, mil gal} = \dfrac{\pi(\text{radius})^2(\text{Depth, ft})(7.48\ \text{gal/ft}^3)}{1,000,000/\text{mil}}$

$$\text{Clarifier, mil gal} = \frac{3.14\,(39.9\ \text{ft})\,(39.9\ \text{ft})\,(15.8\ \text{ft})\,(7.48\ \text{gal/ft}^3)}{1,000,000/\text{mil}} = 0.5908\ \text{mil gal}$$

Total volume = 0.5908 mil gal + 1.15 mil gal = 1.7408 mil gal

Equation: **Target SRT =**

$$\frac{(\text{MLSS mg/L})(\text{Clarifier, Aerator Volume, mil gal})(8.34\ \text{lb/gal})}{(\text{RAS SS mg/L})(x\ \text{mgd})(8.34\ \text{lb/gal}) + (\text{Effluent SS, mg/L})(\text{Flow, mgd})(8.34\ \text{lb/gal})}$$

Substitute values and solve.

$$10 \text{ days SRT} = \frac{(2,722 \text{ mg/L})(1.7408 \text{ mil gal})(8.34 \text{ lb/gal})}{(6,580 \text{ mg/L})(x \text{ mgd})(8.34 \text{ lb/gal}) + (16.5 \text{ mg/L})(3.78 \text{ mgd})(8.34 \text{ lb/gal})}$$

$$10 \text{ days SRT} = \frac{39,518.74 \text{ lb MLSS}}{(6,580 \text{ mg/L})(x \text{ mgd})(8.34 \text{ lb/gal}) + 520.1658 \text{ lb/day}}$$

Rearrange the equation so that x mgd is in the numerator.

$$(6,580 \text{ mg/L})(x \text{ mgd})(8.34 \text{ lb/gal}) + 520.1658 \text{ lb/day} = \frac{39,518.74 \text{ lb MLSS}}{10 \text{ days SRT}}$$

$$(6,580 \text{ mg/L})(x \text{ mgd})(8.34 \text{ lb/gal}) + 520.1658 \text{ lb/day} = 3,951.874 \text{ lb/day}$$

Subtract 520.1658 lb/day from both sides of the equation.

$$(6,580 \text{ mg/L})(x \text{ mgd})(8.34 \text{ lb/gal}) = 3,431.7082 \text{ lb/day}$$

$$x \text{ mgd} = \frac{3,431.7082 \text{ lb/day}}{(6,580 \text{ mg/L})(8.34 \text{ lb/gal})} = \textbf{0.0625 mgd WAS}$$

PUMPING HORSEPOWER, EFFICIENCY, AND COSTING CALCULATIONS

These types of calculations can be used for determining pump size, efficiency, and costing.

84. **Find the whp for the following system: Motor efficiency (ME) is 89.4%; Pump efficiency (PE) is 78.2%; and mhp is 398.**

Equation: **Water horsepower = (mhp)(ME)(PE)**

Water horsepower = (398 mhp)(89.4%/100% ME)(78.2%/100 PE) =

Water horsepower = 278.24 whp, round to **278 whp**

85. Find the motor horsepower (mhp) for a pump with the following parameters:

Motor efficiency (ME): 90.1%
Total head (TH): 37.5 ft
Pump efficiency (PE): 75.9%
Flow: 0.843 mgd

First, convert mgd to gpm.

Gpm = (0.843 mgd)(1,000,000/mil)(1 day/1,440 min) = 585.42 gpm

The equation for determining the mhp with the given data is different than the problem above.

Equation: $\mathbf{mhp} = \dfrac{(\mathbf{Flow, gpm})(\mathbf{TH, ft})}{(3,960)(\mathbf{ME})(\mathbf{PE})}$

Substitute values and solve.

$mhp = \dfrac{(585.42\ \text{gpm})(37.5\ \text{ft})}{(3,960)(90.1\%/100\%\ \text{ME})(75.9\%/100\%\ \text{PE})}$

mhp = **8.11 mhp**

86. If the water horsepower (whp) is 145 and the pump efficiency (PE) is 76.5%, what is the brake horsepower (bhp) for a pump?

Equation: **Brake horsepower = whp/PE**

Brake hp $= \dfrac{145\ \text{whp}}{76.5\%/100\%\ \text{PE}}$ = 189.54 bhp, round to **190 bhp**

87. What is the motor horsepower, if the bhp is 95 and the motor efficiency is 91.3%?

Equation: **Motor hp = bhp/ME**

Motor horsepower $= \dfrac{95\ \text{bhp}}{91.3\%/100\%\ \text{ME}}$ = 104 mhp, round to **100 mhp**

88. **What is the motor horsepower (mhp), if 205 horsepower (hp) is required to run a pump with a motor efficiency of 89.5% and a pump efficiency of 77.4%?** *Note:* **The 205 hp in this problem is called the water horsepower (whp). The whp is the actual energy (horsepower) available to pump water.**

Equation: **Motor horsepower** $= \dfrac{whp}{(ME)(PE)}$

$mhp = \dfrac{205 \; whp}{(89.5\%/100\% \; ME)(77.4\%/100\% \; PE)}$

$mhp = \dfrac{205 \; whp}{(0.895 \; ME)(0.774 \; PE)}$

$mhp = 295.93 \; mhp$, round to **296 mhp**

89. **Given the following data, calculate the cost of running a pump in dollars and cents per day:**

Flow = 2.08 mgd
Total differential head (TDH) = 104 ft
Motor efficiency = 91.2%
Pump efficiency = 73.8%
Cost in kilowatt hours = $0.078

First, convert mgd to gpm.

Number of mgd $= \dfrac{(2.08 \; mgd)(1,000,000/mil)}{(1,440 \; min/day)} = 1,444.4 \; gpm$

Next, determine the horsepower (hp).

Equation: **Motor hp** $= \dfrac{(Flow, \; gpm)(TDH, \; ft)}{(3,960)(Motor \; efficiency)(Pump \; efficiency)}$

Motor hp $= \dfrac{(1,444.4 \; gpm)(104 \; ft)}{(3,960)(91.2\%/100\%)(73.8\%/100\%)} = 56.36 \; hp$

Now, calculate the cost of running the pump in dollars and cents.

Equation: **Cost, $/day = (Motor hp)(24 hr/day)(0.746 kW/hp)(Cost/kW-hr)**

Cost, $/day = (56.36 hp)(24 hr/day)(0.746 kW/hp)($0.078/kW-hr) = **$78.71/day**

90. **Given the following data, calculate the cost of running a pump in dollars and cents per day:**

Flow = 3,450,000 gpd
TDH = 288 ft
Motor efficiency = 89.5%
Pump efficiency = 62.2%
Cost in kilowatt hours = $0.081

First, convert gpd to gpm.

$$\text{Number of gpm} = \frac{(3,450,000 \text{ gpd})}{(1,440 \text{ min/day})} = 2,395.8 \text{ gpm}$$

Next, determine the horsepower (hp).

Equation: $\textbf{Motor hp} = \dfrac{(\textbf{Flow, gpm})(\textbf{TDH, ft})}{(\textbf{3,960})(\textbf{Motor efficiency})(\textbf{Pump efficiency})}$

$$\text{Motor hp} = \frac{(2,395.8 \text{ gpm})(288 \text{ ft})}{(3,960)(89.5\%/100\%)(62.2\%/100\%)} = 313 \text{ hp}$$

Now, calculate the cost of running the pump in dollars.

Equation: **Cost, $/day = (Motor hp)(24 hr/day)(0.746 kW/hp)(Cost/kW-hr)**

Cost, $/day = (313 hp)(24 hr/day)(0.746 kW/hp)($0.081/kW-hr) = **$453.92/day**

91. **Given the following parameters, calculate the bhp for a lift pump.**

Pumping rate = 165 gpm
Pressure differential = 23 psi
Pump efficiency = 71%
Specific gravity (sp gr) of sludge = 1.04

First, convert the pressure differential to total differential head (TDH).

$$\text{TDH, ft} = \frac{(23\,\text{psi})\,(2.31\,\text{ft/psi})}{(1.04\,\text{sp gr})} = 51.09\,\text{ft}$$

$$\text{Equation: } \mathbf{bhp} = \frac{(\text{Pumping rate, gpm})\,(\text{TDH, ft})\,(\text{sp gr})}{(3,960)\,(\text{Percent pump efficiency}/100\%)}$$

$$\text{bhp} = \frac{(165\,\text{gpm})\,(51.09\,\text{ft})\,(1.04\,\text{sp gr})}{(3,960)\,(71\%/100\%)} = \mathbf{3.1\ hp}$$

92. **Given the following data, calculate the brake horsepower for a pump:**

Pumping rate = 375 gpm
Differential pressure = 68 psi
Pump efficiency = 79.5%
Sludge specific gravity = 1.04

$$\text{Equation: } \mathbf{bhp} = \frac{(\text{Flow, gpm})\,(\text{Differential pressure, psi})}{(1,714)\,(\text{Pump efficiency})}$$

$$\text{bhp} = \frac{(375\,\text{gpm})\,(68\,\text{psi})}{(1,714)\,(79.5\%/100\%)} = 18.7\,\text{hp, round to } \mathbf{19\ hp}$$

DOSAGE PROBLEMS

These calculations are used mainly for process control, which requires accurate determination before the chemical is actually applied to a particular process. By keeping accurate records of dosages and thus usage, operators can also plan ordering or costing.

93. **What should the chemical feeder setting be in mL/min for a polymer solution, if the desired dosage is 3.50 mg/L and the treatment plant is treating 2.19 mgd? The specific gravity of the polymer is 1.28.**

First, determine the lb/gal of the polymer solution.

Polymer, lb/gal = (1.28 sp gr)(8.34 lb/gal) = 10.675 lb/gal

Next, find the number of pounds per day of polymer required by using the "pounds" equation.

lb/day, Polymer = (Dosage, mg/L)(mgd)(8.34 lb/gal)

lb/day, Polymer = (3.50 mg/L)(2.19 mgd)(8.34 lb/gal) = 63.926 lb/day

Convert the number of lb/day to number of gal/day.

$$\text{gal/day, Polymer} = \frac{63.926 \text{ lb/day}}{10.675 \text{ lb/gal}} = 5.99 \text{ gal/day}$$

Then, convert gal/day to mL/min.

$$\text{mL/min of Polymer} = \frac{(5.99 \text{ gal/day})(3,785 \text{ mL/gal})}{1,440 \text{ min/day}} = \textbf{15.7 mL/min of Polymer}$$

94. **A wastewater treatment plant is treating 2,845,000 gpd with an 11.7% sodium hypochlorite solution. If the dosage required is 8.85 mg/L and the specific gravity of the hypochlorite is 1.03, how many gpd of sodium hypochlorite are required?**

First, convert gpd to mgd.

$$\text{Number of mgd} = \frac{2,845,000 \text{ gpd}}{1,000,000/\text{mil}} = 2.845 \text{ mgd}$$

Next, determine the lb/gal for the hypochlorite solution.

$$\text{Hypochlorite, lb/gal} = (8.34 \text{ lb/gal})(1.03 \text{ sp gr}) = 8.5902 \text{ lb/gal}$$

Next, using the "pounds equation," calculate the lb day of chlorine needed.

Equation: Chlorine, lb/day = (Dosage, mg/L)(mgd)(8.34 lb/gal)

Chlorine, lb/day = (8.85 mg/L)(2.845 mgd)(8.34 lb/gal) = 209.9866 lb/day

Since the solution is not 100%, divide the percent hypochlorite into the lb/day of chlorine needed.

$$\text{Hypochlorite, lb/day} = \frac{209.9866 \text{ lb/day}}{11.7\%/100\%} = 1,794.757 \text{ lb/day hypochlorite}$$

Lastly, determine the gpd of hypochlorite solution needed.

$$\text{Hypochlorite, gpd} = \frac{1,794.757 \text{ lb/day}}{8.5902 \text{ lb/gal}} = 208.93 \text{ gpd, round to } \textbf{209 gpd Sodium hypochlorite}$$

95. **How many lb/day of chlorine gas are required to treat 1,058,000 gpd, given the following data?**

Chlorine demand = 8.2 mg/L
Chlorine residual = 1.10 mg/L

First, determine the total chlorine dose in mg/L.

Equation: **Chlorine dose = Chlorine demand + Chlorine residual**

Chlorine dose = 8.2 mg/L + 1.10 mg/L = 9.3 mg/L

Next, convert gpd to mgd.

$$\text{Number of mgd} = \frac{1,058,000 \text{ gpd}}{1,000,000/\text{mil}} = 1.058 \text{ mgd}$$

Lastly, calculate the lb/day of chlorine required.

Equation: **Number of lb/day Cl_2 = (Cl_2, mg/L)(Number of mgd)(8.34 lb/gal)**

Substitute values and solve.

Number of lb/day Cl_2 = (9.3 mg/L)(1.058 mgd)(8.34 lb/gal)

Number of lb/day Cl_2 = 82.06 lb/day, round to **82 lb/day Cl_2**

96. **Determine the number of mgd that flows through a wastewater plant's effluent, given the following data:**

SO_2 dosage = **4.7 mg/L**
Pounds of chlorine used = 145 lb/day
Chlorine demand = 5.85 mg/L
Assume SO_2 is 3.5 mg/L higher than the chlorine residual

First, based on the assumption, determine the chlorine residual in mg/L.

Chlorine residual, mg/L = SO_2 dosage − 3.5 mg/L SO_2

Chlorine residual, mg/L = 4.7 mg/L − 3.5 mg/L = 1.2 mg/L

Chlorine dosage, mg/L = Chlorine demand, mg/L + Chlorine residual, mg/L

Chlorine dosage, mg/L = 5.85 mg/L + 1.2 mg/L = 7.05 mg/L

Equation: **Number of lb/day Cl_2 = (Dosage, mg/L)(Number of mgd)(8.34 lb/gal)**

Rearrange the equation to solve for the number of mgd the plant is treating.

$$\text{Number of mgd} = \frac{\text{Number of lb/day}}{(\text{Dosage, mg/L})(8.34 \text{ lb/gal})}$$

Substitute values and solve.

$$\text{Number of mgd} = \frac{145 \text{ lb/day}}{(7.05 \text{ mg/L})(8.34 \text{ lb/gal})} = 2.47 \text{ mgd, round to } \textbf{2.5 mgd}$$

97. **What is the number of lb/day of alum used by a wastewater plant, given the following data?**

Plant's treatment flow = 3,188,000 gpd
Alum dose = 8.65 mg/L
Alum = 48.5% aluminum sulfate

First, convert gpd to mgd.

$$\text{Number of mgd} = \frac{3,188,000 \text{ gpd}}{1,000,000/\text{mil}} = 3.188 \text{ mgd}$$

Next, calculate the number of lb/day of alum required.

Equation: $\textbf{lb/day} = \dfrac{(\text{Alum, mg/L})\,(\text{mgd})\,(8.34\,\text{lb/gal})}{\text{Percent purity}/100\%}$

$\text{Alum, lb/day} = \dfrac{(8.65\,\text{mg/L})\,(3.188\,\text{mgd})\,(8.34\,\text{lb/gal})}{48.5\%/100\%} = \textbf{474 lb/day of Alum}$

CHEMICAL FEED SOLUTION SETTINGS

As above, these calculations are used mainly for process control, which requires accurate determination before the chemical is actually applied to a particular process. Also as above, by keeping accurate records of dosages and thus usage, operators can plan ordering and costing.

98. **What should the chemical feeder be set on in mL/min, if the desired polymer dosage is 75.5 gpd?**

Equation: $\textbf{Number of mL/min} = \dfrac{(\text{Number of gallons used})\,(3,785\,\text{mL/gal})}{1,440\,\text{min/day}}$

Substitute values and solve.

$\text{Polymer, mL/min} = \dfrac{(75.5\,\text{gpd})\,(3,785\,\text{mL/gal})}{1,440\,\text{min/day}} = 198.45 \text{ mL/min, round to } \textbf{198 mL/min}$

99. What should the chemical feed pump be set on in gpd, given the following data?

Plant's treatment flow = 2,073,000 gpd
Alum dose = 17.25 mg/L
Alum = 5.38 lb/gal
(11.093 lb/gal if water is included; assume 48.5% purity of aluminum sulfate, thus
(11.093)(48.5%/100%) = 5.38 lb/gal)

First, convert gpd to mgd.

$$\text{Number of mgd} = \frac{2,073,000 \text{ gpd}}{1,000,000/\text{mil}} = 2.073 \text{ mgd}$$

Next, calculate the number of lb/day of alum required.

Equation: **lb/day = (mg/L)(mgd)(8.34 lb/gal)**

Alum, lb/day = (17.25 mg/L)(2.073 mgd)(8.34 lb/gal) = 298.23 lb/day of dry alum

Now, calculate the amount of liquid alum by dividing the amount of dry alum by 5.38 lb/gal.

$$\text{Alum, gpd} = \frac{298.23 \text{ lb/day}}{5.38 \text{ lb/gal}} = \textbf{55.4 gpd of liquid alum solution}$$

DRY CHEMICAL FEED SETTINGS

As with liquid dosing, accuracy in dosing dry chemicals is important, too. The more accurate the dosage calculation is, the more probability there will be for an operator to control a treatment process and the better the records for future referral.

100. **What is the feed rate of a dry chemical in lb/day, if a sample collection bowl collected 778.24 g in 10.0 minutes?**

Know: 454 grams = 1 pound

First, determine the number of g/min.

Number of grams = 778.24 g/10.0 min = 77.824 g/min

Equation: **Chemical, lb/day** $= \dfrac{(\textbf{Number of g/min})(1,440\ \textbf{min/day})}{454\ \textbf{g/lb}}$

Substitute values and solve.

Chemical, lb/day $= \dfrac{(77.824\ \text{g/min})(1,440\ \text{min/day})}{454\ \text{g/lb}} = 246.84$ lb/day, round to **247 lb/day**

101. **What must have been the setting of a dry chemical feeder in grams/min, if the number of lb/day used was 319.6?**

Equation: **Chemical, lb/day** $= \dfrac{(\textbf{Number of g/min})(1,440\ \textbf{min/day})}{454\ \textbf{g/lb}}$

Rearrange to solve for the feeder setting in g/min.

Number of g/min $= \dfrac{(\text{Chemical, lb/day})(454\ \text{g/lb})}{1,440\ \text{min/day}}$

Substitute values and solve.

Number of g/min $= \dfrac{(319.6\ \text{lb/day})(454\ \text{g/lb})}{1,440\ \text{min/day}} = 100.76$ g/min, round to **101 g/min**

102. Determine the feed rate of dry alum in lb/day, if the drawdown in exactly 5 minutes was 89.4 grams (g) and the flow was 1,679,000 gpd.

First, determine the number of grams used per minute.

Alum, g = 89.4 g/5 min = 17.88 g/min

Equation: **Alum, lb/day** $= \dfrac{(\text{Number of g/min})(1{,}440 \text{ min/day})}{454 \text{ g/lb}}$

Substitute values and solve.

Alum, lb/day $= \dfrac{(17.88 \text{ g/min})(1{,}440 \text{ min/day})}{454 \text{ g/lb}}$ = 56.71 lb/day, round to **56.7 lb/day Alum**

103. **Given the following data, calculate the flocculant feed rate for a belt press in lb/hr and the flocculant dosage in lb/ton:**

Flocculant concentration $= 1.5\%$
Flocculant feed rate $= 1.15$ gpm
Solids loading on belt filter $= 2,750$ lb/hr

Know: $1\% = 10,000$ mg/L, therefore $1.5\% = 15,000$ mg/L

Next, find the number of mgd of flocculant used.

$$\text{Flocculant, mgd} = \frac{(1.15 \text{ gpm})(1,440 \text{ min/day})}{1,000,000/\text{mil}} = 0.001656 \text{ mgd of flocculant}$$

Next, determine the flocculant feed rate in lb/hr.

Equation: $\textbf{Flocculant feed, lb/hr} = \dfrac{(\text{Flocculant, mg/L})(\text{mgd flocculant})(8.34 \text{ lb/gal})}{24 \text{ hr/day}}$

Substitute values and solve.

$$\text{Flocculant feed, lb/hr} = \frac{(15,000 \text{ mg/L})(0.001656 \text{ mgd})(8.34 \text{ lb/gal})}{24 \text{ hr/day}}$$

Flocculant feed, lb/hr $= 8.6319$ lb/hr, round to **8.6 lb/hr**

Next, determine flocculant dosage. Start with converting the solids loading from lb/hr to tons/hr.

$$\text{Solids loading, tons/hr} = \frac{2,750 \text{ lb/hr}}{2,000 \text{ lb/ton}} = 1.375 \text{ tons/hr}$$

Lastly, calculate the flocculant dosage in lb/ton.

$$\text{Flocculant dosage, lb/ton} = \frac{8.6319 \text{ lb/hr}}{1.375 \text{ tons/hr}} = 6.2777 \text{ lb/ton, round to } \textbf{6.3 lb/ton}$$

104. What is the feed rate of a magnesium hydroxide in g/min, if the feed rate is 45.3 lb/day?

Equation: **Chemical, lb/day** $= \dfrac{(\text{Number of g/min})(1{,}440 \text{ min/day})}{454 \text{ g/lb}}$

Rearrange the equation to solve for g/min.

Number of g/min $= \dfrac{(\text{Chemical, lb/day})(454 \text{ g/lb})}{1{,}440 \text{ min/day}}$

Substitute values and solve.

Number of g/min $= \dfrac{(45.3 \text{ lb/day})(454 \text{ g/lb})}{1{,}440 \text{ min/day}} = 14.28$ g/min, round to **14.3 g/min**

105. How many pounds of a dry polymer (80.5% active) are required to make exactly 250 gallons of a solution that is exactly 10%, and what feed rate will be required in mL/min for a dosage of 4.75 mg/L, if the plant is treating 1.88 mgd?

First, convert 250 gallons to be mixed to mil gal.

Number of mil gal $= \dfrac{250 \text{ gallons}}{1{,}000{,}000/\text{mil}} = 0.000250$ mil gal

Know: 1% = 10,000 mg/L, therefore 10% = 100,000 mg/L

Next, calculate the number of lb of dry polymer required.

Equation: **Dry polymer, lb** $= \dfrac{(\text{Dose, mg/L})(\text{mil gal})(8.34 \text{ lb/gal})}{\text{Percent purity}}$

Substitute values and solve.

Dry polymer, lb $= \dfrac{(100{,}000 \text{ mg/L})(0.000250 \text{ mil gal})(8.34 \text{ lb/gal})}{80.5\%/100\%}$

Dry polymer, lb $=$ **259.0 lb**

Next, determine the number of pounds of polymer per gallon in this solution.

$$\text{Dry polymer, lb/gal} = \frac{259.0 \text{ lb}}{250 \text{ gal}} = 1.036 \text{ lb/gal}$$

Now, calculate the mL/min required for a dosage of 4.75 mg/L

Equation: **Dosage desired, mg/L** $= \dfrac{(\text{mL/min})(1{,}440 \text{ min/day})(\text{lb/gal})}{(3{,}785 \text{ mL/gal})(8.34 \text{ lb/gal})(\text{mgd})}$

Rearrange the formula to solve for mL/min.

$$\text{Dry polymer feed, mL/min} = \frac{(\text{Dosage, mg/L})(3{,}785 \text{ mL/gal})(8.34 \text{ lb/gal})(\text{mgd})}{(1{,}440 \text{ min/day})(\text{lb/gal})}$$

Substitute values and solve.

$$\text{Dry polymer feed, mL/min} = \frac{(4.75 \text{ mg/L})(3{,}785 \text{ mL/gal})(8.34 \text{ lb/gal})(1.88 \text{ mgd})}{(1{,}440 \text{ min/day})(1.036 \text{ lb/gal})}$$

Dry polymer feed = 188.96 mL/min, round to **189 mL/min**

SLUDGE PRODUCTION CALCULATIONS

Sludge production calculations are important for costing and disposal purposes. Plants that use processes like digestion or heat treatment produce less sludge compared to plants that use chemical addition to treat wastes because more of the sludge is destroyed.

106. Given the following data, determine the amount of dry solids produced in lb/day:

Flow = 1,742,000 gpd
BOD_5 = 215 mg/L
Influent suspended solids = 316 mg/L
Primary effluent suspended solids = 91 mg/L
Specific gravity (sp gr) = 1.05

First, determine the number of mg/L of suspended solids (SS) removed.

SS removed, mg/L = 316 mg/L, influent − 91 mg/L effluent = 225 mg/L, SS removed

Next, convert gpd to mgd.

$$\text{Number of mgd} = \frac{1{,}742{,}000 \text{ gpd}}{1{,}000{,}000/\text{mil}} = 1.742 \text{ mgd}$$

Equation: **SS removed, lb/day = (SS removed, mg/L)(Number of mgd)(Sludge, lb/gal)**

Equation: **SS removed, lb/day = (SS removed, mg/L)(Number of mgd)(8.34 lb/gal)(SS sp gr)**

Use the latter equation.

SS removed, lb/day = (225 mg/L, SS)(1.742 mgd)(8.34 lb/gal)(1.05 sp gr)

SS removed, lb/day = 3,432.31 lb/day, round to **3,400 lb/day**

107. **What is the amount of dry solids produced in lb/day, if a wastewater treatment plant has an influent flow of 4,282,000 gpd, has primary influent suspended solids of 217 mg/L, and the secondary suspended solids are 89 mg/L?** *Note:* **The suspended solids weigh 9.02 lb/gal.**

First, determine the number of mg/L of suspended solids (SS) removed.

SS removed, mg/L = 217 mg/L, influent − 89 mg/L effluent = 128 mg/L, SS removed

Next, convert gpd to mgd.

$$\text{Number of mgd} = \frac{4{,}282{,}000 \text{ gpd}}{1{,}000{,}000/\text{mil}} = 4.282 \text{ mgd}$$

SS removed, lb/day = (SS removed, mg/L)(Number of mgd)(Sludge, lb/gal)

However, in this problem the suspended solids weight is given, so use 9.02 lb/gal.

SS removed, lb/day = (128 mg/L, SS)(4.282 mgd)(9.02 lb/gal)

SS removed, lb/day = 4,943.83 lb/day, round to **4,900 lb/day**

108. **Given the following data, determine the amount of flow the wastewater plant is treating in gpm.**

Primary effluent suspended solids (SS) = 107 mg/L
Primary effluent suspended solids removed = 3,285 lb/day
SS specific gravity (sp gr) = 1.05

Equation: **SS removed, lb/day = (SS removed, mg/L)(Number of mgd)(8.34 lb/gal)(SS sp gr)**

Rearrange to solve for SS removed in mg/L.

$$\text{Number of mgd} = \frac{\text{SS removed, lb/day}}{(\text{SS removed, mg/L})(8.34 \text{ lb/gal})(\text{SS sp gr})}$$

$$\text{Number of mgd} = \frac{3,285 \text{ lb/day removed}}{(107 \text{ mg/L, SS})(8.34 \text{ lb/gal})(1.05 \text{ sp gr})} = 3.506 \text{ mgd}$$

Next, convert mgd to gpm.

$$\text{Number of gpm} = \frac{(3.506 \text{ mgd})(1,000,000/\text{mil})}{1,440 \text{ min/day}} = 2,434.72 \text{ gpm, round to } \mathbf{2,430 \text{ gpm}}$$

SLUDGE AGE (GOULD) CALCULATIONS

Operators need to understand sludge age calculations because it will help them maintain an appropriate amount of activated sludge in an aeration tank. The age of the sludge refers to the average solids retention time (usually in days) that the solids remain in the aeration tank. The sludge age is controlled by the sludge wasting rate, which affects the sludge yield in the system. This calculation is similar to detention time. See figure E-8 in Appendix E for one type of wastewater plant using an oxidation ditch.

109. **Given the following data, determine the sludge age in days for an oxidation ditch wastewater treatment plant:**

Mixed liquor suspended solids (MLSS) = 2,800 mg/L
Ditch volume = 0.54 mil gal
Solids added = 575 lb/day

First, calculate the amount of solids under aeration.

Know: **Solids under aeration, lb = (Ditch volume, mil gal)(MLSS, mg/L)(8.34 lb/gal)**

Solids under aeration, lb = (0.54 mil gal)(2,800 mg/L)(8.34 lb/gal) = 12,610 lb

Next, determine the sludge age in days.

Know: **Sludge age, days** $= \dfrac{\text{Solids under aeration, lb}}{\text{Solids added, lb/day}}$

Sludge age, days $= \dfrac{12,610\ \text{lb}}{575\ \text{lb/day}} = 21.93$ days, round to **22 days**

110. **Given the following data, determine the sludge age in days for an oxidation ditch at a wastewater treatment plant:**

MLSS = 3,500 mg/L
Ditch volume = 0.63 mil gal
Flow = 1.65 mgd
SS, mg/L = 125 mg/L

Equation: **Sludge age, days** $= \dfrac{(\text{MLSS, mg/L})(\text{Volume, mil gal})(8.34\ \text{lb/gal})}{(\text{SS, mg/L})(\text{Flow, mgd})(8.34\ \text{lb/gal})}$

Sludge age, days $= \dfrac{(3,500\ \text{mg/L})(0.63\ \text{mil gal})(8.34\ \text{lb/gal})}{(125\ \text{mg/L})(1.65\ \text{mgd})(8.34\ \text{lb/gal})} = 10.69$ days, round to **11 days**

ORGANIC LOADING RATE CALCULATIONS

Organic loading rate calculations tell the operator the amount of food entering the plant. These calculations are used for wastewater treatment ponds, rotating biological contactors, or trickling filters. See figures in Appendix E for the types of wastewater plants using these processes.

111. **What is the organic loading rate for a trickling filter that is 121 ft in diameter and 5.1 ft deep in lb BOD$_5$/d/1,000 ft^3, if the primary effluent flow is 1,245 gpm and the BOD$_5$ is 153 mg/L?**

First, determine the volume of the tricking filter in ft^3.

Volume, ft^3 = (0.785)(Diameter)^2(Depth, ft)

Volume, ft^3 = (0.785)(121 ft)(121 ft)(5.1 ft) = 58,615.24 ft^3

Next, factor out 1,000 ft^3 from the volume = (58.615)(1,000 ft^3)

Next, convert gpm to mgd.

$$\text{Number of mgd} = \frac{(1,245 \text{ gpm})(1,440 \text{ min}/\text{day})}{1,000,000/\text{mil}} = 1.7928 \text{ mgd}$$

Next, determine the pounds of BOD$_5$/d/1,000 ft^3 using a modified version of the "pounds" equation.

$$\textbf{Organic loading rate, lb BOD}_5\textbf{/d/1,000 ft}^3 = \frac{(\textbf{BOD}_5, \textbf{mg/L})(\textbf{Flow, mgd})(8.34 \text{ lb/gal})}{\text{Volume of trickling filter, ft}^3/1,000 \text{ ft}^3}$$

$$\text{Organic loading rate, lb BOD}_5\text{/d/1,000 ft}^3 = \frac{(153 \text{ mg/L BOD}_5)(1.7928 \text{ mgd})(8.34 \text{ lb/gal})}{(58.615)(1,000 \text{ ft}^3)}$$

Organic loading rate, lb BOD$_5$/d/1,000 ft^3 = 39.03, round to **39 lb BOD$_5$/d/1,000 ft^3**

112. **What is the organic loading rate for a trickling filter in pounds biochemical oxygen demand per day per 1,000 cubic feet (lb BOD$_5$/d/1,000 ft^3), given the following data?**

Trickling filter radius = 49.9 ft
Trickling filter depth = 5.45 ft
BOD$_5$ = 161 mg/L
Primary effluent flow = 2,882,000 gpd

First, convert gallons per day to mgd.

Number of mgd = (2,882,000 gpd)/(1,000,000/mil) = 2.882 mgd

Next, calculate the volume of the trickling filter.

Know: **Volume, ft^3 = πr^2(Depth)** where r equals the radius

Volume, ft^3 = (3.14)(49.9 ft)(49.9 ft)(5.45 ft) = 42,611.54 ft^3

Next, factor out 1,000 ft^3 from the volume = (42.612)(1,000 ft^3)

$$\text{Organic loading rate, lb BOD}_5\text{/d/1,000 ft}^3 = \frac{(\text{BOD}_5, \text{mg/L})(\text{Flow, mgd})(8.34 \text{ lb/gal})}{(\text{Volume of trickling filter})(1,000 \text{ ft}^3)}$$

$$\text{Organic loading rate, lb BOD}_5\text{/d/1,000 ft}^3 = \frac{(161 \text{ mg/L BOD}_5)(2.882 \text{ mgd})(8.34 \text{ lb/gal})}{(42.612)(1,000 \text{ ft}^3)}$$

Organic loading rate, lb BOD$_5$/d/1,000 ft^3 = 90.81, round to **90.8 lb BOD$_5$/d/1,000 ft^3**

113. **Given the following data on a wastewater treatment pond, calculate the organic loading rate in lb BOD$_5$/d/acre:**

Influent flow = 326,000 gpd
Surface area of pond = 3.62 acre-ft
Influent BOD$_5$ concentration = 227 mg/L

First, convert gallons per day to mgd.

$$\text{Number of mgd} = \frac{326,000 \text{ gpd}}{1,000,000/\text{mil}} = 0.326 \text{ mgd}$$

Next, determine the pounds of BOD_5/d/acre using a modified version of the "pounds" equation.

$$\textbf{Organic loading rate, lb BOD}_5\textbf{/d/acre} = \frac{(BOD_5, \text{mg/L})(\text{Flow, mgd})(8.34 \text{ lb/gal})}{\text{Surface area of pond, acre ft}}$$

$$\text{Organic loading rate, lb BOD}_5\text{/d/acre} = \frac{(227 \text{ mg/L BOD}_5)(0.326 \text{ mgd})(8.34 \text{ lb/gal})}{3.62 \text{ acre ft}}$$

Organic loading rate, lb BOD_5/d/acre = 170.49 lb BOD_5/d/acre, round to **170 lb BOD$_5$/d/acre**

114. What is the organic loading rate for a rotating biological contactor (RBC) in lb BOD$_5$/d/1,000 ft^2, given the following data?

Surface area of RBC = 725,500 ft^2
BOD$_5$ = 135 mg/L
Flow = 3,920,000 gpd

First, convert gallons per day to mgd.

$$\text{Number of mgd} = \frac{3,920,000 \text{ gpd}}{1,000,000/\text{mil}} = 3.92 \text{ mgd}$$

Next, factor out 1,000 ft^2 from the RBC surface area = $(725.5)(1,000 \text{ ft}^2)$

$$\textbf{Organic loading rate, lb BOD}_5\textbf{/d/1,000 ft}^2 = \frac{(BOD_5, \text{mg/L})(\text{Flow, mgd})(8.34 \text{ lb/gal})}{(\text{Surface area of RBC})(1,000 \text{ ft}^2)}$$

$$\text{Organic loading rate, lb BOD}_5\text{/d/1,000 ft}^2 = \frac{(135 \text{ mg/L BOD}_5)(3.92 \text{ mgd})(8.34 \text{ lb/gal})}{(725.5)(1,000 \text{ ft}^2)}$$

Organic loading rate, lb BOD_5/d/1,000 ft^2 = **6.08 lb BOD$_5$/d/1,000 ft^2**

SUSPENDED SOLIDS LOADING CALCULATIONS

Operators use suspended solids loading for evaluating process control.

115. Given the following data, calculate the amount of suspended solids entering a trickling filter in lb/day:

Influent flow = 1,040 gpm
Suspended solids (SS) concentration = 379 mg/L

First, determine the number of mgd.

$$\text{Number of mgd} = \frac{(1{,}040 \text{ gpm})(1{,}440 \text{ min/day})}{1{,}000{,}000/\text{mil}} = 1.4976 \text{ mgd}$$

This problem uses the same equation as a dosage problem.

Equation: **Number of lb/day SS = (SS, mg/L)(Number of mgd)(8.34 lb/gal)**

Substitute values and solve.

Number of lb/day SS = (379 mg/L, SS)(1.4976 mgd)(8.34 lb/gal)

Number of lb/day SS = 4,733.7 lb/day, round to **4,730 lb/day SS**

116. What is the amount of suspended solids entering a trickling filter in lb/day, if the influent flow is 895 gpm and the amount of suspended solids (SS) is 365 mg/L?

First, convert gallons per day to mgd.

$$\text{Number of mgd} = \frac{(895 \text{ gpm})(1{,}440 \text{ min/day})}{1{,}000{,}000/\text{mil}} = 1.2888 \text{ mgd}$$

Then, calculate the lb/day of suspended solids.

Equation: **Number of lb/day SS = (SS, mg/L)(Number of mgd)(8.34 lb/gal)**

Substitute values and solve.

Number of lb/day SS = (365 mg/L, SS)(1.2888 mgd)(8.34 lb/gal)

Number of lb/day SS = 3,923.24 lb/day, round to **3,920 lb/day SS**

117. What are the suspended solids entering a trickling filter in mg/L, if the influent flow to a trickling filter is 825 gpm and the suspended solids loading is 3,115 lb/day?

First, determine the number of mgd.

$$\text{Number of mgd} = \frac{(825 \text{ gpm})(1,440 \text{ min/day})}{1,000,000/\text{mil}} = 1.188 \text{ mgd}$$

Equation: **Number of lb/day SS = (SS, mg/L)(Number of mgd)(8.34 lb/gal)**

Rearrange the equation to solve for mg/L, SS.

$$\text{Number of mg/L, SS} = \frac{\text{Number of lb/day SS}}{(\text{Number of mgd})(8.34 \text{ lb/gal})}$$

Substitute values and solve.

$$\text{Number of mg/L, SS} = \frac{3,115 \text{ lb/day}}{(1.188 \text{ mgd})(8.34 \text{ lb/gal})} = 314.4 \text{ mg/L, round to } \textbf{314 mg/L, SS}$$

118. The flow through a trickling filter is 1,920 gpm. If the influent suspended solids (SS) are 233 mg/L and the effluent suspended solids are 61 mg/L, how many lb/day of suspended solids are removed?

First, determine the amount in mg/L of suspended solids removed by the trickling filter.

SS removed, mg/L = 233 mg/L − 61 mg/L = 172 mg/L

Next, convert gpm to mgd.

$$\text{Number of mgd} = \frac{(1,920 \text{ gpm})(1,440 \text{ min/day})}{1,000,000/\text{mil}} = 2.7648 \text{ mgd}$$

Now, calculate the lb/day of suspended solids removed.

Equation: **SS removed, lb/day = (SS, mg/L)(mgd)(8.34 lb/gal)**

SS removed, lb/day = (172 mg/L, SS)(2.7648 mgd)(8.34 lb/gal)

SS removed, lb/day = 3,966 lb/day, round to **3,970 lb/day SS removed**

119. **If the influent flow entering a trickling filter is 1,145 gpm and the suspended solids loading is 3,075 lb/day, calculate the amount of suspended solids entering a trickling filter in mg/L.**

First, convert gallons per day to mgd.

$$\text{Number of mgd} = \frac{(1,145 \text{ gpm})(1,440 \text{ min/day})}{1,000,000/\text{mil}} = 1.6488 \text{ mgd}$$

Equation: **Number of lb/day SS = (SS, mg/L)(Number of mgd)(8.34 lb/gal)**

Rearrange the equation to solve for mg/L, SS.

$$\text{Number of mg/L, SS} = \frac{\text{Number of lb/day}}{(\text{Number of mgd})(8.34 \text{ lb/gal})}$$

Substitute values and solve.

$$\text{Number of mg/L, SS} = \frac{3,075 \text{ lb/day}}{(1.6488 \text{ mgd})(8.34 \text{ lb/gal})} = 223.6 \text{ mg/L, round to } \textbf{224 mg/L, SS}$$

BIOCHEMICAL OXYGEN DEMAND LOADING CALCULATIONS

Biochemical oxygen demand (BOD_5) is the demand for oxygen made by bacteria as they decompose organic matter in wastewater or in the natural environment. This calculation sometimes is helpful in evaluating treatment pond processes. The BOD_5 is a five-day test. See Figures E-1 and E-6 in Appendix E for two types of wastewater plants using a trickling filter.

120. **Given the following data, calculate the biochemical oxygen demand (BOD_5) loading on a trickling filter in lb/day:**

Influent flow = 2,775,000 gallons per day (gpd)
Influent BOD_5 = 334 mg/L
Effluent BOD_5 = 92 mg/L

First, convert gallons to mgd.

$$\text{Number of mgd} = \frac{2,775,000 \text{ gpd}}{1,000,000/\text{mil}} = 2.775 \text{ mgd}$$

Then, calculate the BOD_5 loading in lb/day.

Equation: **Number of lb/day BOD_5 = (BOD_5, mg/L)(Number of mgd)(8.34 lb/gal)**

Substitute values and solve.

Number of lb/day BOD_5 = (334 mg/L BOD_5)(2.775 mgd)(8.34 lb/gal)

Number of lb/day BOD_5 = 7,729.93 lb/day, round to **7,730 lb/day BOD_5**

121. **What is the influent flow to a trickling filter in gpm, if the BOD$_5$ loading is 2,775 lb/day and the BOD$_5$ removed is 216 mg/L?**

Equation: **Number of lb/day BOD$_5$ = (BOD$_5$, mg/L)(Number of mgd)(8.34 lb/gal)**

Rearrange the equation to solve for mgd.

$$\text{\textbf{Number of mgd}} = \frac{\text{Number of lb/day BOD}_5}{(\text{BOD}_5, \text{mg/L})(8.34 \text{ lb/gal})}$$

Substitute values and solve.

$$\text{Number of mgd flow} = \frac{2,775 \text{ lb/day}}{(216 \text{ mg/L})(8.34 \text{ lb/gal})} = 1.54 \text{ mgd}$$

Lastly, convert mgd to gpm.

$$\text{Number of gpm} = \frac{(1.54 \text{ mgd})(1,000,000/\text{mil})}{1,440 \text{ min/day}} = 1,069 \text{ gpm, round to } \textbf{1,070 gpm}$$

122. **What is the biochemical oxygen demand (BOD$_5$) loading on a trickling filter in lb/day, if the influent flow is 1,125 gpm and the influent BOD$_5$ is 222 mg/L?**

First, determine the number of mgd.

$$\text{Number of mgd} = \frac{(1,125 \text{ gpm})(1,440 \text{ min/day})}{1,000,000/\text{mil}} = 1.62 \text{ mgd}$$

Equation: **Number of lb/day BOD$_5$ = (BOD$_5$, mg/L)(Number of mgd)(8.34 lb/gal)**

Substitute values and solve.

Number of lb/day BOD$_5$ = (222 mg/L BOD$_5$)(1.62 mgd)(8.34 lb/gal)

Number of lb/day BOD$_5$ = 2,999.398 lb/day BOD$_5$, round to **3,000 lb/day BOD$_5$**

123. Calculate the influent flow to a trickling filter in gpm, if the BOD_5 loading is 3,371 lb/day and the BOD_5 is 286 mg/L.

Equation: **Number of lb/day = (BOD_5, mg/L)(Number of mgd)(8.34 lb/gal)**

Rearrange the equation to solve for mgd.

$$(\textbf{Number of mgd}) = \frac{\textbf{Number of lb/day}}{(\textbf{BOD}_5\textbf{,mg/L})(\textbf{8.34 lb/gal})}$$

Substitute values and solve.

$$\text{Number of mgd flow} = \frac{3,371\,\text{lb/day}}{(286\,\text{mg/L})(8.34\,\text{lb/gal})} = 1.4133\,\text{mgd}$$

Lastly, convert mgd to gpm.

$$\text{Number of gpm} = \frac{(1.4133\,\text{mgd})(1,000,000/\text{mil})}{(1,440\,\text{min/day})} = 981.46\,\text{gpm, round to } \textbf{981 gpm}$$

SOLUBLE AND PARTICULATE BIOCHEMICAL OXYGEN DEMAND CALCULATIONS

Biochemical oxygen demand (BOD_5) measures the amount of organic matter that is present in water. Bacteria break down this organic matter by natural decomposition and in the process utilize oxygen. Thus, the more organic matter present in the wastewater, the more demand for oxygen by the bacteria. Operators need to know how to do BOD_5 calculations because there are strict regulations for the amount of BOD_5 that can be discharged to a natural water body from the treated plant. The K value in these problems is the portion of suspended solids in the wastewater that are organic suspended solids. Domestic water usually has about 50 to 70% of the suspended solids as organic suspended solids, which is usually written in decimal form (0.5 to 0.7).

124. What is the particulate BOD_5, given the following data?

Total BOD_5 = 204 mg/L
Soluble BOD_5 = 138 mg/L
K factor = 0.68

Equation: **Total BOD_5 = (Particulate BOD_5)(K factor) + Soluble BOD_5**

Rearrange the equation to solve for particulate BOD_5 by first subtracting soluble BOD_5 from both sides of the equation.

(Particulate BOD_5)(K factor) = Total BOD_5 − Soluble BOD_5

Then, divide both sides by the K factor.

$$\text{Particulate } BOD_5 = \frac{\text{Total } BOD_5 - \text{Soluble } BOD_5}{\text{K factor}}$$

Substitute values and solve.

$$\text{Particulate } BOD_5 = \frac{204 \text{ mg/L } BOD_5 - 138 \text{ mg/L } BOD_5}{0.68 \text{ factor}}$$

Particulate BOD_5 = 97.06 mg/L, round to **97 mg/L Particulate BOD_5**

125. What is the soluble BOD_5, if the total BOD_5 is 234 mg/L, the suspended solids (particulates) are 161 mg/L, and the K factor is 0.52?

Equation: **Total BOD_5 = (Particulate BOD_5)(K factor) + Soluble BOD_5**

Rearrange the equation to solve for soluble BOD_5.

Soluble BOD_5 = Total BOD_5 − (Particulate BOD_5)(K factor)

Substitute values and solve.

Soluble BOD_5 = 234 mg/L BOD_5 − (161 mg/L BOD_5)(0.52 K factor)

Soluble BOD_5 = 234 mg/L BOD_5 − 83.72 mg/L BOD_5

Soluble BOD_5 = 150.28 mg/L, round to **150 mg/L Soluble BOD_5**

126. How many lb/day of soluble BOD_5 enters a rotating biological contactor (RBC) each day, if the flow is 1,450 gpm, total BOD_5 is 224 mg/L, the particulate BOD_5 is 138 mg/L, and the K factor is 0.54?

Equation: **Total BOD_5 = (Particulate BOD_5)(K factor) + Soluble BOD_5**

Rearrange the equation to solve for soluble BOD_5 by subtracting (Particulate BOD_5)(K factor) from each side.

Soluble BOD_5 = Total BOD_5 − (Particulate BOD_5)(K factor)

Substitute values and solve.

Soluble BOD_5 = 224 mg/L BOD_5 − (138 mg/L BOD_5)(0.54 K factor)

Soluble BOD_5 = 224 mg/L BOD_5 − 74.52 mg/L BOD_5

Soluble BOD_5 = 149.48 mg/L Soluble BOD_5

Next, convert gpm to mgd.

$$\text{Number of mgd} = \frac{(1{,}450 \text{ gpm})(1{,}440 \text{ min/day})}{1{,}000{,}000 \text{ /mil}} = 2.088 \text{ mgd}$$

Next, determine the Soluble BOD_5.

Equation: **Soluble BOD_5, lb/day = (Soluble BOD_5, mg/L)(mgd)(8.34 lb/gal)**

Soluble BOD_5, lb/day = (149.48 mg/L)(2.088 mgd)(8.34 lb/gal)

Soluble BOD_5, lb/day = 2,603.03 lb/day, round to **2,600 lb/day Soluble BOD_5**

CHEMICAL OXYGEN DEMAND LOADING CALCULATIONS

Chemical oxygen demand is a measure of the capacity of water to consume oxygen, when organic matter and the oxidation of inorganic matter are decomposed. Because it also measures the decomposition of inorganic matter such as nitrate and ammonia, it is only an indirect measure of the organic matter in water.

127. **If the influent flow to an aeration tank is 2,541,000 gpd and the chemical oxygen demand (COD) loading is 2,083 lb/day, what is the amount of COD entering an aeration tank in mg/L?**

First, convert gpd to mgd.

$$\text{Number of mgd} = \frac{2,541,000 \text{ gpd}}{1,000,000/\text{mil}} = 2.541 \text{ mgd}$$

Next, use the following equation to calculate the COD, mg/L.

Equation: **Number of lb/day COD = (COD, mg/L)(Number of mgd)(8.34 lb/gal)**

Rearrange the equation to solve for mg/L.

$$\text{Number of mg/L COD} = \frac{\text{Number of lb/day COD}}{(\text{Number of mgd})(8.34 \text{ lb/gal})}$$

Substitute values and solve.

$$\text{Number of mg/L COD} = \frac{2,083 \text{ lb/day}}{(2.541 \text{ mgd})(8.34 \text{ lb/gal})} = \textbf{98.3 mg/L COD}$$

128. **Determine the chemical oxygen demand (COD) in lb/day that is applied to an aeration tank, if the flow is 2,026,000 gallons and the COD concentration is 137 mg/L.**

First, convert the gpd to mgd.

$$\text{Number of gpd} = \frac{2,026,000 \text{ gpd}}{1,000,000/\text{mil}} = 2.026 \text{ mgd}$$

Equation: **Number of lb/day COD = (COD, mg/L)(Number of mgd)(8.34 lb/gal)**

Substitute values and solve.

Number of lb/day = (137 mg/L)(2.026 mgd)(8.34 lb/gal)

Number of lb/day = 2,314.87 lb/day, round to **2,310 lb/day**

129. If the COD loading to an aeration tank is 2,347 lb/day and the COD is 117 mg/L, what is the influent flow to the aeration tank in mgd?

Equation: **Number of lb/day COD = (COD, mg/L)(Number of mgd)(8.34 lb/gal)**

Rearrange the equation to solve for mgd.

$$\text{Number of mgd} = \frac{\text{Number of lb/day COD}}{(\text{Number of mg/L COD})(8.34\,\text{lb/gal})}$$

Substitute values and solve.

$$\text{Number of mgd flow} = \frac{2,347\,\text{lb/day}}{(117\,\text{mg/L})(8.34\,\text{lb/gal})} = \textbf{2.41 mgd}$$

130. What is the influent flow to an aeration tank in gpm, if the COD loading is 2,729 lb/day and the COD is 179 mg/L?

First, determine the number of mgd.

Equation: **Number of lb/day COD = (COD, mg/L)(Number of mgd)(8.34 lb/gal)**

Rearrange the equation to solve for mg/L.

$$\text{Number of mgd} = \frac{\text{Number of lb/day COD}}{(\text{Number of mg/L COD})(8.34\,\text{lb/gal})}$$

Substitute values and solve.

$$\text{Number of mgd flow} = \frac{2,729\,\text{lb/day}}{(179\,\text{mg/L})(8.34\,\text{lb/gal})} = 1.828\,\text{mgd}$$

Then, convert mgd to gpm.

$$\text{Number of gpm} = \frac{(1.828\,\text{mgd})(1,000,000/\text{mil})}{1,440\,\text{min/day}} = 1,269.44\,\text{gpm, round to }\textbf{1,270 gpm}$$

HYDRAULIC DIGESTION TIME CALCULATIONS

Hydraulic digestion time tells the operator how long the process will take to complete, and is thus used for planning purposes. See Figures E-2, E-4, E-5, and E-6 in Appendix E for four types of wastewater plants using a digester.

131. **What is the hydraulic digestion time for a digester that is 30.1 ft in radius, has a level of 12.8 ft, and has a sludge flow of 9,980 gallons per day (gpd)?**

First, determine the volume of the digester in gallons.

Volume, gal $= \pi r^2(\text{Depth, ft})(7.48 \text{ gal/ft}^3)$

Volume, gal $= (3.14)(30.1 \text{ ft})(30.1 \text{ ft})(12.8 \text{ ft})(7.48 \text{ gal/ft}^3) = 272{,}379.37 \text{ gal}$

Next, calculate the digestion time in days.

Equation: **Digestion time, days** $= \dfrac{\text{Number of gallons}}{\text{Influent sludge flow, gal/day}}$

Substitute values and solve.

Digestion time, days $= \dfrac{272{,}379.37 \text{ gal}}{9{,}980 \text{ gal/day}} = 27.29$ days, round to **27.3 days**

132. **What is the hydraulic digestion time for a 45.2-ft diameter digester with a level of 13.6 ft and sludge flow of 10,245 gallons per day (gpd)?**

Equation: **Digestion time, days** $= \dfrac{(0.785)(\text{Diameter})^2(\text{Depth, ft})(7.48 \text{ gal/ft}^3)}{\text{Influent sludge flow, gal/day}}$

Substitute values and solve.

Digestion time, days $= \dfrac{(0.785)(45.2 \text{ ft})(45.2 \text{ ft})(13.6 \text{ ft})(7.48 \text{ gal/ft}^3)}{10{,}245 \text{ gpd}}$

Digestion time, days $=$ **15.9 days**

DIGESTER LOADING RATE CALCULATIONS

Digester loading rate calculations tell the operator how much volatile solids are stabilized per cubic foot of digester space. It is used for evaluating process control.

133. **Given the following data, calculate the digester loading in lb volatile solids (VS)/ day/1,000 ft³.**

Digester diameter = 50.0 ft
Sludge level = 11.9 ft
Influent sludge flow = 13,115 gpd
Percent sludge solids = 4.72%
Percent volatile solids = 70.33%
Specific gravity of sludge = 1.04

Equation: **Digester loading, lb VS/day/1,000 ft³ =**

$$\frac{(\text{Flow, gpd})(8.34\ \text{lb/gal})(\text{sp gr})(\text{Percent sludge})(\text{Percent volatile solids})}{(0.785)(\text{Diameter})^2(\text{Sludge level})}$$

Substitute values and solve.

Digester loading, lb VS/day/1,000 ft³ =

$$\frac{(13,115\ \text{gpd})(8.34\ \text{lb/gal})(1.04\ \text{sp gr})(4.72\%/100\%)(70.33\%/100\%)}{(0.785)(50.0\ \text{ft})(50.0\ \text{ft})(11.9\ \text{ft})}$$

Digester loading, lb VS/day/1,000 ft³ = $\dfrac{3,776.16\ \text{lb VS/day}}{23,353.75\ \text{ft}^3}$

Factor out 1,000 ft³ from the denominator and do not divide by 1,000 ft³, as it will become part of the units and not part of the calculation.

Digester loading, lb VS/day/1,000 ft³ = $\dfrac{3,776.16\ \text{lb VS/day}}{(23.35375)(1,000\ \text{ft}^3)}$

Digester loading, lb VS/day/1,000 ft³ = $\dfrac{161.69\ \text{lb VS/day}}{1,000\ \text{ft}^3}$, round to **162 lb VS/day/1,000 ft³**

134. If the sludge flow into a digester is 75,500 gpd, the digester is 49.9 ft in diameter, the sludge level is 18.6 ft, the sludge is 5.23% solids with a specific gravity of 1.06, and the sludge has 69.8% volatile solids, what is the loading on a digester in lb volatile solids (VS)/day/ft³?

Equation: **Digester loading, lb VS/day/ft³** =

$$\frac{(\text{Flow, gpd})(8.34\ \text{lb/gal})(\text{sp gr})(\text{Percent sludge})(\text{Percent volatile solids})}{(0.785)(\text{Diameter})^2(\text{Sludge level})}$$

Substitute values and solve.

$$\text{Digester loading, lb VS/day/ft}^3 = \frac{(75,500\ \text{gpd})(8.34\ \text{lb/gal})(1.06)(5.23\%/100\%)(69.8\%/100\%)}{(0.785)(49.9\ \text{ft})(49.9\ \text{ft})(18.6\ \text{ft})}$$

$$\text{Digester loading, lb VS/day/ft}^3 = \frac{24,365.54\ \text{lb VS/day}}{36,356.64\ \text{ft}^3}$$

$$\text{Digester loading, lb VS/day/ft}^3 = \textbf{0.670 lb VS/day/ft}^3$$

135. Given the following data, calculate the digester loading in lb VS/day/1,000 ft³:

Digester diameter = 39.85 ft
Sludge level = 15.4 ft
Influent sludge flow = 21,460 gpd
Percent sludge solids = 4.78%
Percent volatile solids = 68.7%
Specific gravity of sludge = 1.06

Equation: **Digester loading, lb VS/day/1,000 ft³** =

$$\frac{(\text{Flow, gpd})(8.34\ \text{lb/gal})(\text{sp gr})(\text{Percent sludge})(\text{Percent volatile solids})}{(0.785)(\text{Diameter})^2(\text{Sludge level})}$$

Substitute values and solve.

Digester loading, lb VS/day/1,000 ft^3 =

$$\frac{(21,460 \text{ gpd})\,(8.34 \text{ lb/gal})\,(1.06 \text{ sp gr})\,(4.78\%/100\%)\,(68.7\%/100\%)}{(0.785)\,(39.85 \text{ ft})\,(39.85 \text{ ft})\,(15.4 \text{ ft})}$$

Digester loading, lb VS/day $= \dfrac{6,229.97 \text{ lb VS/day}}{19,197.60 \text{ ft}^3}$

Factor out 1,000 ft^3 from the denominator and do not divide by 1,000 ft^3, as it will become part of the units and not part of the calculation.

Digester loading, lb VS/day $= \dfrac{6,229.97 \text{ lb VS/day}}{(19.1976)\,(1,000 \text{ ft}^3)}$

Digester loading, lb VS/day/1,000 ft^3 $= \dfrac{324.52 \text{ lb VS/day}}{1,000 \text{ ft}^3}$, round to **325 lb VS/day/1,000 ft^3**

MEAN CELL RESIDENCE TIME (SOLIDS RETENTION TIME) CALCULATIONS

The mean cell residence time (MCRT) is the average time the activated-sludge solids are in an activated biosolids system. The MCRT is an important design and operating parameter for operators to use in the activated-sludge process and is normally expressed in days. This calculation is used for operational process control. See Figure E-2 in Appendix E for one type of wastewater plant using the activated sludge process.

136. **Calculate the mean cell residence time (MCRT) in days, given the following data:**

Flow = 3.36 mgd
Aeration tank volume = 525,000 gallons
Clarifier tank volume = 395,000 gallons
Mixed liquor suspended solids (MLSS) = 2,765 mg/L
Waste rate = 27,100 gpd
Waste activated sludge (WAS) = 6,940 mg/L
Effluent TSS = 17.0 mg/L

First, convert the volumes for the tanks to mil gal.

$$\text{Aeration tank, mgd} = \frac{525,000 \text{ gal}}{1,000,000/\text{mil}} = 0.525 \text{ mil gal}$$

$$\text{Clarifier tank, mgd} = \frac{395,000 \text{ gal}}{1,000,000/\text{mil}} = 0.395 \text{ mil gal}$$

Next, convert the waste rate from gpd to mgd.

$$\text{Waste rate, mgd} = \frac{27,100 \text{ gal}}{1,000,000/\text{mil}} = 0.0271 \text{ mgd}$$

Next, calculate the MCRT.

Equation: **MCRT, days =**

$$\frac{(\text{MLSS, mg/L})(\text{Aeration tank mil gal} + \text{Clarifier tank mil gal})(8.34 \text{ lb/gal})}{(\text{WAS, mg/L})(\text{Waste rate, mgd})(8.34 \text{ lb/gal}) + (\text{TSS, mg/L})(\text{Flow, mgd})(8.34 \text{ lb/gal})}$$

$$\text{MCRT, days} = \frac{(2,765 \text{ mg/L MLSS})(0.525 \text{ mil gal} + 0.395 \text{ mil gal})(8.34 \text{ lb/gal})}{(6,940 \text{ mg/L})(0.0271 \text{ mgd})(8.34 \text{ lb/gal}) + (17.0 \text{ mg/L TSS})(3.36 \text{ mgd})(8.34 \text{ lb/gal})}$$

$$\text{MCRT, days} = \frac{(2,765 \text{ mg/L MLSS})(0.92 \text{ mil gal})(8.34 \text{ lb/gal})}{1,568.54 \text{ lb/day} + 476.38 \text{ lb/day}}$$

$$\text{MCRT, days} = \frac{21,215.292 \text{ lb}}{2,044.92 \text{ lb/day}} = \textbf{10.4 days}$$

137. Given the following data, calculate the mean cell residence time (MCRT):

Flow = 3.04 mgd
Aeration tank volume = 393,000 gallons
MLSS = 2,625 mg/L
Clarifier tank volume = 315,000 gallons
Waste rate = 29,400 gpd
Waste activated sludge (WAS) = 7,030 mg/L
Effluent TSS = 16 mg/L

First, convert the volumes for the tanks to mil gal.

$$\text{Aeration tank, mgd} = \frac{393,000 \text{ gal}}{1,000,000/\text{mil}} = 0.393 \text{ mil gal}$$

$$\text{Clarifier tank, mgd} = \frac{315,000 \text{ gal}}{1,000,000/\text{mil}} = 0.315 \text{ mil gal}$$

Next, convert the waste rate from gpd to mgd.

$$\text{Waste rate, mgd} = \frac{29,400 \text{ gal}}{1,000,000/\text{mil}} = 0.0294 \text{ mgd}$$

Next, calculate the MCRT.

Equation: **MCRT, days** =

$$\frac{(\text{MLSS, mg/L})(\text{Aeration tank mil gal} + \text{Clarifier tank mil gal})(8.34 \text{ lb/gal})}{(\text{WAS, mg/L})(\text{Waste rate, mgd})(8.34 \text{ lb/gal}) + (\text{TSS, mg/L})(\text{Flow, mgd})(8.34 \text{ lb/gal})}$$

$$\text{MCRT, days} = \frac{(2,625 \text{ mg/L MLSS})(0.393 \text{ mil gal} + 0.315 \text{ mil gal})(8.34 \text{ lb/gal})}{(7,030 \text{ mg/L})(0.0294 \text{ mgd})(8.34 \text{ lb/gal}) + (16 \text{ mg/L TSS})(3.04 \text{ mgd})(8.34 \text{ lb/gal})}$$

$$\text{MCRT, days} = \frac{(2,625 \text{ mg/L MLSS})(0.708 \text{ mil gal})(8.34 \text{ lb/gal})}{1,723.728 \text{ lb/day} + 405.658 \text{ lb/day}}$$

$$\text{MCRT, days} = \frac{15,499.89 \text{ lb}}{2,129.386 \text{ lb/day}} = \textbf{7.3 days}$$

138. **If the MCRT desired was 7.5 days, what would the waste rate be for the following system in lb/day?**

Flow = 2.85 mgd
Aeration tank (AT) volume = 0.437 mil gal
MLSS = 2,415 mg/L
Clarifier tank (CT) volume = 0.292 mil gal
Effluent TSS = 15 mg/L

Equation: **Waste rate, lb/day** =

$$\frac{\text{MLSS, mg/L}\,[\text{AT, mil gal} + \text{CT, mil gal}]\,(8.34\,\text{lb/gal})}{\text{Desired MCRT}} - (\text{TSS, mg/L})(\text{Flow, mgd})(8.34\,\text{lb/gal})$$

Waste rate, lb/day =

$$\frac{(2{,}415\,\text{mg/L})\,[0.437\,\text{mil gal} + 0.292\,\text{mil gal}]\,(8.34\,\text{lb/gal})}{7.5\,\text{days, Desired MCRT}} - (15\,\text{mg/L TSS})(2.85\,\text{mgd})(8.34\,\text{lb/gal})$$

Waste rate, lb/day = $\dfrac{(2{,}415\,\text{mg/L})\,(0.729\,\text{mil gal})\,(8.34\,\text{lb/gal})}{7.5\,\text{days, Desired MCRT}} - 356.535\,\text{lb/day}$

Waste rate, lb/day = 1,957.715 lb/day − 356.535 lb/day

Waste rate, lb/day = 1,601.18 lb/day, round to **1,600 lb/day**

139. **If the MCRT desired was 8.25 days, what would the waste rate be for the following system in lb/day and gpm?**

Flow = 3.77 mgd
Aeration tank (AT) volume = 0.750 mil gal
MLSS = 2,550 mg/L
Clarifier tank (CT) volume = 0.350 mil gal
Effluent TSS = 17.5 mg/L
Waste activated sludge (WAS) = 6,350 mg/L

Equation: **Waste rate, lb/day** $=$

$$\frac{\text{MLSS, mg/L}\,[\text{AT, mil gal} + \text{CT, mil gal}]\,(8.34\ \text{lb/gal})}{\text{Desired MCRT}} - (\textbf{TSS, mg/L})(\textbf{Flow, mgd})(\textbf{8.34 lb/gal})$$

Waste rate, lb/day $=$

$$\frac{(2{,}550\ \text{mg/L})\,[0.750\ \text{mil gal} + 0.350\ \text{mil gal}]\,(8.34\ \text{lb/gal})}{8.25\ \text{days, Desired MCRT}} - (17.5\ \text{mg/L TSS})(3.77\ \text{mgd})(8.34\ \text{lb/gal})$$

Waste rate, lb/day $=$

$$\frac{(2{,}550\ \text{mg/L})\,(1.10\ \text{mil gal})\,(8.34\ \text{lb/gal})}{8.25\ \text{days, Desired MCRT}} - (17.5\ \text{mg/L TSS})(3.77\ \text{mgd})(8.34\ \text{lb/gal})$$

Waste, lb/day $= 2{,}835.6$ lb/day $- 550.23$ lb/day $= 2{,}285.37$ lb/day, round to **2,290 lb/day**

Since the concentration of WAS is known, the number of mgd waste rate can be calculated.

Equation: Waste, mgd $= \dfrac{\text{Waste, lb/day}}{(\text{WAS, mg/L})\,(8.34\ \text{lb/gal})}$

Substitute values, in this case before rounding, and solve.

Waste, mgd $= \dfrac{2{,}285.37\ \text{lb/day}}{(6{,}350\ \text{mg/L})\,(8.34\ \text{lb/gal})} = 0.04315$ mgd

Lastly, convert mgd to gpm.

Waste, gpm $= \dfrac{(0.04315\ \text{mgd})\,(1{,}000{,}000/\text{mil})}{1{,}440\ \text{min/day}} = \textbf{30.0 gpm}$

DIGESTER VOLATILE SOLIDS LOADING RATIO CALCULATIONS

This calculation compares the volatile solids added to the volatile solids in the digester. It is used for evaluating process control.

140. **Given the following data, calculate the volatile solids (VS) loading ratio on a digester:**

Sludge in digester = 14,575 gallons
VS loading = 1,020 lb/day
Total solids (TS) percentage = 5.02%
VS percentage = 73.4%
Specific gravity of sludge = 1.03

First, convert the number of gallons in the digester to lb.

Sludge, lb = (14,575 gal)(8.34 lb/gal)(1.03 sp gr) = 125,202 lb

Use the following expanded equation with percentages to solve for digester volatile solid ratio.

$$\textbf{Digester VS ratio} = \frac{\text{VS added lb/day}}{(\text{lb VS in digester})(\text{TS \%/100\%})(\text{VS \%/100\%})}$$

Substitute values and solve.

$$\text{Digester VS ratio} = \frac{1,020 \text{ lb/day, VS}}{(125,202 \text{ lb VS})(5.02\%/100\%, \text{TS})(73.4\%/100\%, \text{VS})}$$

Digester VS ratio = **0.221 VS ratio**

141. **What is the ratio of volatile solids loading on a digester, if the digester has 49,200 kg of volatile solids (VS), 1,375 lb/day are pumped into it, percentage total solids (TS) are 4.86%, and percentage volatile solids are 72.7%?**

First, convert the number of kg to lb.

Know: 1 kg = 2.205 lb

Number of lb = (49,200 kg)(2.205 lb/kg) = 108,486 lb

Use expanded equation with percentages.

$$\text{Equation: } \textbf{Digester VS ratio} = \frac{\text{VS added lb/day}}{(\text{lb VS in digester})(\text{TS \%/100\%})(\text{VS \%/100\%})}$$

Substitute values and solve.

$$\text{Digester VS ratio} = \frac{1,375\ \text{lb/day}}{(108,486\ \text{lb})\,(4.86\%/100\%,\ \text{TS})\,(72.7\%/100\%,\ \text{VS})}$$

Digester VS ratio = **0.359 VS ratio**

DIGESTER GAS PRODUCTION PROBLEMS

Operators calculate the amount of gases produced per pound of volatile solids destroyed to determine the effectiveness of the digestion process. Also it is important to know the gas production because in some cases it is used as a fuel for other plant processes.

142. **What is the amount of gas produced in ft³ per lb of volatile solids (VS) destroyed, if a digester produces 5,035 ft³/day of gas and the amount of volatile solids destroyed are 488 lb/day?**

Equation: **Gas produced, ft³/lb VS destroyed** $= \dfrac{\text{Gas production, ft}^3/\text{day}}{\text{VS destroyed, lb/day}}$

Gas produced, ft³/lb VS destroyed $= \dfrac{5,035\ \text{ft}^3/\text{day}}{488\ \text{lb/day}} =$ **10.3 ft³/lb of VS destroyed**

143. **Given the following data, determine the gas produced by a digester in cubic meters (m³) per pound of volatile solids (VS) destroyed:**

Digester gas production = 5,743 ft³/day
Volatile solids destroyed = 461 lb/day
1 cubic meter = 35.3 cubic ft

Equation: **Gas produced, m³/lb VS destroyed** $= \dfrac{\text{Gas production, ft}^3/\text{day}}{(\text{VS destroyed, lb/day})\,(35.3\ \text{ft}^3/\text{m}^3)}$

Gas produced, ft³/lb VS destroyed $= \dfrac{5,743\ \text{ft}^3/\text{day}}{(461\ \text{lb/day})\,(35.3\ \text{ft}^3/\text{m}^3)} =$ **0.353 m³/lb of VS destroyed**

144. What must have been the gas production by a digester in ft³/day, given the following data?

Volatile solids destroyed = 235 kg/day
Gas produced in ft³/lb VS destroyed = 12.6 ft³/lb

First, convert kg/day to lb/day.

Number of lb/day = (235 kg/day)(2.205 lb/kg) = 518.175 lb/day

Equation: **Gas produced, ft³/lb VS destroyed** $= \dfrac{\text{Gas production, ft}^3/\text{day}}{\text{VS destroyed, lb}/\text{day}}$

Rearrange to solve for gas production in ft³/day.

Gas production, ft³/day = (Gas produced, ft³/lb VS destroyed)(VS destroyed, lb/day)

Substitute values and solve.

Gas production, ft³/day = (12.6 ft³/lb)(518.175 lb/day) = 6,529 ft³/day, round to **6,530 ft³/day**

DIGESTER SOLIDS BALANCE

Digester solids balance calculations are used to confirm the many calculations used to evaluate process control and maximize the digester process.

145. Calculate the solids balance for a digester, given the following data:

<u>Sludge Before Digestion</u>
Influent raw sludge = 17,850 lb/day
Percent solids in raw sludge = 6.15%
Percent volatile solids in raw sludge = 67.8%

<u>Digested Sludge</u>
Percent solids = 4.25%
Percent volatile solids (VS) = 51.8%

First, determine the solids entering the digester unit in lb/day.

Equation: **Total solids, lb/day = (Raw sludge, lb/day)(Percent solids)**

Total solids, lb/day = (17,850 lb/day)(6.15%/100%)

Total solids, lb/day = 1,097.775 lb/day, round to **1,100 lb/day Total solids**

Next, calculate the VS entering the digester unit in lb/day.

VS, lb/day = (1,097.775 lb/day)(67.8%/100%) = 744.292 lb/day, round to **744 lb/day VS**

Next, calculate the fixed solids (residue left over after drying at 550°C) entering the digester unit in lb/day.

Equation: **Fixed solids, lb/day = Total solids, lb/day − VS, lb/day**

Fixed solids, lb/day = 1,097.775 lb/day − 744.292 lb/day

Fixed solids, lb/day = 353.483 lb/day, round to **353 lb/day Fixed solids**

Next, determine the amount of water entering the digester unit in lb/day.

Equation: **Water in sludge, lb/day = Sludge, lb/day − Total solids, lb/day**

Water, lb/day = 17,850 lb/day − 1,097.775 lb/day

Water, lb/day = 16,752.225 lb/day, round to **16,800 lb/day Water**

Next, calculate the volatile solids reduction (VSR).

Equation: $$\textbf{Percent VSR} = \frac{(\text{In} - \text{Out})(100\%)}{\text{In} - (\text{In})(\text{Out})}$$

First, convert percentage to decimal form.

Decimal form for 67.8% = 67.8%/100% = 0.678

Decimal form for 51.8% = 51.8%/100% = 0.518

Percent VSR $= \dfrac{(0.678 - 0.518)\,(100\%)}{0.678 - (0.678)\,(0.518)}$

Reduce and simplify.

Percent VSR $= \dfrac{(0.16)\,(100\%)}{0.678 - 0.3512} = \dfrac{16\%}{0.3268} = 48.96\%$, round to **49.0% VSR**

Next, determine the gas produced by the digester in lb/day.

Know: Pounds VSR = Pounds gas produced

Equation: **Gas produced, lb/day = (Effluent VS, lb/day)(Percent VSR)**

Gas produced, lb/day = (744.292 lb/day VS)(48.96%/100% VSR)

Gas produced, lb/day = 364.40 lb/day, round to **364 lb/day Gas production**

Next, calculate the VS in the digested sludge in lb/day.

Equation: **VS in digested sludge, lb/day = Influent VS, lb/day − Destroyed VS, lb/day**

VS in digested sludge, lb/day = 744.292 lb/day VS − 364.40 lb/day Destroyed VS

VS in digested sludge, lb/day = 379.892, lb/day, round to **380 lb/day VS in digested sludge**

Next, calculate the total solids in digested sludge (lb/day).

Equation: **Total digested solids, lb/day** $= \dfrac{\text{VS digested, lb/day}}{\text{Percent digested VS}}$

Total digested solids, lb/day $= \dfrac{379.892 \text{ lb/day}}{51.8\%/100\% \text{ digested VS}}$

Total digested solids, lb/day = 733.38 lb/day, round to **733 lb/day Total digested solids**

Next, calculate the fixed solids in the digested sludge in lb/day.

Equation: **Fixed solids, lb/day = Total digested solids, lb/day − VS digested, lb/day**

Fixed solids, lb/day = 733.38 lb/day − 379.891 lb/day

Fixed solids, lb/day = 353.489 lb/day, round to **353 lb/day Fixed solids**

Next, calculate the digested sludge in lb/day.

Equation: **Digested sludge, lb/day** $= \dfrac{\text{Total digested solids, lb/day}}{\text{Digested sludge percent solids}}$

Digested sludge, lb/day $= \dfrac{733.38 \text{ lb/day}}{4.25\%/100\% \text{ Digested solids}}$

Digested sludge, lb/day = 17,256 lb/day, round to **17,300 lb/day**

Next, calculate the lb/day of water in the digested sludge.

Equation: **Water in digested sludge, lb/day = Sludge, lb/day − Total solids, lb/day**

Water in digested sludge, lb/day = 17,256 lb/day − 733.38 lb/day

Water in digested sludge, lb/day = 16,522.62 lb/day, round to **16,500 lb/day Water**

Finally do a comparison analysis of the results.

COMPARISON ANALYSES OF DIGESTER SOLIDS MASS BALANCE			
Sludge Entering the Digester		**Digested Sludge Exiting Digester**	
Total Solids	1,097.775 lb	Total Solids	733.38 lb
Volatile Solids	(744.292 lb)	Volatile Solids	(379.892 lb)
Fixed Solids	(353.483 lb)	Fixed Solids	(353.489 lb)
Water in Sludge	16,752.225 lb	Water in Sludge	16,522.62 lb
		Gas Produced	364.40 lb
Final rounded total	**17,850 lb**	Final rounded total	**17,620 lb**

Note: Numbers in parenthesis are not added.

146. **Calculate the solids balance for a digester, given the following data:**

<u>Sludge Before Digestion</u>
Influent raw sludge = 23,220 lb/day
Percent solids in raw sludge = 6.35%
Percent volatile solids in raw sludge = 68.5%

<u>Digested Sludge</u>
Percent solids = 4.37
Percent volatile solids (VS) = 53.0%

First, determine the solids entering the digester unit in lb/day.

Equation: **Total solids, lb/day = (Raw sludge, lb/day)(Percent solids)**

Total solids, lb/day = (23,220 lb/day)(6.35%/100%)

Total solids, lb/day = 1,474.47 lb/day, round to **1,470 lb/day Total solids**

Next, calculate the VS entering the digester unit in lb/day.

VS, lb/day = (1,474.47 lb/day)(68.5%/100%) = 1,010.01 lb/day, round to **1,010 lb/day VS**

Next, calculate the fixed solids (residue left over after drying at 550°C) entering the digester unit in lb/day.

Equation: **Fixed solids, lb/day = Total solids, lb/day − VS, lb/day**

Fixed solids, lb/day = 1,474.47 lb/day − 1,010.01 lb/day

Fixed solids, lb/day = 464.46 lb/day, round to **464 lb/day Fixed solids**

Next, determine the amount of water entering the digester unit in lb/day.

Equation: **Water in sludge, lb/day = Sludge, lb/day − Total solids, lb/day**

Water, lb/day = 23,220 lb/day − 1,474.47 lb/day

Water, lb/day = 21,745.53 lb/day, round to **21,700 lb/day Water**

Next, calculate the volatile solids reduction (VSR).

Equation: **Percent VSR** $= \dfrac{(\text{In} - \text{Out})(100\%)}{\text{In} - (\text{In})(\text{Out})}$

First, convert percentage to decimal form.

Decimal form for 68.5% = 68.5%/100% = 0.685

Decimal form for 53.0% = 53.0%/100% = 0.530

Percent VSR $= \dfrac{(0.685 - 0.530)(100\%)}{0.685 - (0.685)(0.530)}$

Reduce and simplify.

Percent VSR $= \dfrac{(0.155)(100\%)}{0.685 - 0.36305} = \dfrac{15.5\%}{0.32195} = 48.14\%$, round to **48.1% VSR**

Next, determine the gas produced by the digester in lb/day.

Know: Pounds VSR = Pounds gas produced

Equation: **Gas produced, lb/day = (Effluent VS, lb/day)(Percent VSR)**

Gas produced, lb/day = (1,010.01 lb/day VS)(48.14%/100% VSR)

Gas produced, lb/day = 486.22 lb/day, round to **486 lb/day Gas production**

Next, calculate the VS in the digested sludge in lb/day.

Equation: **VS in digested sludge, lb/day = Influent VS, lb/day − Destroyed VS, lb/day**

VS in digested sludge, lb/day = 1,010.01 lb/day VS − 486.22 lb/day Destroyed VS

VS in digested sludge, lb/day = 523.79, lb/day, round to **524 lb/day VS in digested sludge**

Next, calculate the total solids in digested sludge (lb/day).

Equation: **Total digested solids, lb/day** $= \dfrac{\text{VS digested, lb/day}}{\text{Percent digested VS}}$

Total digested solids, lb/day $= \dfrac{525.79\ \text{lb/day}}{53.0\%/100\%\ \text{digested VS}}$

Total digested solids, lb/day = 988.28 lb/day, round to **988 lb/day Total digested solids**

Next, calculate the fixed solids in the digested sludge in lb/day.

Equation: **Fixed solids, lb/day** $=$ **Total digested solids, lb/day** $-$ **VS digested, lb/day**

Fixed solids, lb/day = 988.28 lb/day $-$ 523.79 lb/day

Fixed solids, lb/day = 464.49 lb/day, round to **464 lb/day Fixed solids**

Next, calculate the digested sludge in lb/day.

Equation: **Digested sludge, lb/day** $= \dfrac{\text{Total digested solids, lb/day}}{\text{Digested sludge percent solids}}$

Digested sludge, lb/day $= \dfrac{988.28\ \text{lb/day}}{4.37/100\%\ \text{Digested solids}}$

Digested sludge, lb/day = 22,615.10 lb/day, round to **22,600 lb/day**

Next, calculate the lb/day of water in the digested sludge.

Equation: **Water in digested sludge, lb/day = Sludge, lb/day − Total solids, lb/day**

Water in digested sludge, lb/day = 22,615.10 lb/day − 988.28 lb/day

Water in digested sludge, lb/day = 21,626.82 lb/day, round to **21,600 lb/day Water**

Finally do a comparison analysis of the results.

COMPARISON ANALYSES OF DIGESTER SOLIDS MASS BALANCE			
Sludge Entering the Digester		**Digested Sludge Exiting Digester**	
Total Solids	1,474.47 lb	Total Solids	988.28 lb
Volatile Solids	(1,010.01 lb)	Volatile Solids	(523.79 lb)
Fixed Solids	(464.46 lb)	Fixed Solids	(464.49 lb)
Water in Sludge	21,745.53 lb	Water in Sludge	21,626.82 lb
		Gas Produced	486.22 lb
Final rounded total	**23,220 lb**	Final rounded total	**23,101 lb**

Note: Numbers in parenthesis are not added.

SOLIDS (MASS) BALANCE CALCULATIONS

Solids balance calculations are used to confirm the many calculations used to evaluate process control and maximize the wastewater treatment process.

Note: **Systems within 10% are considered in balance.**

147. Given the following data, calculate whether a gravity thickener's sludge blanket will increase, remain the same, or decrease, and how much that change will be in lb/day solids, if it increases or decreases:

Pumped sludge to thickener = 135 gpm
Pumped sludge from thickener = 51 gpm
Thickener's effluent suspended solids (SS) = 782 mg/L
Primary sludge solids = 3.20%
Thickened sludge solids = 7.36%

First, calculate the thickener's influent and effluent solids in lb/day.

Equation: **Influent solids, lb/day** =

(Influent flow gpm)(1,440 min/day)(Percent sludge solids)(8.34 lb/gal)

Influent solids, lb/day = (135 gpm)(1,440 min/day)(3.20%/100%)(8.34 lb/gal)

Influent solids, lb/day = 51,881.47 lb/day

Effluent solids, lb/day = (51 gpm)(1,440 min/day)(7.36%/100%)(8.34 lb/gal)

Effluent solids, lb/day = 45,079.23 lb/day

Next, calculate the flow leaving the thickener.

Effluent flow = 135 gpm − 51 gpm = 84 gpm

Next, convert thickened sludge solids to percent.

Know: 1% = 10,000 mg/L

Percent sludge solids = (782 mg/L)(1%/10,000 mg/L) = 0.0782%

Now, calculate the sludge solids that are leaving the thickener.

Sludge solids, lb/day = (84 gpm)(1,440 min/day)(0.0782%/100%)(8.34 lb/gal)

Sludge solids, lb/day = 788.887 lb/day

Now, calculate whether the sludge blanket will increase, remain the same, or decrease.

Total solids, lb/day = 51,881.47 lb/day − 45,079.23 lb/day − 788.887 lb/day

Total solids, lb/day = 6,013.353 lb/day, round to **6,000 lb/day**

The sludge blanket will increase because it is positive.

148. **Given the following data, calculate the mass balance for the following activated biosolids system that has primary treatment using BOD$_5$ removal. Is there a problem with this system? If so, discuss.**

Plant flow = 2.83 mgd
Plant influent BOD$_5$ = 236 mg/L
Plant influent TSS = 249 mg/L
Effluent flow = 2.76 mgd
Effluent system activated biosolids BOD$_5$ = 24 mg/L
Effluent system activated biosolids TSS = 18 mg/L
Waste concentration, TSS = 6,980 mg/L
Waste flow = 0.0510 mgd

Given: This activated biosolids system has lb solids/lb BOD$_5$ = 0.71 lb solids/lb BOD$_5$

First, calculate the BOD$_5$ influent and effluent in lb/day.

Influent BOD$_5$, lb/day = (236 mg/L)(2.83 mgd)(8.34 lb/gal) = 5,570.12 lb/day

Effluent BOD$_5$, lb/day = (24 mg/L)(2.83 mgd)(8.34 lb/gal) = 566.4528 lb/day

Then, determine the difference in lb/day BOD_5 removal.

BOD_5 removed, lb/day = 5,570.12 lb/day − 566.4528 lb/day = 5,003.67 lb/day

Next, determine the solids produced in lb/day.

Equation: **Solids produced, lb/day = (BOD_5 removed, lb/day)(0.71 lb solids/lb BOD_5)**

Solids produced, lb/day = (5,003.67 lb/day)(0.71 lb/lb BOD_5) = 3,552.61 lb/day

Next, calculate the solids and sludge removed.

Effluent solids, lb/day = (18 mg/L)(2.76 mgd)(8.34 lb/gal) = 414.33 lb/day

Sludge removed, lb/day = (6,980 mg/L)(0.0510 mgd)(8.34 lb/gal) = 2,968.87 lb/day

Next, calculate the total solids removed.

Total solids removed, lb/day = 414.33 lb/day + 2,968.87 lb/day = 3,383.20 lb/day

Now, determine the difference in lb/day.

Balance difference, lb/day = 3,552.61 lb/day − 3,383.20 lb/day = 169.41 lb/day

Lastly, calculate the percent difference.

Equation: **Percent mass balance** $= \dfrac{(\text{Solids produced, lb/day} + \text{Solids removed, lb/day})(100\%)}{\text{Solids produced, lb/day}}$

Percent difference $= \dfrac{(3,552.61 \text{ lb/day} - 3,383.20 \text{ lb/day})(100\%)}{3,552.61 \text{ lb/day}} = \textbf{4.8\%}$

This system is in balance.

149. **Calculate the mass balance for the following conventional biological system. Is there a problem with this system? If so, discuss.**

Plant influent waste flow = 2.63 mgd
Influent BOD_5 = 197 mg/L
Influent TSS = 274 mg/L
Effluent flow = 2.46 mgd
Effluent BOD_5 = 22 mg/L
Effluent TSS = 44 mg/L
Waste concentration, TSS = 8,060 mg/L
Waste flow = 0.0359 mgd

Given: This conventional activated biosolids system without primary at this plant has a lb solids/lb BOD_5 = 0.84 lb solids/lb BOD_5

First, calculate the BOD_5 influent and then the effluent in lb/day.

Influent BOD_5, lb/day = (197 mg/L)(2.63 mgd)(8.34 lb/gal) = 4,321.04 lb/day

Effluent BOD_5, lb/day = (22 mg/L)(2.63 mgd)(8.34 lb/gal) = 482.55 lb/day

BOD_5 removed, lb/day = 4,321.04 lb/day − 482.55 lb/day = 3,838.49 lb/day

Next, determine the solids produced in lb/day.

Equation: **Solids produced, lb/day = (BOD_5 removed, lb/day)(0.84 lb solids/lb BOD_5)**

Solids produced, lb/day = (3,838.49 lb/day)(0.84 lb/lb BOD_5) = 3,224.33 lb/day

Next, calculate the solids and sludge removed.

Effluent solids, lb/day = (44 mg/L)(2.46 mgd)(8.34 lb/gal) = 902.72 lb/day

Effluent sludge, lb/day = (8,060 mg/L)(0.0359 mgd)(8.34 lb/gal) = 2,413.21 lb/day

Next, calculate the total solids removed.

Total solids removed, lb/day = 902.72 lb/day + 2,413.21 lb/day = 3,315.93 lb/day

Now, calculate the percent mass balance of the system.

$$\text{Percent mass balance} = \frac{(3{,}224.33 \text{ lb/day} - 3{,}315.93 \text{ lb/day})(100\%)}{3{,}224.33 \text{ lb/day}} = -2.8\%$$

This system is within balance.

150. **Calculate the mass balance for the following conventional biological system. Is there a problem with this system? If so, discuss.**

Influent waste flow = 2.09 mgd
Influent BOD$_5$ = 204 mg/L
Influent TSS = 247 mg/L
Effluent flow = 2.01 mgd
Effluent BOD$_5$ = 33 mg/L
Effluent TSS = 41 mg/L
Waste flow = 0.0261 mgd
Waste TSS = 8,486 mg/L

Given: This activated biosolids without primary, extended air system wastewater plant has a lb solids/lb BOD$_5$ = 0.66 lb solids/lb BOD$_5$

First, calculate the BOD$_5$ influent and then the effluent in lb/day.

Influent BOD$_5$, lb/day = (204 mg/L)(2.09 mgd)(8.34 lb/gal) = 3,555.84 lb/day

Effluent BOD$_5$, lb/day = (33 mg/L)(2.09 mgd)(8.34 lb/gal) = 575.21 lb/day

BOD$_5$ removed, lb/day = 3,555.84 lb/day − 575.21 lb/day = 2,980.63 lb/day

Next, determine the solids produced in lb/day.

Equation: **Solids produced, lb/day = (BOD$_5$ removed, lb/day)(0.66 lb solids/lb BOD$_5$)**

Solids produced, lb/day = (2,980.63 lb/day)(0.66 lb/lb BOD$_5$) = 1,967.22 lb/day

Next, calculate the solids and sludge removed.

Effluent solids, lb/day = (41 mg/L)(2.01 mgd)(8.34 lb/gal) = 687.30 lb/day

Effluent sludge, lb/day = (8,486 mg/L)(0.0261 mgd)(8.34 lb/gal) = 1,847.18 lb/day

Next, calculate the total solids removed.

Total solids removed, lb/day = 687.30 lb/day + 1,847.18 lb/day = 2,534.48 lb/day

Now, calculate the percent mass balance of the system.

Equation: $\textbf{Percent mass balance} = \dfrac{(\text{Solids produced, lb/day} + \text{Solids removed, lb/day})\,(100\%)}{\text{Solids produced, lb/day}}$

$\text{Percent mass balance} = \dfrac{(1,967.22\ \text{lb/day} - 2,534.48\ \text{lb/day})\,(100\%)}{1,967.22\ \text{lb/day}} = \textbf{--28.8\%}$

This system is *not* in balance, since more solids are being removed than produced.

Now, calculate the waste rate in gpd based on the mass balance results.

Equation: $\textbf{Waste rate, gpd} = \dfrac{(\text{Solids produced, lb/day})\,(1,000,000/\text{mil})}{(\text{Waste TSS, mg/L})\,(8.34\ \text{lb/gal})}$

$\text{Waste rate, gpd} = \dfrac{(1,967.22\ \text{lb/day})\,(1,000,000/\text{mil})}{(8,486\ \text{mg/L TSS})\,(8.34\ \text{lb/gal})} = 27,796\ \text{gpd, round to}\ \textbf{27,800 gpd}$

Current waste rate is 0.0261 mil gal, which is 26,100 gpd. This is less than the calculated waste rate of 27,796 gpd.

Analyzing these results indicates that less wasting is required because more solids are being removed than added based on the mass balance results. The waste rate result indicates more solids need to be wasted. These conflicting results indicate the system is out of balance. In addition, problems with this system could be accentuated or be caused by improper sampling and by laboratory analytical error(s).

VOLATILE-ACIDS-TO-ALKALINITY RATIO PROBLEMS

The first phase of anaerobic digestion is acid fermentation, which is dependent upon new volatile solids entering the digester. The second stage is methane fermentation. These two processes need to be in delicate balance with each other for the anaerobic digestion process to proceed properly. Different treatment plants have different ratios, but typically the ratio is less than 0.1.

151. **What is the ratio of volatile acids to alkalinity, if the alkalinity in an anaerobic digester is 1,587 mg/L and the volatile acid concentration of the sludge is 198 mg/L?**

Equation: **Ratio = Volatile acids/Alkalinity**

Substitute values and solve.

$$\text{Ratio} = \frac{198 \text{ mg/L}}{1,587 \text{ mg/L}} = \textbf{0.125 Volatile-acids-to-alkalinity ratio}$$

152. **What must have been the volatile acid concentration in an anaerobic digester, if the alkalinity was 1,726 mg/L and the ratio of volatile acids to alkalinity was 0.107?**

Equation: **Volatile acids = (Alkalinity)(Ratio)**

Substitute values and solve.

Volatile acids = (1,726 mg/L)(0.107) = 184.68 mg/L, round to **185 mg/L**

153. **What is the amount of volatile solids in lb/day that must have been added to a digester, if the ratio of volatile solids (VS) added to volatile solids already in a digester is 0.096 and the amount of VS already in the digester is 21,380 lb?**

Equation: **Digester VS ratio** $= \dfrac{\text{VS added lb/day}}{\text{lb VS in digester}}$

Rearrange the problem to solve for volatile solids added.

VS added lb/day $=$ (Digester VS ratio)(lb VS in digester)

Substitute values and solve.

VS added lb/day $=$ (0.096 VS ratio)(21,380 lb VS in digester)

VS added lb/day $=$ 2,052.48 lb/day, round to **2,100 lb/day VS added**

LIME NEUTRALIZATION PROBLEMS

When the sludge in an anaerobic digester becomes acidic, it is called a sour digester. A sour digester occurs when the volatile acid-to-alkalinity ratio increases above 0.8. It is not always possible to wait for a digester to naturally correct itself because of the digester's capacity or time constraints. Under these circumstances, it is necessary to neutralize the acid conditions in the digester with lime. The following problems show how operators calculate the appropriate dosage of lime. The lime dosage is based on the amount of volatile acids in the sludge and is in a 1-to-1 ratio that is 1 mg/L of lime will neutralize 1 mg/L of volatile acid.

154. **Given the following data, determine the amount of lime in lb that are needed to neutralize a sour digester:**

Digester diameter $=$ **38.5 ft**
Digester fluid level $=$ **13.4 ft**
Volatile acids $=$ **2,615 mg/L**

Know: 1 mg/L of lime will neutralize 1 mg/L of volatile acids

First, calculate the volume of the digester in gallons.

Number of gallons $= (0.785)(38.5 \text{ ft})(38.5 \text{ ft})(13.4 \text{ ft})(7.48 \text{ gal/ft}^3) = 116,627$ gal

Next, convert gallons to mil gal.

Number of mil gal $= (116,627 \text{ gal}) / (1,000,000/\text{mil}) = 0.116627$ mil gal

Equation: **Number of lb of Lime = (Volatile acids, mg/L)(mil gal)(8.34 lb/gal)**

Lime, lb $= (2,615 \text{ mg/L})(0.116627 \text{ mil gal})(8.34 \text{ lb/gal}) = 2,543.53$ lb, round to **2,540 lb of Lime**

155. How many pounds of lime are required to neutralize a sour digester that is 45.1 ft in diameter, has a sludge level of 12.9 ft, and has a volatile acid content of 2,720 mg/L?

Know: 1 mg/L of lime will neutralize 1 mg/L of volatile acids

First, calculate the volume of the digester in gallons.

Number of gallons $= (0.785)(45.1 \text{ ft})(45.1 \text{ ft})(12.9 \text{ ft})(7.48 \text{ gal/ft}^3) = 154,069$ gal

Next, convert gallons to mil gal.

Number of mil gal $= \dfrac{154,069 \text{ gal}}{1,000,000/\text{mil}} = 0.154069$ mil gal

Equation: **Number of lb of Lime = (Volatile acids, mg/L)(mil gal)(8.34 lb/gal)**

Lime, lb $= (2,720 \text{ mg/L})(0.154069 \text{ mil gal})(8.34 \text{ lb/gal}) = 3,495.02$ lb, round to **3,500 lb of Lime**

156. **What must have been the concentration of volatile acids in mg/L for a sour digester with a radius of 24.8 ft and a sludge level of 13.5 ft, if the number of pounds of lime to neutralize the volatile acids was 1,483 lb?**

Know: 1 mg/L of lime will neutralize 1 mg/L of volatile acids

First, determine the sludge volume in gallons that is in the digester.

Number of gallons $= \pi(r)^2(\text{Height})(7.48 \text{ gal/ft}^3)$

Number of gallons $= 3.14(24.8 \text{ ft})(24.8 \text{ ft})(13.5 \text{ ft})(7.48 \text{ gal/ft}^3) = 195{,}015 \text{ gal}$

First, convert the digester's volume from gallons to mil gal.

Number of mil gal $= \dfrac{195{,}015 \text{ gal}}{1{,}000{,}000/\text{mil}} = 0.195 \text{ mil gal}$

Equation: **Number of lb of Lime = (Volatile acids, mg/L)(mil gal)(8.34 lb/gal)**

Rearrange the equation to solve for the concentration of volatile acids.

Volatile acids, mg/L $= \dfrac{\text{Number of lb, lime}}{(\text{mil gal})(8.34 \text{ lb/gal})}$

Substitute values and solve.

Volatile acids, mg/L $= \dfrac{1{,}483 \text{ lb, lime}}{(0.195 \text{ mil gal})(8.34 \text{ lb/gal})}$

Volatile acids, mg/L $= 911.89$ mg/L, round to **912 mg/L Volatile acids**

POPULATION LOADING CALCULATIONS

These calculations are used for wastewater treatment ponds. They are based on the number of people per acre of pond, and it is a helpful tool in evaluating process control of ponds.

157. Given the following data, calculate the population loading in people per acre on 4 ponds:

Pond 1 = 7.64 acres
Pond 2 = 5.07 acres
Pond 3 = 4.51 acres
Pond 4 = 3.86 acres
Services = 9,250
Average persons per service = 3.13

First, add the area in acres for each pond to get the total acres.

Total area of ponds = 7.64 acres + 5.07 acres + 4.51 acres + 3.86 acres = 21.08 acres

Next, determine the number of people served.

Number of people = (9,250 services)(3.13 people/service) = 28,952 people served

Next, using the following equation determine the population loading.

$$\text{Population loading, people/acre} = \frac{\text{Number of people served}}{\text{Area of pond(s), acres}}$$

Substitute values and solve.

$$\text{Population loading, people/acre} = \frac{28,952 \text{ people served}}{21.08 \text{ acres}}$$

Population loading, people/acre = 1,373 people/acre, round to **1,370 people/acre**

158. What is the population loading in people/acre, if there are 7,285 services with 3.24 people per service, given the following data?

Wastewater pond 1 = 11.07 acres
Wastewater pond 2 = 6.89 acres
Wastewater pond 3 = 7.25 acres

First, add the area in acres for each pond to get the total acres.

Total area of ponds = 11.07 acres + 6.89 acres + 7.25 acres = 25.21 acres

Next, determine the number of people served.

Number of people = (7,285 services)(3.24 people/service) = 23,603 people served

Next, using the following equation, determine the population loading.

$$\text{Population loading, people/acre} = \frac{\text{Number of people served}}{\text{Area of pond(s), acres}}$$

Substitute values and solve.

$$\text{Population loading, people/acre} = \frac{23,603 \text{ people served}}{25.21 \text{ acres}}$$

Population loading, people/acre = 936.26 people/acre, round to **936 people/acre**

POPULATION EQUIVALENT CALCULATIONS

Wastewater discharge from industries or commercial sources usually has a higher organic content than domestic wastewaters. Operators use population equivalent calculations to compare domestic wastewater to wastewater from these former sources. This is important in determining the loading that will be placed on a wastewater system when a new industry wants to connect to a system. What is needed is the flow from this industry in mgd and the BOD_5 concentration in mg/L. Domestic wastewater systems usually contain a range of 0.17 to 0.20 pounds of BOD_5 per day, which the wastewater plant should have already determined. Also, population equivalent calculations are required for designing proper size wastewater treatment plants, pump stations, and pipe sizes because the volumetric flow that is expected to be treated and pumped needs to be estimated.

159. Given the following data, determine the population equivalent:

Average wastewater flow for the day $= 3.95$ ft^3/s
BOD$_5$/person $= 0.24$ lb/day
BOD$_5$ concentration in the wastewater $= 2,708$ mg/L

Know: 1 ft^3/s $= 0.6463$ mgd

First, convert ft^3/s to mgd.

Number of mgd $= (3.95$ ft^3/s$)(0.6463$ mgd/ ft^3/s$) = 2.553$ mgd

Use the following equation to solve this problem.

$$\textbf{Number of people} = \frac{(\text{BOD}_5, \text{mg/L})(\text{mgd})(8.34 \text{ lb/gal})}{\text{lb/day of BOD}_5/\text{person}}$$

$$\text{Number of people} = \frac{(2,708 \text{ mg/L BOD}_5)(2.553 \text{ mgd})(8.34 \text{ lb/gal})}{0.24 \text{ lb/day}}$$

Number of people $= 240,245$ people, round to **240,000 people**

160. A wastewater treatment plant has an influent flow of 4,546,000 gpd. If the BOD$_5$ is 2,910 mg/L and the average BOD$_5$ per person is 0.25 lb/day, what is the population equivalent that this plant is currently treating?

First, convert the flow in gpd to mgd.

$$\text{Number of mgd} = \frac{4,546,000 \text{ gpd}}{1,000,000/\text{mil}} = 4.546 \text{ mgd}$$

Next, use the following equation to solve this problem.

$$\textbf{Number of people} = \frac{(\text{BOD}_5, \text{mg/L})(\text{mgd})(8.34 \text{ lb/gal})}{\text{lb/day of BOD}_5/\text{person}}$$

$$\text{Number of people} = \frac{(2,910 \text{ mg/L BOD}_5)(4.546 \text{ mgd})(8.34 \text{ lb/gal})}{0.25 \text{ lb/day}}$$

Number of people $= 441,315$ people, round to **440,000 people**

SOLIDS UNDER AERATION

Solids under aeration calculations are used by operators for evaluating process control.

161. **An aeration tank is 50.1 ft in diameter and has a working level of 15.6 ft. Calculate the number of pounds of suspended solids (SS) contained in an aeration tank, if the concentration of SS is 2,808 mg/L and the specific gravity is 1.03.**

First, calculate the number of mil gal in the tank.

Equation: $\text{Number of mil gal} = \dfrac{(0.785)\,(\text{Diameter, ft})^2\,(\text{Depth, ft})\,(7.48 \text{ gal/ft}^3)}{1,000,000/\text{mil}}$

$\text{Number of mil gal} = \dfrac{(0.785)\,(50.1 \text{ ft})\,(50.1 \text{ ft})\,(15.6 \text{ ft})\,(7.48 \text{ gal/ft}^3)}{1,000,000/\text{mil}} = 0.23 \text{ mil gal}$

Next, determine the number of lb/gal for the SS.

Number of lb/gal, SS = (8.34 lb/gal)(1.03 sp gr) = 8.59 lb/gal

Next, determine the pounds of SS under aeration using a modified version of the "pounds" equation.

Equation: **Number of lb SS = (SS, mg/L)(Number of mil gal)(SS lb/gal)**

Substitute values and solve.

Number of lb SS = (2,808 mg/L, SS)(0.23 mgd)(8.59 lb/gal)

Number of lb SS = 5,547.77 lb SS, round to **5,550 lb SS**

162. **How many pounds of mixed liquor suspended solids (MLSS) are being aerated, if the aeration tank is 301 ft by 42.5 ft with a level of 16.1 ft, the concentration of MLSS is 2,276 mg/L, and the specific gravity of the MLSS is 1.02?**

First, determine how many gallons are in the aeration tank.

Equation: **Number of gallons** $= \dfrac{(\text{Length, ft})\,(\text{Width, ft})\,(\text{Height, ft})\,(7.48\ \text{gal/ft}^3)}{1,000,000/\text{mil}}$

Number of gallons $= \dfrac{(301\ \text{ft})\,(42.5\ \text{ft})\,(16.1,\ \text{ft})\,(7.48\ \text{gal/ft}^3)}{1,000,000/\text{mil}} = 1.5406\ \text{mil gal}$

Next, determine the lb/gal for the MLSS.

Number of lb/gal, MLSS $= (8.34\ \text{lb/gal})(1.02\ \text{sp gr}) = 8.5068\ \text{lb/gal}$

Next, determine the pounds of MLSS under aeration using a modified version of the "pounds" equation.

Equation: **Number of lb MLSS $=$ (MLSS, mg/L)(Number of mil gal)(MLSS lb/gal)**

Substitute values and solve.

Number of lb MLSS $= (2{,}276\ \text{mg/L MLSS})(1.5406\ \text{mil gal})(8.5068\ \text{lb/gal})$

Number of lb MLSS $= 29{,}828$ lb SS, round to **29,800 lb MLSS**

SUSPENDED SOLIDS REMOVAL

The suspended solids removal calculations are used by operators as a sign for the efficiency of the treatment process in question. Typically, the suspended solids removed from wastewater systems ranges from 100 to 350 mg/L.

163. **Find the number of pounds of mixed liquor suspended solids (MLSS) being aerated, if the aeration tank is 44.8 ft in diameter, with the sludge height of 14.7 ft, the concentration of MLSS is 2,015 mg/L, and the specific gravity of the MLSS is 1.05.**

First, determine how many gallons are in the aeration tank.

Number of gallons = (0.785)(Diameter)^2(Height, ft)(7.48 gal/ft^3)

Number of gallons = (0.785)(44.8 ft)(44.8 ft)(14.7 ft)(7.48 gal/ft^3) = 173,239 gal

Next, convert gallons to mil gal.

$$\text{Number of mgd} = \frac{173,239 \text{ gal}}{1,000,000/\text{mil}} = 0.173239 \text{ mgd}$$

Next, determine the lb/gal for the MLSS.

Number of lb/gal, MLSS = (8.34 lb/gal)(1.05 sp gr) = 8.757 lb/gal

Next, determine the pounds of MLSS under aeration using a modified version of the "pounds" equation because the MLSS weighs more than water.

Equation: **Number of lb MLSS = (MLSS, mg/L)(Number of mil gal)(MLSS lb/gal)**

Substitute values and solve.

Number of lb MLSS = (2,015 mg/L MLSS)(0.173239 mil gal)(8.757 lb/gal)

Number of lb MLSS = 3,056.86 lb, round to **3,060 lb MLSS**

164. Given the following data, calculate the quantity of suspended solids (SS) in lb/day that was removed from a primary clarifier:

Average influent flow for the day = **2.86 ft³/s**
Suspended solids = **119 mg/L**
Specific gravity of the SS = **1.06**

First, convert ft³/s to mgd.

Number of mgd = (2.86 ft³/s)(0.6463 mgd/ ft³/s) = 1.848 mgd

Next, determine the lb/gal for the SS.

Number of lb/gal, MLSS = (8.34 lb/gal)(1.06 sp gr) = 8.84 lb/gal

Next, determine the pounds of SS under aeration using a modified version of the "pounds" equation because the SS weighs more than water.

Equation: **Number of lb/day SS** = **(SS, mg/L)(Number of mgd)(SS lb/gal)**

Substitute values and solve.

Number of lb/day SS removed = (119 mg/L, SS)(1.848 mgd)(8.84 lb/gal)

Number of lb/day SS removed = 1,944 lb/day, round to **1,940 lb/day SS removed**

165. What must have been the average influent concentration of suspended solids (SS) in mg/L, if a wastewater treatment plant's clarifier had a flow of 1,275 gpm and removed 2,630 lb/day of SS?

First, convert gpm to mgd.

$$\text{Number of mgd} = \frac{(1,275 \text{ gpm})(1,440 \text{ min/day})}{1,000,000/\text{mil}} = 1.836 \text{ mgd}$$

Equation: **Number of lb/day SS = (SS, mg/L)(Number of mgd)(8.34 lb/gal)**

Rearrange the equation, then substitute values and solve.

$$\textbf{SS removed, mg/L} = \frac{\text{Number of lb/day SS}}{(\text{Number of mgd})(8.34 \text{ lb/gal})}$$

$$\text{SS removed, mg/L} = \frac{2,630 \text{ lb/day}}{(1.836 \text{ mgd})(8.34 \text{ lb/gal})} = 171.76 \text{ mg/L, round to } \textbf{172 mg/L, SS removed}$$

BIOCHEMICAL OXYGEN DEMAND REMOVAL CALCULATIONS

The biochemical oxygen demand (BOD_5) removal calculations are used to inform operators about the efficiency of the treatment process for a pond or trickling filter. The BOD_5 is an empirical test that informs the operator on the relative oxygen requirements of a wastewater, and is an indicator of how much food is in the wastewater. The BOD_5 is a five-day test.

166. Given the following data, determine the BOD_5 removal in lb/day from a trickling filter.

Plant influent flow = 1,875 gpm
Influent BOD_5 concentration = 326 mg/L
Effluent BOD_5 concentration = 115 mg/L

First, determine the amount of BOD_5 removed in mg/L by subtracting the influent BOD_5 from the effluent BOD_5.

$$BOD_5 \text{ removed, mg/L} = \text{Influent } BOD_5\text{, mg/L} - \text{Effluent } BOD_5\text{, mg/L}$$

$$BOD_5 \text{ removed, mg/L} = 326 \text{ mg/L} - 115 \text{ mg/L} = 211 \text{ mg/L } BOD_5 \text{ removed}$$

Next, convert gpm to mgd.

$$\text{Number of mgd} = \frac{(1,875 \text{ gpm})(1,440 \text{ min/day})}{1,000,000/\text{mil}} = 2.7 \text{ mgd}$$

Next, solve for the amount of BOD_5 removed in lb/day by using the "pounds" formula.

Equation: **Number of lb/day BOD_5 removed = (BOD_5, mg/L)(Number of mgd)(8.34 lb/gal)**

Substitute values and solve.

Number of lb/day BOD_5 removed = (211 mg/L BOD_5)(2.7 mgd)(8.34 lb/gal)

Number of lb/day BOD_5 removed = 4,751.298 lb/day, round to **4,750 lb/day BOD_5 removed**

167. **What must have been the daily flow to a trickling filter in mgd, if the influent BOD_5 was 305 mg/L, the effluent BOD_5 was 106 mg/L, and the BOD_5 removed was 3,410 lb/day?**

First, the amount of BOD_5 removed must be determined by subtracting the influent BOD_5 from the effluent BOD_5.

BOD_5 removed, mg/L = 305 mg/L − 106 mg/L = 199 mg/L BOD_5 removed

Next, solve for the amount of BOD_5 removed in lb/day by using the "pounds" formula.

Equation: **Number of lb/day BOD_5 removed = (BOD_5, mg/L)(Number of mgd)(8.34 lb/gal)**

Rearrange the equation, then substitute values and solve.

$$\textbf{Number of mgd} = \frac{\textbf{Number of lb/day } BOD_5 \textbf{ removed}}{(\textbf{mg/L } BOD_5 \textbf{ removed})(\textbf{8.34 lb/gal})}$$

Substitute values and solve.

$$\text{Number of mgd} = \frac{3,410 \text{ lb/day}}{(199 \text{ mg/L } BOD_5 \text{ removed})(8.34 \text{ lb/gal})} = \textbf{2.05 mgd}$$

FOOD-TO-MICROORGANISM RATIO CALCULATIONS

A properly operated activated sludge process has a balance between the food entering the system and the microorganisms in the aeration tank. The best ratio varies because it depends on the activated sludge process and the characteristics of the wastewater being treated. It measures the pounds of food coming in divided by the pounds of microorganisms present. The ratio is a process control number because it helps the operator determine the proper number of microorganisms for the system in question.

Note: The day in mgd flow can be dropped in these types of problems.

168. **Given the following data on an aeration tank, calculate the current food-to-microorganism (F/M) ratio.**

Primary effluent flow = 2,050 gpm
Aeration tank radius = 30.2 ft
Average height of wastewater = 9.87 ft
Mixed liquor volatile suspended solids (MLVSS) = 2,587 mg/L
BOD_5 = 252 mg/L

First, calculate the aeration tank volume in gallons.

Volume, gal = $\pi(r)^2$(Height)(7.48 gal/ft^3)

Substitute values and solve.

Volume, gal = 3.14(30.2 ft)(30.2 ft)(9.87 ft)(7.48 gal/ft^3) = 211,428 gal

Now, convert the volume of wastewater in the aeration tank to mil gal.

$$\text{Number of mil gal} = \frac{211,428 \text{ gal}}{1,000,000/\text{mil}} = 0.211428 \text{ mil gal}$$

Next, convert the primary effluent flow from gpm to mgd.

$$\text{Number of mgd} = \frac{(2,050 \text{ gpm})(1,440 \text{ min}/\text{day})}{1,000,000/\text{mil}} = 2.952 \text{ mgd}$$

Next, write the equation: $\mathbf{F/M} = \dfrac{(\text{BOD}_5, \text{mg/L})(\text{Flow, mgd})(8.34\ \text{lb/gal})}{(\text{mg/L MLVSS})(\text{Volume in tank, mil gal})(8.34\ \text{lb/gal})}$

The 8.34 lb/gal cancels, leaving the following equation.

$$\text{F/M} = \dfrac{(\text{mg/L BOD}_5)(\text{Flow, mgd})}{(\text{mg/L MLVSS})(\text{Volume of tank, mil gal})}$$

Substitute values and solve.

$$\text{F/M} = \dfrac{(252\ \text{mg/L BOD}_5)(2.952\ \text{mgd})}{(2{,}587\ \text{mg/L MLVSS})(0.211428\ \text{mil gal})} = \mathbf{1.36\ F/M\ ratio}$$

169. **What is the food-to-microorganism (F/M) ratio for an aeration tank that has a radius of 27.9 ft, liquid level of 14.8 ft, if the primary effluent flow averages 2,120 gpm, the MLVSS is 2,850 mg/L, and the BOD$_5$ is 281 mg/L?**

First, calculate the number of gallons in the aeration tank.

Number of gallons = $\pi(\text{radius})^2(\text{Height, ft})(7.48\ \text{gal/ft}^3)$

Number of gallons = $(3.14)(27.9\ \text{ft})(27.9\ \text{ft})(14.8\ \text{ft})(7.48\ \text{gal/ft}^3) = 270{,}584\ \text{gal}$

Next, convert the number of gallons in the aeration tank to mil gal.

$$\text{Number of mil gal} = \dfrac{270{,}584\ \text{gal}}{1{,}000{,}000/\text{mil}} = 0.270584\ \text{mil gal}$$

Next, convert the effluent flow in gpm to mgd.

$$\text{Number of mgd} = \dfrac{(2{,}120\ \text{gpm})(1{,}440\ \text{min/day})}{1{,}000{,}000/\text{mil}} = 3.0528\ \text{mgd}$$

Next, write the equation.

$$\mathbf{F/M} = \dfrac{(\text{BOD}_5, \text{mg/L})(\text{Flow, mgd})}{(\text{mg/L MLVSS})(\text{Volume of tank, mil gal})}$$

Substitute values and solve.

$$\text{F/M} = \dfrac{(281\ \text{mg/L BOD}_5)(3.0528\ \text{mgd})}{(2{,}850\ \text{mg/L MLVSS})(0.270584\ \text{mil gal})} = \mathbf{1.11\ F/M\ ratio}$$

WASTE RATES USING FOOD/MICROORGANISM RATIO CALCULATIONS

These calculations are used in determining the wasting rate from aeration tanks. Because these calculations concentrate on organics, it is important to also use the mean cell residence time determinations, as it also includes inorganics. In addition, because the F/M ratio varies from day to day, the seven-day moving average should be used when estimating wasting rates with the F/M ratio calculation.

170. **Calculate the required waste rate from an aeration tank in mgd and gpm, given the following data:**

Aeration tank volume = 0.806 mil gal
Desired COD lb/MLVSS lb = 0.19
Primary effluent flow = 2.95 mgd
Primary effluent COD = 128 mg/L
Mixed liquor volatile suspended solids (MLVSS) = 3,672 mg/L
Waste volatile solids (WVS) concentration = 4,125 mg/L

First, find the existing MLVSS in pounds.

Equation: **Existing MLVSS, lb = (MLVSS, mg/L)(mil gal)(8.34 lb/gal)**

Existing MLVSS, lb = (3,672 mg/L)(0.806 mil gal)(8.34 lb/gal) = 24,683 lb MLVSS

Next, determine the desired MLVSS in pounds.

Equation: $\textbf{Desired MLVSS, lb} = \dfrac{\textbf{(Primary effluent COD, mg/L) (mgd) (8.34 lb/gal)}}{\textbf{Desired COD lb/MLVSS lb}}$

Desired MLVSS, lb = $\dfrac{(128 \text{ mg/L}) (2.95 \text{ mgd}) (8.34 \text{ lb/gal})}{0.19 \text{ COD lb/MLVSS lb}}$ = 16,575 lb MLVSS

Next, subtract the existing MLVSS from the desired MLVSS to find the waste in pounds.

Waste, lb = 24,683 lb − 16,575 lb = 8,108 lb

Next, calculate the waste rate in mgd.

Equation: **Waste rate, mgd** $= \dfrac{\text{Waste, lb}}{(\text{WVS concentration, mg/L})\,(8.34\ \text{lb/gal})}$

Waste rate, mgd $= \dfrac{8{,}108\ \text{lb}}{(4{,}125\ \text{mg/L})\,(8.34\ \text{lb/gal})} = 0.23568$ mgd, round to **0.24 mgd**

Lastly, calculate the waste rate in gpm.

Waste rate, gpm $= \dfrac{(0.23568\ \text{mgd})\,(1{,}000{,}000\ \text{gpd/mgd})}{1{,}440\ \text{min/day}} = 163.67$ gpm, round to **160 gpm**

---

171. Calculate the required waste rate from an aeration tank in mgd and gpm, given the following data:

Aeration tank volume = 1.245 mil gal
Desired COD lb/MLVSS lb = 0.200
Primary effluent flow = 4.82 mgd
Primary effluent COD = 158 mg/L
MLVSS = 3,790 mg/L
Waste volatile solids (WVS) concentration = 4,285 mg/L

First, find the existing MLVSS in pounds.

Equation: **Existing MLVSS, lb = (MLVSS, mg/L)(Aeration tank, mil gal)(8.34 lb/gal)**

Existing MLVSS, lb = (3,790 mg/L)(1.245 mil gal)(8.34 lb/gal) = 39,353 lb MLVSS

Next, determine the desired MLVSS in pounds.

Equation: **Desired MLVSS, lb** $= \dfrac{(\text{Primary effluent COD, mg/L})\,(\text{mgd})\,(8.34\ \text{lb/gal})}{\text{Desired COD lb/MLVSS lb}}$

Desired MLVSS, lb $= \dfrac{(158\ \text{mg/L})\,(4.82\ \text{mgd})\,(8.34\ \text{lb/gal})}{0.200\ \text{COD lb/MLVSS lb}} = 31{,}757$ lb MLVSS

Next, subtract the existing MLVSS from the desired MLVSS to find the waste in pounds.

Waste, lb = 39,353 lb − 31,757 lb = 7,596 lb

Next, calculate the waste rate in mgd.

Equation: **Waste rate, mgd** $= \dfrac{\text{Waste, lb}}{(\text{WVS concentration, mg/L})(8.34 \text{ lb/gal})}$

Waste rate, mgd $= \dfrac{7,596 \text{ lb}}{(4,285 \text{ mg/L})(8.34 \text{ lb/gal})} = 0.21255$ mgd, round to **0.213 mgd**

Lastly, calculate the waste rate in gpm.

Waste rate, gpm $= \dfrac{(0.21255 \text{ mgd})(1,000,000 \text{ gpd/mgd})}{1,440 \text{ min/day}} = 147.6$ gpm, round to **148 gpm**

SEED SLUDGE PROBLEMS

This calculation is required for determining how much seed sludge in gallons to use for starting a new digester.

172. Given the following data, determine the seed sludge required in gallons:

Digester diameter = 48.8 ft
Liquid level in digester = 16.2 ft
Requires 14.5% seed sludge

First, determine the number of gallons in the digester.

Volume, gal = (0.785)(Diameter)²(Depth, ft)(7.48 gal/ft³)

Volume, gal = (0.785)(48.8 ft)(48.8 ft)(16.2 ft)(7.48 gal/ft³) = 226,530 gal

Next, use the following equation.

Seed sludge, gal = (Capacity of digester) $\dfrac{\text{(Percent seed sludge required)}}{100\%}$

Seed sludge, gal = (226,530 gallons)$\dfrac{(14.5\%)}{100\%}$

Seed sludge, gal = 32,846.85 gal, round to **32,800 gal of seed sludge**

173. **A digester with a radius of 24.6 ft and a sludge level of 12.7 ft has a seed sludge requirement of 15.0% of the digester capacity. How many gallons of seed sludge will be needed?**

First, determine the number of gallons in the digester.

Number of gallons = π(radius)^2(Height, ft)(7.48 gal/ft^3)

Number of gallons = (3.14)(24.6 ft)(24.6 ft)(12.7 ft)(7.48 gal/ft^3) = 180,512 gal

Next, use the following equation.

Seed sludge, gal = (Capacity of digester)$\dfrac{\text{(Percent seed sludge required)}}{100\%}$

Seed sludge, gal = (180,512 gallons)$\dfrac{(15.0\%)}{100\%}$ = **27,100 gal of seed sludge**

GRAVITY THICKENER SOLIDS LOADING PROBLEMS

Gravity thickeners use large tanks that separate suspended solid and mineral matter from the liquid by gravity. The gravity thickener concentrates the sludge to reduce the load on processes that follow (conditioning, dewatering, and digestion) and produces a clear liquid, which is decanted. Flocculants are used to speed up the settling process. Operators can calculate the solids loading in lb/d/ft^2 or the hydraulic loading in gal/day/ft^2. The hydraulic loading calculation is used by operators to determine if the process is being overloaded or underloaded. See Figures E-4, E-5, and E-6 in Appendix E for three types of wastewater plants using the thickening process.

174. **Given the following data, determine the solids loading on a gravity thickener in lb/d/ft²:**

Gravity thickener radius = 29.3 ft
Influent flow = 31.0 gpm
Percent solids = 3.77%

First, determine the area of the gravity thickener.

Know: Area = π(radius)² or πr^2

Area = (3.14)(29.3 ft)(29.3 ft) = 2,695.66 ft²

Equation: **Solids loading, lb/d/ft²** $= \dfrac{\text{(Flow, gpm)}\,(1{,}440\ \text{min/day})\,(8.34\ \text{lb/gal})\,(\text{Percent solids})}{\text{(Gravity thickener area)}\,(100\%)}$

Substitute values and solve.

Solids loading, lb/d/ft² $= \dfrac{(31.0\ \text{gpm})\,(1{,}440\ \text{min/day})\,(8.34\ \text{lb/gal})\,(3.77\%)}{(2{,}695.66\ \text{ft}^2)\,(100\%)}$

Solids loading, lb/d/ft² = **5.21 lb/d/ft²**

175. **What is the solids loading on the gravity thickener, if the percent solids is 4.01%, the influent flow is 38.5 gpm, and the gravity thickener has a radius of 27.25 ft?**

First, convert the gpm to gpd.

Number of gpd = (38.5 gpm)(1,440 min/day) = 55,440 gpd

Know: Area of gravity thickener = πr^2 where $\pi = 3.14$

Equation: **Solids loading, lb/d/ft²** $= \dfrac{\text{(Flow, gpd)}\,(8.34\ \text{lb/gal})\,(\text{Percent solids})}{\text{(Gravity thickener area)}\,(100\%)}$

Substitute values and solve.

Solids loading, lb/d/ft² $= \dfrac{(55{,}440\ \text{gpd})\,(8.34\ \text{lb/gal})\,(4.01\%)}{(3.14)\,(27.25\ \text{ft})\,(27.25\ \text{ft})\,(100\%)}$

Solids loading, lb/d/ft² = **7.95 lb/d/ft²**

DISSOLVED AIR FLOTATION: THICKENER SOLIDS LOADING PROBLEMS

The dissolved air flotation technique is used to thicken sludge. These types of calculations are used for evaluating process control.

176. **Given the following data, determine the solids loading in lb/d/ft² on a dissolved air flotation (DAF) thickener unit:**

DAF unit length = **70.2 ft**
DAF unit width = **29.9 ft**
Waste-activated sludge (WAS) = **7,983 mg/L**
Sludge flow = **121 gpm**
Sludge specific gravity = **1.05 lb/gal**

First, determine the area of the DAF.

DAF area, ft² = (Length, ft)(Width, ft)

DAF area, ft² = (70.2 ft)(29.9 ft) = 2,098.98 ft²

Next, convert gpm to mgd.

$$\text{Number of gpd} = \frac{(121 \text{ gpm})(1,440 \text{ min}/\text{day})}{1,000,000/\text{mil}} = 0.17424 \text{ mgd}$$

Next, determine the lb/gal for the sludge using the specific gravity.

Sludge, lb/gal = (8.34 lb/gal)(1.05 sp gr) = 8.757 lb/gal

Equation: **Solids loading, lb/d/ft²** $= \dfrac{(\text{WAS, mg/L})(\text{mgd})(\text{Sludge, lb/gal})}{\text{DAF area, ft}^2}$

$$\text{Solids loading, lb/d/ft}^2 = \frac{(7,983 \text{ mg/L, WAS})(0.17424 \text{ mgd})(8.757 \text{ lb/gal, Sludge})}{2,098.98 \text{ ft}^2 \text{ DAF}}$$

Solids loading, lb/d/ft² = **5.80 lb/d/ft²**

177. What are the solids loading for a dissolved air flotation (DAF) unit in lb/hr/ft² that is 60.1 ft by 24.1 ft, where sludge flow averages 109 gpm, with a waste-activated sludge (WAS) concentration of 6,779 mg/L, and the sludge has a specific gravity of 1.03?

First, determine the area of the DAF unit in ft².

DAF area, ft² $=$ (60.1 ft)(24.1 ft) $=$ 1,448.41 ft²

Next, convert gpm to mgd.

$$\text{Number of mgd} = \frac{(109 \text{ gpm})\,(1,440 \text{ min}/\text{day})}{1,000,000/\text{mil}} = 0.15696 \text{ mgd}$$

Next, calculate the weight of the sludge in lb/gal.

Sludge, lb/gal $=$ (8.34 lb/gal)(1.03 sp gr) $=$ 8.59 lb/gal

Next, calculate the solids loading.

Equation: **Solids loading, lb/hr/ft²** $= \dfrac{(\text{WAS, mg/L})\,(\text{mgd})\,(\text{Sludge, lb/gal})}{(\text{DAF area, ft}^2)\,(24 \text{ hr}/\text{day})}$

Solids loading, lb/hr/ft² $= \dfrac{(6,779 \text{ mg/L, WAS})\,(0.15696 \text{ mgd})\,(8.59 \text{ lb/gal, Sludge})}{(1,448.41 \text{ ft}^2 \text{ DAF})\,(24 \text{ hr}/\text{day})}$

Solids loading, lb/hr/ft² $=$ **0.263 lb/hr/ft²**

DISSOLVED AIR FLOTATION: AIR-TO-SOLIDS RATIO CALCULATIONS

Air-to-solids ratio calculations are used to determine the efficiency of the process, as the air flotation thickener and the solids in the system must be in balance. Typically the ratio ranges from 0.01 to 0.1.

178. What is the air-to-solids ratio for a dissolved air flotation (DAF) unit that has an air flow rate equal to 9.7 ft³/min, a solids concentration of 0.72%, and a flow of 161,000 gpd?

Know: Air = 0.0807 lb/ft³ at standard temperature, pressure, and average composition

Equation: **Air-to-solids ratio** $= \dfrac{(\text{Air flow, ft}^3/\text{min})\,(\text{Air, lb/ft}^3)}{(\text{gpm})\,(\text{Percent solids}/100\%)\,(8.34\ \text{lb/gal})}$

Because equation requires gpm, first convert gpd to gpm.

Number of gpm $= \dfrac{161,000\ \text{gpd}}{1,440\ \text{min/day}} = 111.8\ \text{gpm}$

Now, using the air-to-solids equation, Substitute values and solve.

Air-to-solids ratio $= \dfrac{(9.7\ \text{ft}^3/\text{min})\,(0.0807\ \text{lb/ft}^3)}{(111.8\ \text{gpm})\,(0.72\%/100\%)\,(8.34\ \text{lb/gal})} =$ **0.12 Air-to-solids ratio**

179. Given the following data, determine the air-to-solids ratio for a DAF unit:

DAF influent flow = 0.148 mgd
Air flow = 7.25 ft³/min
Solids concentration = 0.745%
Solids specific gravity = 1.04

Know: Air = 0.0807 lb/ft³ at standard temperature, pressure, and average composition

Equation: **Air-to-solids ratio** $= \dfrac{(\text{Air flow, ft}^3/\text{min})\,(\text{Air, lb/ft}^3)}{(\text{gpm})\,(\text{Percent solids}/100\%)\,(\text{Solids, lb/gal})}$

Since equation requires gpm, first convert mgd to gpm.

Number of gpm $= \dfrac{(0.148\ \text{mgd})\,(1,000,000/\text{mil})}{1,440\ \text{min/day}} = 102.78\ \text{gpm}$

Then, the number of lb/gal is required because the solids weigh more than water.

Solids, lb/gal = (8.34 lb/gal)(1.04 sp gr) = 8.6736 lb/gal

Now, using the air-to-solids equation, Substitute values and solve.

Air-to-solids ratio $= \dfrac{(7.25\ \text{ft}^3/\text{min})\,(0.0807\ \text{lb/ft}^3)}{(102.78\ \text{gpm})\,(0.745\%/100\%)\,(8.6736\ \text{lb/gal})} =$ **0.0881 Air-to-solids ratio**

DISSOLVED AIR FLOTATION: AIR RATE FLOW CALCULATIONS

Operators use air rate flow calculations for evaluating process control.

180. **If a DAF unit receives air at an average rate of 0.227 m³/min, how many lb/hr of air does it receive?** *Note:* **1 cubic meter = 35.3 cubic ft.**

First, convert m³/min to ft³/min.

Air flow, ft³/min = (0.227 m³/min)(35.3 ft³/m³) = 8.0131 ft³/min

Know: Air = 0.0807 lb/ft³ at standard temperature, pressure, and average composition

Equation: **Air, lb/hr = (Air flow, ft³/min)(60 min/hr)(0.0807 lb/ft³, Air)**

Substitute values and solve.

Air, lb/hr = (8.0131 ft³/min)(60 min/hr)(0.0807 lb/ft³)

Air, lb/hr = **38.8 lb/hr of Air**

181. **If a DAF unit receives air at an average rate of 0.219 m³/min, how many lb/day of air does it receive?**

First, convert m³/min to ft³/min.

Air flow, ft³/min = (0.219 m³/min)(35.3 ft³/m³) = 7.7307 ft³/min

Know: Air = 0.0807 lb/ft³ at standard temperature, pressure, and average composition

Equation: **Air, lb/day = (Air flow, ft³/min)(1,440 min/day)(0.0807 lb/ft³, Air)**

Substitute values and solve.

Air, lb/day = $(7.7307 \text{ ft}^3/\text{min})(1{,}440 \text{ min}/\text{day})(0.0807 \text{ lb/ft}^3)$

Air, lb/day = 898.37 lb/day, round to **898 lb/day of Air**

CENTRIFUGE THICKENING PROBLEMS

Centrifuges are used to dewater sludge, usually after applying gravity thickening. They apply forces that are a thousand times greater than gravity. Polymers may be applied to the influent of the centrifuge to facilitate solids thickening.

182. **What is the solids loading on a scroll centrifuge in lb/hr, if it has a waste activated sludge (WAS) flow of 98,400 gpd and a solids concentration of 6,990 mg/L?**

First, convert the solids concentration from mg/L to percent.

$$\text{Percent solids} = \frac{6{,}990 \text{ mg/L}}{10{,}000 \text{ mg/L}/1\%} = 0.699\%$$

$$\text{Solids loading, lb/hr} = \frac{(98{,}400 \text{ gpd})(8.34 \text{ lb/gal})(0.699\%/100\%)}{24 \text{ hr/day}} = \textbf{239 lb/hr}$$

183. **Given the following data, what is the feed time in minutes for a basket centrifuge thickener?**

Basket centrifuge thickener capacity = 24 ft³
Sludge flow rate = 42,900 gpd
Solids concentration = 6,552 mg/L
Percent solids = 5.94%

Know: 1% = 10,000 ppm or mg/L

First, convert the solids concentration in mg/L to percent.

$$\text{Percent solids} = \frac{6,552 \text{ mg/L}}{10,000 \text{ mg/L/1\%}} = 0.6552\%$$

Next, convert gpd to gpm.

$$\text{Number of gpm} = \frac{42,900 \text{ gpd}}{1,440 \text{ min/day}} = 29.8 \text{ gpm}$$

Now, calculate the feed time in minutes.

Equation: **Feed time, min** $= \dfrac{(\text{Capacity, ft}^3)(\text{Solids, \%/100\%})(7.48 \text{ gal/ft}^3)(8.34 \text{ lb/gal})}{(\text{Flow, gpm})(\text{Solids concentration, \%/100\%})(8.34 \text{ lb/gal})}$

Simplify equation by canceling out the 8.34 lb/gal and the 100%.

Feed time, min $= \dfrac{(\text{Capacity, ft}^3)(\text{Solids, \%})(7.48 \text{ gal/ft}^3)}{(\text{Flow, gpm})(\text{Solids concentration, \%})}$

$$\text{Feed time, min} = \frac{(24 \text{ ft}^3)(5.94\%)(7.48 \text{ gal/ft}^3)}{(29.8 \text{ gpm})(0.6552\%)} = 54.6 \text{ min, round to } \mathbf{55 \text{ min}}$$

SAND DRYING BED PROBLEMS

By knowing how thick the sludge was when applied and later measuring the thickness of the dried sludge, an operator can use these calculations to determine the efficiency of the drying bed process.

184. A drying bed is 312 ft long and 51.8 ft wide. If on average 3.46 in. of sludge that has 5.02% solids were applied to the drying bed and the drying and removal cycle on average takes 21.5 days, how many gallons of sludge were applied on average for each cycle, and what is the solids loading rate in lb/yr/ft^2?

First, convert 3.46 inches to feet.

Number of feet = 3.46 in./12 in./ft = 0.28833 ft

Next, determine the volume in ft^3 sent to the drying bed.

Volume, ft^3 = (312 ft)(51.8 ft)(0.28833 ft) = 4,659.928 ft^3

Next, calculate the volume in gallons sent to the sand drying beds.

Number of gal = (4,659.928 ft^3)(7.48 gal/ft^3) = 34,856.26 gal, round to **34,900 gal**

Next, calculate the number of lb.

Number of lb = (4,659.928 ft^3)(7.48 gal/ft^2)(8.34 lb/gal) = 290,701 lb

Lastly, calculate the solids loading rate.

Equation: **Solids loading rate, lb/yr/ft^2** $= \dfrac{\dfrac{(\text{lb})(365 \text{ days})(\text{Percent solids})}{(\text{Drying cycle})(\text{yr})(100\%)}}{\text{Drying bed area, ft}^2}$

Substitute values and solve.

Solids loading rate, lb/yr/ft^2 $= \dfrac{\dfrac{(290,701 \text{ lb})(365 \text{ days})(5.02\% \text{ solids})}{(21.5 \text{ days})(\text{yr})(100\%)}}{(312 \text{ ft})(51.8 \text{ ft})}$

Solids loading rate, lb/yr/ft^2 $= \dfrac{(13,520.977 \text{ lb/day})(365 \text{ days/yr})(0.0502)}{16,161.6 \text{ ft}^2}$

Solids loading rate, lb/yr/ft^2 = **15.3 lb/yr/ft^2**

185. Given the following data, what is the total amount of sludge in gallons applied to these drying beds?

Drying bed 1 = 283 ft by 42.0 ft
Drying bed 2 = 275 ft by 41.9 ft
Drying bed 1 had 3.49 in. of sludge applied to it
Drying bed 2 had 3.27 in. of sludge applied to it

First, convert the inches of sludge applied to each drying bed to feet.

Drying bed 1, ft = 3.49 in./12 in./ft = 0.2908 ft

Drying bed 2, ft = 3.27 in./12 in./ft = 0.2725 ft

Next, determine the volume in ft^3 sent to drying beds 1 and 2.

Drying bed 1, ft^3 = (283 ft)(42.0 ft)(0.2908 ft) = 3,456 ft^3

Drying bed 2, ft^3 = (275 ft)(41.9 ft)(0.2725 ft) = 3,140 ft^3

Lastly, calculate the volume in gallons sent to both sand drying beds.

Number of gal = (3,456 ft^3 + 3,140 ft^3)(7.48 gal/ft^3)

Number of gal = (6,596 ft^3)(7.48 gal/ft^3) = 49,338.08 gal, round to **49,300 gal**

DEWATERING CALCULATIONS

This section contains several types of dewatering problems. The more water that is removed from sludge, the less cost associated with further processing or disposal. The problems are important to the operator because they are helpful in evaluating process control or in informing the operator of process efficiency. See Figure E-12 in Appendix E for one type of sludge process using dewatering.

186. **What is the sludge feed rate for a belt filter press to process 10,780 lb/day of sludge, if it operates only 9.75 hr/day?**

$$\text{Sludge feed rate, lb/hr} = \frac{10,780 \text{ lb/day}}{9.75 \text{ hr/day}} = 1,105.64 \text{ lb/hr, round to } \mathbf{1,110 \text{ lb/hr}}$$

187. **A vacuum filter has a wet cake flow of 4,090 lb/hr and a filter that is 13.2 by 29.8 ft. Calculate the filter yield in lb/hr/ft², if the percent solids are 17.4%.**

First, determine the area of the filter.

Area of filter, ft² = (13.2 ft)(29.8 ft) = 393.36 ft²

Equation: **Filter yield, lb/hr/ft²** $= \dfrac{(\text{Wet cake flow, lb/hr})(\text{Percent solids/100\%})}{\text{Area, ft}^2}$

Substitute values and solve.

$$\text{Filter yield, lb/hr/ft}^2 = \frac{(4,090 \text{ lb/hr})(17.4\%/100\%)}{393.36 \text{ ft}^2} = 1.809 \text{ lb/hr/ft}^2, \text{ round to } \mathbf{1.81 \text{ lb/hr/ft}^2}$$

SETTLEABLE SOLIDS CALCULATIONS

These tests are performed on samples from either the clarifier's influent or effluent or from a sedimentation tank. They are used to determine the percent and thus the efficiency of settleable solids. Calculations based on these tests follow.

188. **A 2,000.0 mL of activated sludge was collected in a graduated cylinder. What is the percent of settleable solids, if after exactly 30 minutes the sludge solids that settled totaled 319 mL?**

Equation: **Percent settleable solids** $= \dfrac{(\text{Settled sludge, mL})(100\%)}{\text{Sample size, mL}}$

Substitute values and solve.

$$\text{Percent settleable solids} = \frac{(319 \text{ mL})(100\%)}{2,000.0 \text{ mL}} = 15.95\%, \text{ round to } \mathbf{16.0\% \text{ Settled solids}}$$

189. Given the following data, calculate the percent settleable solids:

Activated sludge sample = 0.980 gallon poured into a large graduated cylinder
Settling time is exactly = 30 minutes
Sludge solids in graduated cylinder = 802 mL

First, convert the sample in gallons to mL.

Number of mL = (0.980 gal)(3,785 mL/gal) = 3,709.3 mL

Equation: **Percent settleable solids** $= \dfrac{(\text{Settled sludge, mL})(100\%)}{\text{Sample size, mL}}$

Substitute values and solve.

Percent settleable solids $= \dfrac{(802 \text{ mL})(100\%)}{3,709.3 \text{ mL}} = 21.62\%$, round to **21.6% Settled solids**

COMPOSTING CALCULATIONS

Composting is an aerobic biological process. This process decomposes organic matter to a stable end product. The optimum moisture content for composting ranges from 50 to 60% water. Several different composting calculations follow.

190. What is the percent moisture content of a compost blend, given the following data?

9,450 lb of sludge was added and mixed with 5,380 lb of compost
Compost = 61.8% solids
Added sludge = 28.1% solids

Since percent moisture needs to be solved, first determine the percent moisture of both the compost and sludge.

Compost percent moisture = 100% − 61.8% solids = 38.2% moisture content

Sludge percent moisture = 100% − 28.1% solids = 71.9% moisture content

Equation: **Mixture's % moisture** $=$

$$\frac{[(\text{Sludge, lb})(\%\text{ moisture}) + (\text{Compost, lb})(\%\text{ moisture})]100\%}{\text{Sludge, lb} + \text{Compost, lb}}$$

Substitute values and solve.

Mixture's percent moisture $= \dfrac{[(9,450\text{ lb})(71.9\%/100\%) + (5,380\text{ lb})(38.2\%/100\%)]100\%}{9,450\text{ lb} + 5,380\text{ lb}}$

Mixture's percent moisture $= \dfrac{(6,794.55\text{ lb} + 2,055.16\text{ lb})100\%}{14,830\text{ lb}} = \mathbf{59.7\%}$

191. **Given the following data, calculate the amount of compost in lb/day that needs to be blended with a dewatered sludge to make a mixture that has a moisture content of 45.0%.**

Dewatered digester primary sludge $=$ 7,850 lb/day
Compost solids $=$ 72.4%
Dewatered sludge solids $=$ 39.5%

First, determine the moisture content of the dewatered sludge and the compost.

Dewatered sludge moisture $= 100\% - 39.5\% = 60.5\%$

Compost percent moisture $= 100\% - 72.4\% = 27.6\%$

Equation: **Mixture's % moisture** $=$

$$\frac{[(\text{Sludge, lb/day})(\%\text{ moisture}) + (\text{Compost, lb/day})(\%\text{ moisture})]100\%}{\text{Sludge, lb/day} + \text{Compost, lb/day}}$$

Substitute values and solve.

45.0% moisture content $= \dfrac{[(7,850\text{ lb/day})(60.5\%/100\%) + (x\text{ lb/day})(27.6\%/100\%)]100\%}{7,850\text{ lb/day} + x\text{ lb/day}}$

Solve for x.

Divide both sides of the equation by 100%.

$$0.450 = \frac{(7{,}850\ \text{lb/day})\,(60.5\%/100\%) + (x\ \text{lb/day})\,(27.6\%/100\%)}{7{,}850\ \text{lb/day} + x\ \text{lb/day}}$$

Simplify terms in the numerator.

$$0.450 = \frac{4{,}749.25\ \text{lb/day} + 0.276\,x\ \text{lb/day}}{7{,}850\ \text{lb/day} + x\ \text{lb/day}}$$

Multiply both sides of the equation by $(7{,}850\ \text{lb/day} + x\ \text{lb/day})$

$$0.450(7{,}850\ \text{lb/day} + x\ \text{lb/day}) = 4{,}749.25\ \text{lb/day} + 0.276x\ \text{lb/day}$$

Multiply terms on the left side of the equation.

$$3{,}532.5\ \text{lb/day} + 0.450x\ \text{lb/day} = 4{,}749.25\ \text{lb/day} + 0.276x\ \text{lb/day}$$

Subtract $0.276x$ and $3{,}532.5$ lb/day from both sides of the equation.

$$0.450x\ \text{lb/day} - 0.276x\ \text{lb/day} = 4{,}749.25\ \text{lb/day} - 3{,}532.5\ \text{lb/day}$$

Simplify both sides of the equation.

$$0.174x\ \text{lb/day} = 1{,}216.75\ \text{lb/day}$$

$x = 6{,}992.82$ lb/day, round to **6,990 lb/day of Compost required**

192. **Given the following data for blending compost (BC) with wood chips, calculate the percent of the blended compost:**

Bulk density of sludge = 1,690 lb/yd³
Sludge volume = 22.5 yd³
Sludge solids content = 20.3%
Density of wood chips = 625 lb/yd³
Wood chip solids = 51.8%
Mix ratio (MR) of wood chips to sludge = exactly 3 to 1

Equation: **Percent solids BC** $=$

$$\frac{(\text{Sludge, yd}^3)(\text{lb/yd}^3)(\% \text{ solids, sludge}) + (\text{Sludge, yd}^3)(\text{MR})(\text{lb/yd}^3)(\% \text{ solids, chips})(100\%)}{(\text{Sludge, yd}^3)(\text{lb/yd}^3) + (\text{Sludge, yd}^3)(\text{Mix ratio})(\text{lb/yd}^3)}$$

Substitute values and solve.

Percent solids BC $=$

$$\frac{[(22.5 \text{ yd}^3)(1,690 \text{ lb/yd}^3)(20.3\%/100\%) + (22.5 \text{ yd}^3)(3)(625 \text{ lb/yd}^3)(51.8\%/100\%)](100\%)}{(22.5 \text{ yd}^3)(1,690 \text{ lb/yd}^3) + (22.5 \text{ yd}^3)(3)(625 \text{ lb/yd}^3)}$$

Percent solids BC $= \dfrac{(7,719.075 + 21,853.125)(100\%)}{38,025 + 42,187.5} = \dfrac{(29,572.2)(100\%)}{80,212.5}$

Percent solids BC $= 36.87\%$, round to **36.9% Solids in blended compost**

193. **Given the following parameters calculate the amount of wet compost a composting site can process in lb/day and tons/day:**

Compost site capacity $= $ 12,750 yd^3
Average compost cycle $= $ 23.8 days
Bulk density of compost averages 918 lb/yd^3

Equation: **Compost cycle, days** $= \dfrac{(\text{Site capacity, yd}^3)(\text{Density of compost, lb/yd}^3)}{x \text{ Wet compost lb/day}}$

Substitute values.

$23.8 \text{ days} = \dfrac{(12,750 \text{ yd}^3)(918 \text{ lb/yd}^3)}{x \text{ lb/day}}$

Solve for x.

$x \text{ lb/day} = \dfrac{(12,750 \text{ yd}^3)(918 \text{ lb/yd}^3)}{(23.8 \text{ days})} = 491,786 \text{ lb/day}$, round to **492,000 lb/day Capacity**

Lastly, calculate the capacity in tons.

Site capacity, tons $= \dfrac{491,786 \text{ lb/day}}{2,000 \text{ lb/ton}} = 245.893 \text{ tons}$, round to **246 tons**

194. If 7,260 lb of dewatered biosolids are mixed with 3,180 lb of compost each day, the dewatered biosolids (DB) have a solids content of 27.3%, and the compost has a moisture content of 32.7%, what is the moisture content of the mixture?

First, calculate the moisture content in the biosolids.

DB percent moisture content = 100% − 27.3% = 72.7%

Next, calculate the percent moisture in mixture.

Equation: **Percent moisture in mixture =**

$$\frac{[(DB, \text{lb/day})(\text{Percent moisture DB}) + (\text{Compost lb/day})(\text{Percent moisture compost})]\,100\%}{DB, \text{lb/day} + \text{Compost, lb/day}}$$

$$\text{Percent moisture in mixture} = \frac{[(7,260\ \text{lb/day})(72.7\%/100\%) + (3,180\ \text{lb/day})(32.7\%)]\,100\%}{7,260\ \text{lb/day} + 3,180,\ \text{lb/day}}$$

$$\text{Percent moisture in mixture} = \frac{(5,278.02\ \text{lb/day} + 1,039.86\ \text{lb/day})(100\%)}{10,440\ \text{lb/day}}$$

Percent moisture in mixture = 60.52%, round to **60.5% Moisture content of mixture**

195. Calculate this particular system's composting cycle time in days for the following set of data:

Site available capacity = 7,250 yd³
Bulk density of wet sludge = 1,695 lb/yd³
Bulk density of wet compost = 975 lb/yd³
Sludge solids content = 21.7%
Bulk density of wet wood chips = 735 lb/yd³
Dry solids = 20,220 lb/day
Mix ratio (MR) of wood chips to sludge = 3.25 to 1

Equation: **Cycle time, days =**

$$\frac{(\text{Capacity, yd}^3)(\text{Bulk density of wet compost lb/yd}^3)}{\dfrac{\text{Dry solids, lb/day}}{\text{Percent solids}} + \dfrac{(\text{Dry solids, lb/day})}{\text{Percent solids}}\,(\text{MR})\dfrac{(\text{Bulk density of wood chips, lb/yd}^3)}{(\text{Bulk density of wet sludge lb/yd}^3)}}$$

Substitute values and solve.

$$\text{Cycle time, days} = \cfrac{(7,250 \text{ yd}^3)(975 \text{ lb/yd}^3)}{\cfrac{20,220 \text{ lb/day}}{21.7\%/100\%} + \cfrac{(20,220 \text{ lb/day})}{21.7\%/100\%} (3.25) \cfrac{(735 \text{ lb/yd}^3)}{1,695 \text{ lb/yd}^3}}$$

Reduce and simplify.

$$\text{Cycle time, days} = \cfrac{7,068,750 \text{ lb}}{\cfrac{20,220 \text{ lb/day}}{0.217} + \cfrac{(20,220 \text{ lb/day})(1.4093)}{0.217}}$$

Reduce again.

$$\text{Cycle time , days} = \cfrac{7,068,750 \text{ lb}}{93,179.724 \text{ lb/day} + 131,318.18 \text{ lb/day}}$$

$$\text{Cycle time, days} = \cfrac{7,068,750 \text{ lb}}{224,497.90 \text{ lb/day}}$$

Cycle time, days = 31.49 days, round to **31.5 days**

196. **What is the capacity for a compost site in processing dry sludge in lb/day and tons/day, given the following data?**

Site capacity = 11,100 yd³
Average compost cycle = 28.2 days
Bulk density of wet sludge = 1,690 lb/yd³
Bulk density of wet compost = 965 lb/yd³
Sludge solids content = 18.3%
Bulk density of wet wood chips = 625 lb/yd³
Mix ratio (MR) of wood chips to sludge = exactly 3 to 1

Equation: **Cycle time, days =**

$$\cfrac{(\text{Capacity, yd}^3)(\text{Bulk density of wet compost lb/yd}^3)}{\cfrac{x \text{ Dry solids, lb/day}}{\text{Percent solids}} + \cfrac{(x \text{ Dry solids, lb/day})}{\text{Percent solids}} (\text{MR}) \cfrac{(\text{Bulk density of wood chips, lb/yd}^3)}{(\text{Bulk density of wet sludge lb/yd}^3)}}$$

Substitute values and solve.

$$28.2 \text{ days} = \cfrac{(11,100 \text{ yd}^3)(965 \text{ lb/yd}^3)}{\cfrac{x \text{ Dry solids, lb/day}}{18.3\%/100\%} + \cfrac{(x \text{ Dry solids, lb/day})}{18.3\%/100\%} (3) \cfrac{(625 \text{ lb/yd}^3)}{1,690 \text{ lb/yd}^3}}$$

Reduce and simplify.

$$28.2 \text{ days} = \cfrac{10,711,500 \text{ lb}}{\cfrac{x \text{ Dry solids, lb/day}}{0.183} + \cfrac{(x \text{ Dry solids, lb/day})(1.1095)}{0.183}}$$

Reduce again by dividing 0.183 into $1x$ and $1.1095\ x$.

$$28.2\ \text{days} = \frac{10{,}711{,}500\ \text{lb}}{5.464x\ \text{lb/day} + 6.063x, \text{lb/day}}$$

$$28.2\ \text{days} = \frac{10{,}711{,}500\ \text{lb}}{11.527x}$$

Solve for x.

$$11.527x = \frac{10{,}711{,}500\ \text{lb}}{28.2\ \text{days}}$$

$$x = \frac{10{,}711{,}500\ \text{lb}}{(28.2\ \text{days})\,(11.527)}$$

$x = 32{,}952$ lb/day, round to **33,000 lb/day Dry sludge**

$$\text{Dry sludge, tons/day} = \frac{32{,}952\ \text{lb/day}}{2{,}000\ \text{lb/ton}} = 16.476\ \text{tons/day, round to } \textbf{16.5 tons/day Dry sludge}$$

BIOSOLIDS DISPOSAL CALCULATIONS

Biosolids disposal is a beneficial and an environmentally safe reuse of the stable waste products in the wastewater treatment process. Biosolids are increasingly being land applied because they no longer can be dumped into the ocean, and landfill space is in shorter supply. Various factors must be considered when applying biosolids to land, such as the amount of nitrogen available for plant growth, nitrogen loading rate, phosphorus, or metals in the biosolids. See Figure E-12 in Appendix E for one type of sludge process using land application.

197. **Determine the number of acres used in the land application of biosolids, given the following data:**

Hydraulic loading rate = 0.60 in./day
Flow = 1,235,000 gpd

First, determine the number of gallons per acre-inch.

Number of gals/acre-in. = (43,560 ft²/acre)(1 ft/12 in.)(7.48 gal/ft³) = 27,152 gal/acre-in.

Equation: **Hydraulic loading rate, in./day** $= \dfrac{\text{Flow, gpd}}{(27,152 \text{ gal/acre in.})(\text{Area, acres})}$

Rearrange the equation and solve for area in acres.

$$\text{Area, acres} = \dfrac{\text{Flow, gpd}}{(27,152 \text{ gal/acre in.})(\text{Hydraulic loading rate, in./day})}$$

Substitute values and solve.

$$\text{Area, acres} = \dfrac{1,235,000 \text{ gpd}}{(27,152 \text{ gal/acre in.})(0.60 \text{ in./day})}$$

Area, acres $= 75.8$ acres, round to **76 acres**

198. Calculate the flow in gpd for a 115-acre land application site, if the hydraulic loading rate is 0.38.

Know: 27,152 gal/acre-in.

Equation: **Hydraulic loading rate, in./day** $= \dfrac{\text{Flow, gpd}}{(27,152 \text{ gal/acre in.})(\text{Area, acres})}$

Rearrange the equation and solve for flow in gpd.

Flow, gpd = (Area, acres)(27,152 gal/acre-in.)(Hydraulic loading rate, in./day)

Substitute values and solve.

Flow, gpd = (115 acres)(27,152 gal/acre-in.)(0.38 in./day)

Flow, gpd $= 1,186,542.4$ gal, round to **1,200,00 gpd**

199. What is the plant available nitrogen (PAN) in lb per dry ton, given the following data?

Organic nitrogen (N) in biosolids $= 23,400$ mg/kg
Ammonia nitrogen (N) $= 11,900$ mg/kg

Biosolids volatilization rate (VR) = 0.50
Mineralization rate (MR) from aerated lagoon = 0.30

Equation: **PAN, lb/dry ton** =

[(Organic N, mg/kg)(MR) + (Ammonia N, mg/kg)(VR)](0.002 lb/dry ton)

PAN, lb/dry ton = [(23,400 mg/kg)(0.30) + (11,900 mg/kg)(0.50)](0.002 lb/dry ton)

PAN, lb/dry ton = (7,020 + 5,950)(0.002 lb/dry ton)

PAN, lb/dry ton = (12,970)(0.002 lb/dry ton) = 25.94 lb/dry ton, round to **26 lb/dry ton**

200. **If the plant available nitrogen (PAN) of the biosolids is 32.5 lb/dry ton, how many dry tons per acre of PAN should be applied for a crop that requires 195 lb of nitrogen per acre?**

Equation: **PAN, dry tons/acre** $= \dfrac{\text{Plant nitrogen required, lb/acre}}{\text{PAN, lb/dry ton}}$

PAN, dry tons/acre $= \dfrac{195 \text{ lb/acre}}{32.5 \text{ lb/dry ton}} =$ **6.00 dry tons/acre PAN**

201. **How many lb/acre/year of plant available nitrogen (PAN) will be applied to a wastewater land application field, given the following parameters?**

Nitrate (NO_3) = 13.4 mg/L
Nitrite (NO_2) = 0.59 mg/L
Total Kjeldahl nitrogen (TKN)* = 61.8 mg/L
Ammonia (NH_3) = 20.7 mg/L
Applying 17.5 in/acre/year
Mineralization rate (MR) = 0.30
Volatilization rate (VR) = 0.50

First, determine the PAN applied per year to each acre.

Equation: **PAN, mg/L = [MR(TKN − NH$_3$)] + [1 − VR(NH$_3$)] + (NO$_3$ + NO$_2$)**

PAN, mg/L = [0.30(61.8 mg/L − 20.7 mg/L)] + [1 − 0.50(20.7 mg/L)] + (13.4 mg/L + 0.59 mg/L)

PAN, mg/L = 0.30(41.1 mg/L) + 10.35 mg/L + 13.99 mg/L

PAN, mg/L = 12.33 mg/L + 10.35 mg/L + 13.99 mg/L = 36.67 mg/L

Next, convert the hydraulic loading rate from inches to mil gal/acre/yr.

$$\text{Flow, mil gal/acre/yr} = \frac{(43,560 \text{ ft}^2/\text{acre})(1 \text{ ft}/12 \text{ in.})(7.48 \text{ gal/ft}^3)(17.5 \text{ in.}/\text{acre/yr})(1 \text{ acre})}{1,000,000/\text{mil}}$$

Flow, mil gal/acre/yr = 0.475167 mil gal/acre/yr

Lastly, solve for PAN in lb/acre/yr.

PAN, lb/acre/yr = (36.67 mg/L)(0.475167 mil gal/acre/yr)(8.34 lb/gal)

PAN, lb/acre/yr = 145.32 lb/acre/yr, round to **150 lb/acre/yr PAN**

Note: TKN = organic nitrogen plus ammonia

202. **Given the parameters for the following land application site, calculate the concentration of total nitrogen; nitrogen loading rate, and phosphorus loading rate in lb/day, lb/yr, and lb/acre/yr, and COD growing season loading rate in lb/day and lb/acre/day:**

Land application site = 166 acres
Growing season = 196 days/yr
Average hydraulic loading rate = 0.172 mgd
Total Kjeldahl nitrogen (TKN) = 18.7 mg/L
Nitrate (NO$_3$) = 0.31 mg/L
Nitrite (NO$_2$) = 0.11 mg/L
Ammonia (NH$_3$) = 5.6 mg/L
Total phosphate = 3.22 mg/L
COD = 284 mg/L

First, calculate the total nitrogen.

Total nitrogen (N) = Nitrate, mg/L + Nitrite, mg/L + TKN, mg/L

Total nitrogen (N) = 0.31 mg/L + 0.11 mg/L + 18.7 mg/L = 19.12 mg/L, round to **19.1mg/L N**

Now, determine the total nitrogen loading rate.

Nitrogen loading rate, lb/day = (Total Nitrogen, mg/L)(mgd)(8.34 lb/gal)

Nitrogen loading rate, lb/day = (19.12 mg/L)(0.172 mgd)(8.34 lb/gal)

Nitrogen loading rate, lb/day = 27.427 lb/day, round to **27.4 lb/day Nitrogen**

Nitrogen loading rate, lb/yr = (27.427 lb/day)(196 days/yr)

Nitrogen loading rate, lb/yr = 5,375.692 lb/yr, round to **5,380 lb/yr Nitrogen**

Nitrogen loading rate, lb/acre/yr = $\dfrac{5,375.692 \text{ lb/yr}}{166 \text{ acres}}$

Nitrogen loading rate, lb/acre/yr = 32.384 lb/acre/yr, round to **32.4 lb/acre/yr Nitrogen**

Phosphorus (P) loading rate, lb/day = (Total P, mg/L)(mgd)(8.34 lb/gal)

Phosphorus loading rate, lb/day = (3.22 mg/L)(0.172 mgd)(8.34 lb/gal)

Phosphorus loading rate, lb/day = 4.619 lb/day, round to **4.62 lb/day Phosphorus**

Phosphorus loading rate, lb/yr = (4.619 lb/day)(196 days/yr)

Phosphorus loading rate, lb/yr = 905.324 lb/yr, round to **905 lb/yr Phosphorus**

Phosphorus loading rate, lb/acre/yr = $\dfrac{905.324 \text{ lb/yr}}{166 \text{ acres}}$

Phosphorus loading rate, lb/acre/yr = 5.454 lb/acre/yr, round to **5.45 lb/acre/yr Phosphorus**

COD loading rate, lb/day = (COD, mg/L)(mgd)(8.34 lb/gal)

COD loading rate, lb/day = (284 mg/L)(0.172 mgd)(8.34 lb/gal)

COD loading rate, lb/day = 407.39 lb/day, round to **407 lb/day COD**

COD loading rate, lb/acre/yr = $\dfrac{407.39 \text{ lb/day}}{166 \text{ acres}}$

COD loading rate, lb/acre/yr = **2.45 lb/acre/day COD**

203. Given the following data, calculate the sodium absorption ratio (SAR) for a wastewater:

Sodium (Na$^+$) = 133 mg/L
Calcium (Ca^{2+}) = 37.9 mg/L
Magnesium (Mg^{2+}) = 20.3 mg/L
Equivalent weight of sodium = 22.99
Equivalent weight of calcium = 20.04
Equivalent weight of magnesium = 12.15

Note: See next section, Laboratory Calculations, on how to calculate equivalent weights.

First, determine the milliequivalents (meq) for sodium, calcium, and magnesium.

Sodium, meq = $\dfrac{133 \text{ mg/L}}{22.99}$ = 5.785 meq of sodium

Calcium, meq = $\dfrac{37.9 \text{ mg/L}}{20.04}$ = 1.891 meq of calcium

Magnesium, meq = $\dfrac{20.3 \text{ mg/L}}{12.15}$ = 1.671 meq of magnesium

Now, Solve for SAR.

Equation: **SAR** $= \dfrac{\textbf{Na}^+}{[(0.5)(\textbf{Ca}^{2+} + \textbf{Mg}^2 +)]^{1/2}}$

Substitute values and solve.

$$SAR = \frac{5.785 \text{ meq Na}^+}{[(0.5)(1.891 \text{ meq Ca}^{2+} + 1.671 \text{ meq Mg}^{2+})]^{1/2}}$$

$$SAR = \frac{5.785 \text{ meq Na}^+}{[(0.5)(3.562)]^{1/2}}$$

$$SAR = \frac{5.785 \text{ meq Na}^+}{[1.781]^{1/2}}$$

Calculate the square root of 1.781.

$$SAR = \frac{5.785 \text{ meq Na}^+}{1.3345} = \textbf{4.33 SAR}$$

CHEMISTRY AND LABORATORY PROBLEMS

Operators should have a thorough understanding of many laboratory calculations for they help in evaluating plant processes and efficiencies. Following are a few examples.

204. **What is the normality (N) of an HCl solution if 1.25 equivalents are dissolved in 4.5 liters of solution?**

Equation: $\textbf{Normality} = \dfrac{\text{Number of equivalents of solute}}{\text{Liters of solution}}$

$\text{Normality} = \dfrac{1.25 \text{ equivalents}}{4.5 \text{ liters}} = \textbf{0.28 N HCl}$

205. **What is the normality (N) of a potassium permanganate solution if 200.5 grams (g) are dissolved in 2.00 liters of deionized (D.I.) water?**

Know: $KMnO_4 = 158.034$ g/equivalent (g-eq)

First, determine the number of equivalents that were dissolved in the D.I. water.

Equation: $\textbf{Equivalents} = \dfrac{\text{Grams in solution}}{\text{Equivalent weight}}$

In this case, it would be as follows:

$$KMnO_4 \text{ equivalents} = \frac{\text{Grams, } KMnO_4}{\text{Equivalent weight of } KMnO_4}$$

$$KMnO_4 \text{ equivalents} = \frac{200.5g \ KMnO_4}{158.034 \text{ g-eq of } KMnO_4} = 1.269 \text{ g-eq}$$

Next, calculate the normality of the solution.

$$\text{Normality, } KMnO_4 = \frac{\text{Number of equivalents } KMnO_4}{\text{Liters of solution}}$$

$$\text{Normality, } KMnO_4 = \frac{1.269 \text{ g-eq } KMnO_4}{2.00 \text{ liters}} = \textbf{0.634 N } KMnO_4$$

206. What is the molarity of a calcium carbonate ($CaCO_3$) solution, if 345.8 grams (g) of $CaCO_3$ is dissolved in 3.28 liters of deionized water?

Given the following grams/mole atomic weights:

Ca = 40.08 g/mole
C = 12.011 g/mole
O = 15.999 g/mole

First, determine the number of g/mole for **$CaCO_3$**.

Equation: **$CaCO_3$, g/mole = Ca, g/mole + C, g/mole + 3(O, g/mole)**

$CaCO_3$, g/mole = 40.08 g/mole + 12.011 g/mole + 3(15.999 g/mole)

$CaCO_3$, g/mole = 40.08 g/mole + 12.011 g/mole + 47.997 g/mole) = 100.088 g/mole

Next, determine the number of moles that are in 345.8 grams of $CaCO_3$.

$$\text{Moles, } CaCO_3 = \frac{345.8 \text{ g}}{100.088 \text{ g/mole}} = 3.455 \text{ moles of } CaCO_3$$

Now, calculate the molarity.

Equation: **Molarity** $= \dfrac{\text{Moles solute}}{\text{Liters solution}}$

Molarity of solution $= \dfrac{3.455 \text{ moles of solute}}{3.28 \text{ liters}} = $ **1.05 Molarity CaCO$_3$ solution**

EQUIVALENT WEIGHTS

Chemical reactions such as oxidation-reduction and neutralization will always occur between substances in equal numbers of gram-equivalents. (Appendix C contains the atomic weights of all the naturally occurring substances known, along with some manmade substances, in Table C-1. Table C-2 contains some common substances that are encountered in the water industry. These tables allow you to look up the atomic or formula weights and determine what the equivalent weights are of different chemical species.) In oxidation-reduction reactions, the number of electrons gained by the oxidizer will be equal to the number of electrons lost by the reducing agent. The same holds true with neutralization reactions, where one H^+ will neutralize one OH^- molecule.

Example 1: The reaction between NaOH and HCl will react in such a way that one equivalent weight of NaOH will react with exactly one equivalent weight of HCl. How do you find the equivalent weight? Look at Table C-3 for NaOH. It shows NaOH to be 39.997 or 39.997 grams/ formula weight (fw) or 39.997 equivalent weight (eq-wt) or if dealing in milliequivalents, 39.997 meq-wt, or most chemists use 39.997 eq or 39.997 meq. Thus, for NaOH the equivalent weight is simply the formula weight. The same can be said of HCl. The formula weight equals the eq-wt. In Table C-3, HCl is 36.461 grams/formula. Thus one gram formula weight of NaOH will react with one gram formula weight of HCl. That is, 39.997 grams of NaOH will react exactly with 36.461 grams of HCl.

Example 2: The reaction between NaOH and H$_2$S. This reaction will be different because to neutralize one molecule of H$_2$S will require two molecules of NaOH. Reactions must balance.

$2NaOH + H_2S \rightarrow Na_2S + 2H_2O$

What this means is that it takes 2 eq-wt of NaOH to neutralize 1 eq-wt of H$_2$S. Most chemists will write or say 1 eq-wt of NaOH will neutralize ½ eq-wt of H$_2$S. Chemists also write as g-eq.

Example 3: The reaction of hydrogen (H^+) with electrons

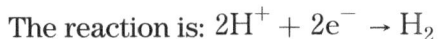

The reaction is: $2H^+ + 2e^- \rightarrow H_2$

This one is a little tricky. In order to form hydrogen gas, we have to have two atoms of hydrogen and they require one electron each. Thus the ratio of reactants is 2:2, which reduces to 1:1. This means that it takes one eq-wt of hydrogen. Table C-3 shows this to be 1.008 g/fw.

Example 4: The reaction of calcium ions with electrons.

$$Ca^{2+} + 2e^- \rightarrow Ca^0$$

The zero indicates neutral and is usually not written.

In this reaction it still takes two electrons to neutralize one calcium. Thus it is a ratio of 1:2 or, as chemists write it, ½:1. This means that it only takes ½ grams/formula weight of calcium. In Table C-1, the atomic weight of calcium is 40.08. Thus, the eq-wt is half this, or 20.04 g/eq or 20.04 mg/meq.

207. What is the equivalent weight in grams of Na_2CO_3 in the following reaction?

$$2HCl + Na_2CO_3 \rightarrow 2NaCl + H_2O + CO_2$$

Since it takes two molecules HCl to neutralize one molecule of Na_2CO_3, it follows that

Two eq of HCl will react with one eq of Na_2CO_3 or written as 2 g-eq HCl to 1 g-eq Na_2CO_3. Usually the ratio is written 1 g-eq HCl to 1/2 g-eq Na_2CO_3. Thus, the g-eq weight of Na_2CO_3 will need to be divided by 2.

The ratio is 2:1 or 1:½.

Looking up the formula weight of Na_2CO_3 in Table C-3 shows it to be 105.989 g/fw.

Now, solve the problem.

$$Na_2CO_3, \text{g-eq} = \frac{105.989 \text{ g/fw}}{2 \text{ eq/fw}} = \textbf{52.9945 g-eq of } Na_2CO_3$$

208. What is the equivalent weight in grams of H_2SO_4 in its reaction with NaOH?

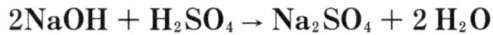

$$2NaOH + H_2SO_4 \rightarrow Na_2SO_4 + 2\,H_2O$$

Two eq-wt of NaOH will react with one of H_2SO_4, which is a ratio of 2:1 or 1:½. Keep NaOH at 1 equivalent in the ratio, as the problem above. Then we will need to divide by 2 (½) to find the number of grams/equivalent in H_2SO_4. By keeping the highest number in the ratio at one, you will see what needs to be done to the chemical species in question, that is, ½, ⅓, ¼, etc., division by 2, 3, and 4 respectively.

Now, solve the problem. Look up in Table C-3 for the g/fw of H_2SO_4. It is 98.07 g/fw

$$H_2SO_4,\ \text{g-eq} = \frac{98.07\ \text{g/fw}}{2\ \text{eq/fw}} = \textbf{49.035 g-eq } \mathbf{H_2SO_4}$$

209. How many grams of NaOH will react with 20.45 g of H_2SO_4?

Know: 1 g-eq of NaOH reacts with ½ g-eq of H_2SO_4, which equals 49.035 g/g-eq

Know: The formula weight of NaOH equals the g-eq weight (39.997 g)

First, determine the number of equivalents in 20.45 g of H_2SO_4.

$$\text{Number of g-eq in 20.45 g of } H_2SO_4 = \frac{20.45\ \text{g}}{49.035\ \text{g/g eq}} = 0.4170\ \text{g-eq } H_2SO_4$$

Thus, 0.4170 g-eq of NaOH are required.

Know: eq-wt or g-eq of NaOH equals 39.997 g/g-eq

Number of NaOH g = (0.4170 g-eq)(39.997 g/g-eq) = **16.68 g of NaOH**

210. What is the percent Ba in barium sulfate ($BaSO_4$)? *Note:* **round atomic weights to nearest 0.01.**

The equation for calculating the % Ba in $BaSO_4$ is:

$$\% \ Ba = \frac{(\text{Molecular Wt of Ba})(100)}{\text{Molecular Wt of } BaSO_4}$$

Next, look up in Table C-1, Appendix C, the atomic weights for potassium (K), phosphorus (P), and oxygen (O). Arrange the data in tabular form, as shown below. Multiply the atomic weight by the corresponding number of atoms for each element. Then, add the formula weights to find the number of g/fw of $BaSO_4$.

First, determine the molecular weight of each of the elements in the compound.

Element	Number of Atoms		Atomic Wt		Molecular Wt
Ba	1	×	137.33	=	137.33
S	1	×	32.06	=	32.06
O	4	×	16.00	=	64.00
			Molecular Wt of $BaSO_4$	=	233.39

The molecular wt of Ba in $BaSO_4$ is 233.40.

Substitute in above formula.

$$\% \ Ba = \frac{(137.33)(100)}{233.39} = \textbf{58.84\% Ba}$$

211. What is the concentration of polymer in mg/L, if 29.2 mL of a 0.375 grams/liter polymer solution is added to 1,000 mL of deionized water?

Equation: $\textbf{Polymer, mg/L} = \dfrac{(\text{Stock, mL})(1,000 \ \text{mg/gram})(\text{Concentration in grams/liter})}{\text{Sample size, mL}}$

$$\text{Polymer, mg/L} = \frac{(29.2 \ \text{mL})(1,000 \ \text{mg/gram})(0.375 \ \text{grams/liter})}{1,000 \ \text{mL}} = \textbf{10.95 mg/L}$$

212. A 0.1% stock solution (1,000 ppm or 1,000 mg/L) is required for doing jar tests. If the alum has a specific gravity of 1.32 and is 48.5% aluminum sulfate, how many milliliters of alum are required to make exactly 1,000 mL of stock solution?

First, find the number of lb/gal of alum.

Alum, lb/gal = (sp gr)(8.34 lb/gal) = (1.32 sp gr)(8.34 lb/gal) = 11.0088 lb/gal

Next, determine the number of grams/mL.

$$\text{Number grams/mL, Alum} = \frac{(11.0088 \text{ lb/gal})(48.5\% \text{ Al}_2\text{SO}_4, \text{Purity})(454 \text{ grams/lb})}{(3,785 \text{ mL/gal})(100\%)}$$

Number grams/mL, Alum = 0.6404 grams/mL

Convert grams/mL to milligrams/mL

Number mg/mL = (0.6404 grams/mL)(1,000 mg/g) = 640.4 mg/mL

Next, convert mL to liters by multiplying by 1,000.

Number mg/L = (640.4 mg/mL)(1,000 mL/liter) = 640,400 mg/liter

Next, determine the number of mL required.

Equation: $\mathbf{C_1V_1 = C_2V_2}$

(640,400 mg/mL)(x, mL) = (1,000 mg/liter)(1,000 mL)

$$x, \text{mL} = \frac{(1,000 \text{ mg/liter})(1,000 \text{ mL})}{640,400 \text{ mg/liter}} = \mathbf{1.56 \text{ mL, Alum}}$$

Now, using a pipette, add 1.56 mL of the 48.5% alum solution to a clean, dry 1,000-mL flask. Dilute the alum to the 1,000-mL mark with deionized water. Add a magnetic stir bar and place the flask on a magnetic stirrer. Turn the magnetic stirrer on and mix this solution with the bar as vigorously as possible for at least 10 minutes.

Thus, every 1 mL of stock alum solution that is added to a 1,000-mL jar of raw water will be equivalent to adding a 1-mg/L dose (because of second dilution with raw water; 1,000 mg/L/ 1,000 mg/L raw water sample = 1 mg/L). If 10 mL of this stock solution were added to the 1,000-mL raw water sample, it would be a dose of 10 mg/L. If you are using the 2-liter square jars, simply double the mL added for each mg/L dosage increase desired. Another way is to feed the alum neat by using a micropipette; pipette 0.00156 mL of alum into a 1,000-mL raw water sample.

213. **A 0.1% stock polymer solution (1,000 ppm or 1,000 mg/L) is desired for performing a jar test. If the polymer has a specific gravity of 1.24 and is 100% polymer, how many milliliters of polymer are required to make exactly 1,000 mL stock solution?**

First, find the number of lb/gal of polymer.

Polymer, lb/gal = (sp gr)(8.34 lb/gal) = (1.24 sp gr)(8.34 lb/gal) = 10.34 lb/gal

Next, determine the number of grams/mL.

$$\text{Number of grams/mL, Polymer} = \frac{(10.34\ \text{lb/gal})(63.0\%\ \text{Polymer})(454\ \text{grams/lb})}{(3,785\ \text{mL/gal})(100\%)}$$

Number of grams/mL, Polymer = 0.781 grams/mL

Convert grams/mL to milligrams/mL.

Number of mg/mL = (0.781 grams/mL)(1,000 mg/g) = 781 mg/mL

Next, convert mL to liters by multiplying by 1,000.

Number mg/L = (781 mg/mL)(1,000 mL/liter) = 781,000 mg/liter

Next, determine the number of mL required.

Equation: $C_1V_1 = C_2V_2$

(781 mg/mL)(x, mL) = (1,000 mg/liter)(1,000 mL)

$$x,\ \text{mL} = \frac{(1,000\ \text{mg/liter})(1,000\ \text{mL})}{781,000\ \text{mg/Liter}} = \textbf{1.28 mL, Polymer}$$

Now, using a pipette, add 1.28 mL of the 100% polymer solution to a clean, dry 1,000-mL flask. Dilute the polymer to the 1,000-mL mark with deionized water. Add a magnetic stir bar and place the flask on a magnetic stirrer. Turn the magnetic stirrer on and mix this solution with the bar as vigorously as possible for at least 10 minutes.

Thus, every 1 mL of polymer solution that is added to 1,000-mL sample of raw water will add a 1-mg/L dose (because of second dilution with raw water; 1,000 mg/L/1,000 mg/L raw water sample = 1 mg/L). If 10 mL of this stock solution were added to the 1,000-mL raw water sample, it would be a dose of 10 mg/L. If you are using the 2-liter square jars, simply double the mL added for each mg/L dosage increase desired. Another way is to feed the polymer neat by using a micropipette; pipette 0.00128 mL of polymer into a 1,000 mL raw water sample.

214. How many milliliters of 1.35 Normal (N) sulfuric acid (H_2SO_4) solution will neutralize 8.95 g NaOH?

Know: 1 gram-equivalent (g-eq) of H_2SO_4 will completely react with 1 g-eq of NaOH

The equivalent weight of NaOH = Formula weight = 39.997 grams (Table C-3, Appendix C)

Gram-equivalents of NaOH = Gram-equivalents of H_2SO_4

Gram-equivalents of NaOH = (Liters, H_2SO_4)(Normality H_2SO_4)

$$\frac{8.95 \text{ g NaOH}}{39.997 \text{ g/g eq}} = (\text{Liters, } H_2SO_4)(1.35 \text{ N } H_2SO_4)$$

Solve for liters H_2SO_4.

$$\text{Liters, } H_2SO_4 = \frac{8.95 \text{ g NaOH}}{(39.997 \text{ g/g eq})(1.35 \text{ N } H_2SO_4)} = 0.1658 \text{ L}$$

Finally, convert liters to milliliters.

Number of mL = (0.1658 L)(1,000 mL/L) = 165.8 mL, round to **166 mL H_2SO_4**

215. Calculate the unseeded BOD$_5$ in mg/L, given the following data:

Start of test bottle dissolved oxygen (DO) = 8.4 mg/L
Bottle was incubated for 5 days in the dark at 20°C
After 5 days DO = 3.0 mg/L
Sample size = 125 mL
Total volume = 300 mL

Equation: $\mathbf{BOD_5}$ **unseeded, mg/L** $= \dfrac{(\text{Initial DO, mg/L} - \text{Final DO, mg/L})(\text{Total volume, mL})}{\text{Sample volume, mL}}$

$\text{BOD}_5 \text{ unseeded, mg/L} = \dfrac{(8.4 \text{ mg/L} - 3.0 \text{ mg/L})(300 \text{ mL})}{125 \text{ mL}} = 12.96 \text{ mg/L, round to } \mathbf{13 \text{ mg/L BOD}_5}$

216. Calculate the seeded BOD$_5$ in mg/L, given the following data:

Sample size = 120 mL
Initial DO = 8.2 mg/L
Final DO = 2.9 mg/L
BOD$_5$ of seed stock = 75 mg/L
Seed stock = 5.0 mL
Total volume = 300 mL

First, calculate the seed correction in mg/L.

Equation: **Seed correction, mg/L** $= \dfrac{(\text{BOD}_5 \text{ of seed stock, mg/L})(\text{Seed stock, mg/L})}{\text{Total volume, mL}}$

$\text{Seed correction, mg/L} = \dfrac{(75 \text{ mg/L})(5.0 \text{ mg/L})}{300 \text{ mL}} = 1.25 \text{ mg/L}$

Next, calculate the BOD$_5$ seeded in mg/L.

Equation: **BOD$_5$ seeded, mg/L** $=$

$\dfrac{(\text{Initial DO, mg/L} - \text{Final DO, mg/L} - \text{Seed correction, mg/L})(\text{Total volume, mL})}{\text{Sample volume, mL}}$

$\text{BOD}_5 \text{ seeded, mg/L} = \dfrac{(8.2 \text{ mg/L} - 2.9 \text{ mg/L} - 1.25 \text{ mg/L})(300 \text{ mL})}{120 \text{ mL}}$

$\text{BOD}_5 \text{ seeded, mg/L} = \dfrac{(4.05 \text{ mg/L})(300 \text{ mL})}{120 \text{ mL}} = 10.125 \text{ mg/L, round to } \mathbf{10 \text{ mg/L}}$

217. **Calculate the percent total solids (TS) and percent volatile solids (VS) for a biosolids sample, given the following data:**

	Biosolids Sample, g	Dried Sample, g	Burnt Sample (ash), g
Sample and dish wt	89.88 g	29.57 g	26.43 g
Weight of dish	25.06 g	25.06 g	25.06 g

First, determine the total solids by subtracting weight of the dish from the weight of the dried sample.

Total solids, g = 29.57 g − 25.06 g = 4.51 g

Next, determine the original weight of the biosolids (sample).

Equation: weight of biosolids, g = Sample and dish wt, g − wt of dish, g

Weight of biosolids, g = 89.88 g − 25.06 g = 64.82 g

Now, calculate the percent of total solids.

Equation: **Percent total solids** $= \dfrac{\text{(Weight of total solids, g)}\,(100\%)}{\text{Weight of biosolids sample, g}}$

Percent total solids $= \dfrac{(4.51 \text{ g})\,(100\%)}{64.82 \text{ g}} = 6.9577\%$, round to **6.96% TS**

Lastly, calculate the percent VS.

First, find the solids lost in burning.

Equation: **Solids (ash), g = Sample and dish dried, g − Burnt sample, g**

Solids (ash), g = 29.57 g − 26.43 g = 3.14 g

Equation: **Percent VS** $= \dfrac{\text{(Solids lost, g)}\,(100\%)}{\text{Weight of total solids, g}}$

Percent VS $= \dfrac{(3.14 \text{ g})\,(100\%)}{4.51 \text{ g}} =$ **69.6% VS**

218. Calculate the percent solids and percent volatile solids (VS) for a biosolids sample, given the following data:

	Biosolids Sample, g	Dried Sample, g	Burnt Sample (ash), g
Sample and dish wt	96.71 g	29.48 g	26.67 g
Weight of dish	25.02 g	25.02 g	25.02 g

First, determine the total solids by subtracting weight of the dish from the weight of the dried sample.

Total solids, g = 29.48 g − 25.02 g = 4.46 g

Next, determine the original weight of the biosolids (sample).

Equation: weight of biosolids, g = Sample and dish wt, g – wt of dish, g

Weight of biosolids, g = 96.71 g − 25.02 g = 71.69 g

Now, calculate the percent of total solids.

Equation: **Percent total solids** $= \dfrac{(\text{Weight of total solids, g})\,(100\%)}{\text{Weight of biosolids sample, g}}$

Percent total solids $= \dfrac{(4.46\text{ g})\,(100\%)}{71.69\text{ g}} = 6.2212\%$, round to **6.22%**

Lastly, calculate the percent VS.

First, find the solids lost in burning.

Solids (ash), g = Sample and dish dried, g − Burnt sample, g

Solids (ash), g = 29.48 g − 26.67 g = 2.81 g

Equation: **Percent VS** $= \dfrac{(\text{Solids (ash), g})\,(100\%)}{\text{Weight of total solids, g}}$

Percent VS $= \dfrac{(2.81\text{ g})\,(100\%)}{4.46\text{ g}} = \mathbf{63.0\%}$

219. **Calculate the number of mg/L of suspended solids (SS) and percent of volatile suspended solids (VSS) in a primary effluent sample, given the following data:**

	Dried Sample, g	Burnt Sample (ash), g
Sample and dish wt	25.7183	25.7102
Weight of dish	25.7049	25.7049
Volume of sample, mL	100.0 mL	

First, determine the amount of suspended solids in grams that were in the 50 mL sample.

$$SS, g = 25.7183\ g - 25.7102\ g = 0.0081\ g\ SS$$

Next, calculate the amount of SS in the primary effluent.

$$SS, mg/L = \frac{(0.0081\ g)(1{,}000\ mg)(10)}{(100\ mL)(1\ g)(10)} = \frac{81\ mg}{1{,}000\ mL}$$

Note: 10 is a multiplier to convert 100 mL to 1,000 mL.

$$SS, mg/L = \frac{(81\ mg)(1{,}000\ mL)}{(1{,}000\ mL)(1\ L)} = \textbf{81 mg/L}$$

Next, determine the weight of the VSS.

$$VSS, g = 25.7102\ g - 25.7049\ g = 0.0053\ g\ VSS$$

Now, calculate the percent VSS.

$$Percent\ VSS = \frac{(Weight\ of\ VS, g)(100\%)}{Weight\ of\ SS}$$

$$Percent\ VSS = \frac{(0.0053\ g\ VSS)(100\%)}{0.0081\ g\ SS} = \textbf{65.43\%}$$

220. A composite sample is being collected at a wastewater plant. If eight samples totaling 4,000 mL are required and the average flow rate is 2.222 mgd, what will be the proportioning factor and what will be the number of mL for each sample time?

Time	Flow, mgd
0600 Hours	2.15
0900 Hours	2.48
1200 Hours	2.75
1500 Hours	2.26
1800 Hours	2.34
2100 Hours	2.86
2400 Hours	2.05
0300 Hours	0.89
Total Flow	17.78
Average Flow	2.222

First, calculate the proportioning factor.

Equation: Proportioning factor $= \dfrac{\text{Total mL required}}{(\text{Number of samples})(\text{Average flow, mgd})}$

Substitute values and solve.

Proportioning factor $= \dfrac{4,000 \text{ mL}}{(8 \text{ samples})(2.222 \text{ mgd})} = $ **225 mL**

Next, calculate the volumes needed for each sample time.

Sample volume $=$ (mgd)(Proportioning factor)

Substitute the known values for each time frame and round samples to nearest 10 mL.

Sample volume 0600 Hours	=	(2.15)(225)	=	483.75 mL, round to **480 mL**
Sample volume 0900 Hours	=	(2.48)(225)	=	558.00 mL, round to **560 mL**
Sample volume 1200 Hours	=	(2.75)(225)	=	618.75 mL, round to **620 mL**
Sample volume 1500 Hours	=	(2.26)(225)	=	508.50 mL, round to **510 mL**
Sample volume 1800 Hours	=	(2.34)(225)	=	526.50 mL, round to **530 mL**
Sample volume 2100 Hours	=	(2.86)(225)	=	643.50 mL, round to **640 mL**
Sample volume 2400 Hours	=	(2.05)(225)	=	461.25 mL, round to **460 mL**
Sample volume 0300 Hours	=	(0.89)(225)	=	200.25 mL, round to **200 mL**

Total (cross check) = **4,000 mL**

BASIC ELECTRICITY PROBLEMS

Operators should have a basic understanding of electrical calculations, and they must always exercise safety in dealing with electricity at wastewater treatment plants or anywhere.

221. What is the voltage (E) on a circuit, if the current is 12 amperes (I or amps) and the resistance (R) is 18.3 ohms?

Equation: **Voltage = (Amps)(Resistance, ohms) or E = (I)(R)**

Substitute values and solve.

Voltage = (12 amps)(18.3 ohms) = 219.6 volts, round to **220 volts**

222. **What is the resistance on a wire, if the amperes are 12.6 and the voltage is 110.2 volts?**

Equation: **Voltage = (Amps)(Resistance, ohms)**

Rearrange the equation to solve for the resistance in ohms.

Resistance, ohms = Voltage/Amps

Substitute values and solve.

Resistance, ohms = 110.2 volts/12.6 amps = 8.746 ohms, round to **8.75 ohms**

223. **The valve on a secondary clarifier is controlled by a 4- to 20-mA signal from the SCADA system. If the valve is 82.5% open and has a 0% to 100% range, what must the signal be in milliamp (mA) from the SCADA system?**

Equation: **Current process reading** $= \dfrac{(\text{Live signal, mA} - 4\,\text{mA offset})(\text{Maximum capacity})}{16\,\text{milliamp span}}$

Substitute values and solve.

$82.5\% = \dfrac{(\text{Live signal, mA} - 4\,\text{mA offset})(100\%\,\text{capacity})}{16\,\text{mA}}$

Rearrange the formula as follows and solve.

Live signal, mA $= \dfrac{(82.5\%)(16\,\text{mA})}{100\%} + 4\,\text{mA}$

Live signal, mA = **17.2 mA**

224. What would the SCADA reading be on the board in milliamps (mA) for a 4-mA to 20-mA signal, if a digester tank has a level capacity of 29.8 ft and it currently has 22.7 ft of sludge water in the tank?

Equation: **Current process reading** $= \dfrac{(\text{Live signal, mA} - 4 \text{ mA offset})(\text{Maximum capacity})}{16 \text{ milliamp span}}$

Substitute values and solve.

$22.7 \text{ ft} = \dfrac{(\text{Live signal, mA} - 4 \text{ mA offset})(29.8 \text{ ft Maximum level})}{16 \text{ mA}}$

Rearrange the formula as follows.

Live signal, mA $= \dfrac{(22.7 \text{ ft})(16 \text{ mA})}{29.8 \text{ ft}} + 4\text{mA}$

Live signal, mA $= 16.188$ mA, round to **16.2 mA**

225. The SCADA system at a wastewater plant uses a 4-mA to 20-mA signal to monitor digester tank levels. If the readout on a SCADA board reads 17.5 mA, what is the height in feet of the sludge water in a tank with a capacity of 29.3 ft?

Know: 4 mA = 0 ft in the digester tank and 20 mA = 29.3 ft in the tank

Equation: **Current process reading** $= \dfrac{(\text{Live signal, mA} - 4 \text{ mA offset})(\text{Maximum capacity})}{16 \text{ milliamp span}}$

Substitute values and solve.

Polymer level, ft $= \dfrac{(17.5 \text{ mA} - 4 \text{ mA offset})(29.3 \text{ ft Maximum level})}{16 \text{ mA}}$

Polymer level, ft $= $ **24.7 ft**

KILOWATT DETERMINATIONS

As above, operators should have a basic understanding of kilowatt calculations, and they must always exercise safety in dealing with electricity at wastewater treatment plants or anywhere.

226. How many kilowatts will it take to operate a 250-hp pump, assuming the startup energy is 2.0 times?

kW = (# of hp)(0.746 kW/hp)(Startup energy)

kW = (250 hp)(0.746 kW/hp)(2.0) = 373 kW, round to **370 kW**

227. Calculate the total kilowatts needed to operate the following facility, if everything is operating:

Automatic screen	8	hp
Chemical pumps	6	hp
Sludge pump	20	hp
Trickling filter	45	hp
Lighting	7.5	hp
Instrumentation	0.5	hp
First, add the total hp:	87.0	hp

Formula: kW = (# of hp)(0.746 kW/hp)

kW = (87 hp)(0.746 kW/hp) = 64.902 kW, round to **65 kW**

228. Calculate the kilowatts needed to run a single-phase alternating current (AC) motor, given the following data:

Volts = 440
Amps = 35
Power factor = 0.82

Equation: $\textbf{Kilowatts} = \dfrac{(\text{Volts})(\text{Amps})(\text{Power factor})}{1{,}000}$

Substitute values and solve.

$\text{Kilowatts} = \dfrac{(440 \text{ V})(35 \text{ A})(0.82)}{1{,}000} = 12.628 \text{ kW, round to } \textbf{13 kW}$

229. Calculate the kilowatts needed to run a two-phase alternating current (AC) motor, given the following data:

Volts = 460
Amps = 29
Power factor = 0.84

Equation: $\textbf{Kilowatts} = \dfrac{(\text{Volts})(\text{Amps})(\text{Power factor})}{1{,}000}$

Substitute values and solve.

$\text{Kilowatts} = \dfrac{(460 \text{ V})(29 \text{ A})(0.84)}{1{,}000} = 11.2056 \text{ kW, round to } \textbf{11 kW}$

230. Calculate the kilowatts needed to run a three-phase alternating current (AC) motor, given the following data:

Volts = 440
Amps = 31
Power factor = 0.80

Equation: $\textbf{Kilowatts} = \dfrac{(1.73)(\text{Volts})(\text{Amps})(\text{Power factor})}{1{,}000}$

Substitute values and solve.

$\text{Kilowatts} = \dfrac{(1.73)(440 \text{ V})(31 \text{ A})(0.80)}{1{,}000} = 18.878 \text{ kW, round to } \textbf{19 kW}$

231. What is the power cost in dollars and cents for a 120-mhp pump, if it operates an average of 8 hr each day and the cost is $0.07862/kW-hr? Assume 30-day month.

First, determine the number of hours the pump operates each month.

Pump operating time, hr/month = (30 days/mo)(8 hr/day) = 240 hr/month

Next, determine the kW for the pump.

Know: 746 watts per hp

$$\text{Number of kW} = \frac{(120 \text{ mhp})(746 \text{ watts/hp})}{1,000 \text{ watts/kW}} = 89.52 \text{ kW}$$

Now, calculate the cost to run the pump for one month.

Cost/month = (240 hr/mo)(89.52 kW)($0.07862/kW-hr) = **$1,689.13**

1. A total of 4,330 gpd of sludge (primary sludge) with 5.24% solids content and weighing 8.38 lb/gal is pumped to a wastewater thickener tank. Determine the amount of sludge that should flow from the thickener tank in gpd, if the sludge (secondary sludge) is further concentrated to 6.98% solids and weighs 8.44 lb/gal. Note: For convenience, primary sludge is abbreviated to 1° sludge and secondary sludge to 2° sludge.

2. What percent hypochlorite solution would result, if 175 gallons of a 13% solution were mixed with 405 gallons of a 5.7% solution? Assume both solutions have the same density.

3. Given the following data, calculate the percent solids content that results when a primary sludge is mixed with a secondary sludge.

Average of primary sludge = 2,860 gpd
Average of secondary sludge = 2,550 gpd
Primary sludge = 6.61% solids
Secondary sludge = 3.48% solids
Primary sludge = 8.37 lb/gal
Secondary sludge = 8.43 lb/gal

4. An 18-inch sewage pipeline is flowing at a velocity of 1.10 ft/s and the depth of the sewage averages 6.7 inches. Determine the flow in the pipeline in ft³/s and gpm.

5. Calculate the solids loading rate in lb of solids/d/ft^2, given the following data:

Secondary clarifier radius = 36.0 ft
Primary effluent flow = 1,750,000 gpd
Return of activated sludge is 0.475 mgd
Mixed liquor suspended solids (MLSS) = 2,645 mg/L
Specific gravity of the solids is 1.04

6. Given the following parameters, calculate how long a primary sludge pump should operate in minutes per hour:

Plant flow = 1,500 gpm
Sludge pump = 62 gpm
Influent suspended solids (SS) = 282 mg/L
Effluent SS = 112 mg/L
Sludge = 4.45% solids

7. What is the biosolids retention time (BRT) in days for a digester that is 70.3 ft in diameter, has a working level of 20.75 ft, and receives an average flow of 14.5 gpm?

8. Determine the waste activated sludge (WAS) in mgd, given the following data.

Influent flow = 2.59 mgd
Clarifier radius = 42.5 ft
Clarifier depth = 15.3 ft
Aerator = 0.840 mil gal
MLSS = 2,835 mg/L
Return activated sludge (RAS) SS = 6,970 mg/L
Secondary effluent SS = 17.2 mg/L
Target solids retention time (SRT) = 10 days exactly

9. Given the following data, calculate the cost of running a pump in dollars and cents per day:

Flow = 1.39 mgd
Total differential head (TDH) = 214 ft
Motor efficiency = 88.6%
Pump efficiency = 71.3%
Cost in kilowatt hours = $0.07568

10. A wastewater treatment plant is treating 2,065,000 gpd with a 12.2% sodium hypochlorite solution. If the dosage required is 9.65 mg/L and the specific gravity of the hypochlorite is 1.03, how many gpd of sodium hypochlorite are required?

11. Determine the number of mgd that flows through a wastewater plant's effluent, given the following data:

SO_2 dosage = 4.6 mg/L
Pounds of chlorine used = 173 lb/day
Chlorine demand = 6.35 mg/L
Assume SO_2 is exactly 3.25 mg/L higher than the chlorine residual

12. Given the following data, calculate the flocculant feed rate for a belt press in lb/hr and the flocculant dosage in lb/ton:

Flocculant concentration = 1.25%
Flocculant feed rate = 1.20 gpm
Solids loading on belt filter = 2,830 lb/hr

13. Given the following data, determine the sludge age in days for an oxidation ditch wastewater treatment plant:

MLSS = 2,690 mg/L
Ditch volume = 0.634 mil gal
Solids added = 845 lb/day

14. Calculate the mean cell residence time (MCRT), given the following data:

Flow = 2.92 mgd
Aeration tank volume = 460,000 gallons
Clarifier tank volume = 308,000 gallons
MLSS = 2,885 mg/L
Waste rate = 25,600 gpd
Waste activated sludge (WAS) = 7,050 mg/L
Effluent TSS = 15.5 mg/L

15. Given the following data, calculate whether a gravity thickener's sludge blanket will increase, remain the same, or decrease, and how much that change will be in lb/day solids, if it increases or decreases:

Pumped sludge to thickener = 110 gpm
Pumped sludge from thickener = 53 gpm
Thickener's effluent suspended solids (SS) = 757 mg/L
Primary sludge solids = 3.45%
Thickened sludge solids = 7.18%

16. What must have been the concentration of volatile acids in mg/L for a sour digester with a radius of 29.7 ft and a sludge level of 16.3 ft, if the number of pounds of lime to neutralize the volatile acids was 2,930 lb?

17. Calculate the required waste rate from an aeration tank in mgd and gpm, given the following data:

Aeration tank volume = 0.698 mil gal
Desired COD lb/MLVSS lb = 0.18
Primary effluent flow = 3.10 mgd
Primary effluent COD = 136 mg/L
Mixed liquor volatile suspended solids (MLVSS) = 3,755 mg/L
Waste volatile solids (WVS) concentration = 4,075 mg/L

18. Given the following data, determine the solids loading in lb/d/ft² on a dissolved air flotation (DAF) thickener unit:

DAF unit length = 79.9 ft
DAF unit width = 32.1 ft
Waste-activated sludge (WAS) = 7,828 mg/L
Sludge flow = 113 gpm
Sludge specific gravity = 1.04 lb/gal

19. What is the air-to-solids ratio for a dissolved air flotation (DAF) unit that has an air flow rate equal to 10.3 ft³/min, a solids concentration of 0.78%, and a flow of 144,000 gpd?

20. A drying bed is 289 feet long and 50.0 feet wide. If on average 2.87 inches of sludge that has 5.36% solids were applied to the drying bed and the drying and removal cycle on average takes 23.3 days, how many gallons of sludge were applied on average for each cycle and what is the solids loading rate in lb/yr/ft²?

21. Given the following data, calculate the amount of compost in lb/day that needs to be blended with a dewatered sludge to make a mixture that has a moisture content of 42.0%.

Dewatered digester primary sludge = 9,150 lb/day
Dewatered sludge solids = 38.4%
Compost solids = 70.6%

22. Determine the number of acres used in the following land application of biosolids, given the following data:

Hydraulic loading rate = 0.52 in./day
Flow = 1,620,000 gpd

23. A 0.1% stock solution (1,000 ppm or 1,000 mg/L) is required for doing jar tests. If the alum has a specific gravity of 1.293 and is 48.2% aluminum sulfate, how many milliliters of alum are required to make exactly 1,000 mL of stock solution?

24. Calculate the seeded BOD_5 in mg/L, given the following data:

Sample size = 150 mL
Initial dissolved oxygen (DO) = 8.8 mg/L
Final DO = 3.1 mg/L
BOD_5 of seed stock = 75 mg/L
Seed stock = 4.0 mL
Total volume = 300 mL

25. The valve on a secondary clarifier is controlled by a 4- to 20-mA signal from the SCADA system. If the valve is 43.2% open and has a 0% to 100% range, what must the signal be in mA from the SCADA system?

1. A total of 4,330 gpd of sludge (primary sludge) with 5.24% solids content and weighing 8.38 lb/gal is pumped to a wastewater thickener tank. Determine the amount of sludge that should flow from the thickener tank in gpd, if the sludge (secondary sludge) is further concentrated to 6.98% solids and weighs 8.44 lb/gal? Note: For convenience, primary sludge is abbreviated to 1° sludge and secondary sludge to 2° sludge.

Equation: **$(x$ gpd$)(2°$ sludge lb/gal$)(\%$ $2°$ sludge$) =$**

$(1°$ sludge, gpd$)(1°$ sludge lb/gal$)(\%$ $2°$ sludge$)$

Substitute values and solve.

$(x$ gpd$)(8.44$ lb/gal$)(6.98\%/100\%) = (4{,}330$ gpd$)(8.38$ lb/gal$)(5.24\%/100\%)$

$$x \text{ gpd} = \frac{(4{,}330 \text{ gpd})\,(8.38 \text{ lb/gal})\,(5.24\%/100\%)}{(8.44 \text{ lb/gal})\,(6.98\%/100\%)} = 3{,}312 \text{ gpd, round to } \textbf{3,310 gpd}$$

2. **What percent hypochlorite solution would result, if 175 gallons of a 13% solution were mixed with 405 gallons of a 5.7% solution? Assume both solutions have the same density.**

First, find the total volume that would result from mixing these two solutions.

Total Volume = 175 gal + 405 gal = 580 gal

Then, write the equation:

$$(\textbf{Concentration}_1)(\textbf{Volume}_1) + (\textbf{Concentration}_2)(\textbf{Volume}_2) = (\textbf{Concentration}_3)(\textbf{Volume}_3)$$

Condensed as $C_1V_1 + C_2V_2 = C_3V_3$, where C_1 and C_2 = % Concentration of the two solutions before being mixed, V_1 and V_2 = Volume of the two solutions before being mixed, and C_3 and V_3 = the resulting % Concentration and Volume, respectively.

Substitute values and solve.

$$\frac{(13\%)(175\text{ gal})}{100\%} + \frac{(5.7\%)(405\text{ gal})}{100\%} = \frac{C_3(580\text{ gal})}{100\%}$$

$$22.75\text{ gal} + 23.085\text{ gal} = \frac{C_3(580\text{ gal})}{100\%}$$

Solving for C_3.

$$C_3 = \frac{(22.75\text{ gal} + 23.085\text{ gal})(100\%)}{580\text{ gal}} = \frac{(45.835\text{ gal})(100\%)}{580\text{ gal}}$$

$C_3 = 0.079$, **7.9% Final solution**

3. Given the following data, calculate the percent solids content that results when a primary sludge is mixed with a secondary sludge.

Average of primary sludge = 2,860 gpd
Average of secondary sludge = 2,550 gpd
Primary sludge = 6.61% solids
Secondary sludge = 3.48% solids
Primary sludge = 8.37 lb/gal
Secondary sludge = 8.43 lb/gal

Equation: % **Sludge mixture** =

$$\frac{[(\text{Sludge 1 gal})(\text{lb/gal})(\text{Avail \%}/100\%) + (\text{Sludge 2 gal})(\text{lb/gal})(\text{Avail \%}/100\%)]100\%}{(\text{Sludge 1, gal})(\text{lb/gal}) + (\text{Sludge 2, gal})(\text{lb/gal})}$$

Where Avail = Available

% Mixture strength =

$$\frac{[(2,860 \text{ gal})(8.37 \text{ lb/gal})(6.61\%/100\%) + (2,550 \text{ gal})(8.43 \text{ lb/gal})(3.48\%/100\%)]100\%}{(2,860 \text{ gal})(8.37 \text{ lb/gal}) + (2,550 \text{ gal})(8.43 \text{ lb/gal})}$$

% Mixture strength = $\dfrac{(1,582.315 \text{ lb} + 748.078 \text{ lb})(100\%)}{23,938.2 \text{ lb} + 21,496.5 \text{ lb}} = \dfrac{(2,330.393 \text{ lb})(100\%)}{45,434.7 \text{ lb}}$

% Mixture strength = **5.13% Mixture strength**

4. An 18-inch sewage pipeline is flowing at a velocity of 1.10 ft/s and the depth of the sewage averages 6.7 inches. Determine the flow in the pipeline in ft³/s and gpm.

First, divide the depth of sewage flow by the diameter of the pipe. Converting inches to feet is not necessary.

Ratio = depth/Diameter = 6.7 in./12 in. = 0.5583, round to 0.56

Next, determine the factor that needs to be used.

In Appendix G look up 0.56 under the column d/D. The number immediately to the right will be the factor that needs to be used. In this case it is 0.4526. This will be the number used rather than 0.785.

Next, convert the pipe's diameter from inches to feet.

$$\text{Number of feet} = \frac{18 \text{ in.}}{12 \text{ in./ft}} = 1.5 \text{ ft}$$

Equation: **Flow, ft³/s = (Area, ft²)(Velocity, ft/s)**

Where the area = (Factor)(Diameter)²

Substitute values and solve.

Flow, ft³/s = (0.4526)(1.5 ft)(1.5 ft)(1.10 ft/s) = 1.120 ft³/s, round to **1.1 ft³/s**

5. Calculate the solids loading rate in lb of solids/d/ft^2, given the following data:

Secondary clarifier radius = 36.0 ft
Primary effluent flow = 1,750,000 gpd
Return of activated sludge is 0.475 mgd
Mixed liquor suspended solids (MLSS) = 2,645 mg/L
Specific gravity of the solids is 1.04

First, convert the primary effluent flow in gallons per day to mgd.

$$\text{mgd} = \frac{1,750,000 \text{ gpd}}{1,000,000/\text{mil}} = 1.75 \text{ mgd}$$

Next, determine the total flow.

Total flow = Primary flow + Return of activated sludge

Total flow = 1.75 mgd + 0.475 mgd = 2.225 mgd

Next, calculate the area of the clarifier.

Area = πr^2

Area = (3.14)(36.0 ft)(36.0 ft) = 4,069.44 ft^2

Next, determine the lb/gal of solids.

Solids, lb/gal = (8.34 lb/gal)(1.04 sp gr) = 3.6736 lb/gal

Next, calculate the solids loading rate.

Equation: **Solids loading rate** $= \dfrac{(\text{MLSS, mg/L})\,(\text{mgd})\,(8.34 \text{ lb/gal})}{\text{Area, ft}^2}$

Solids loading rate $= \dfrac{(2,645 \text{ mg/L})\,(2.225 \text{ mgd})\,(8.6736 \text{ lb/gal})}{4,069.44 \text{ ft}^2} =$

Solids loading rate = 12.54 lb of solids/d/ft^2, round to **12.5 lb of solids/d/ft^2**

6. **Given the following parameters, calculate how long a primary sludge pump should operate in minutes per hour:**

Plant flow = 1,500 gpm
Sludge pump = 62 gpm
Influent suspended solids (SS) = 282 mg/L
Effluent SS = 112 mg/L
Sludge = 4.45% solids

First, convert gpm to mgd.

$$\text{Number of mgd} = \frac{(1{,}500 \text{ gpm})(1{,}440 \text{ min/day})}{1{,}000{,}000/\text{mil}} = 2.16 \text{ mgd}$$

Equation: Operating time, min/hr =

$$\frac{(\text{Flow, mgd})(\text{Influent SS, mg/L} - \text{Effluent SS, mg/L})(100\%)}{(\text{Sludge pump, gpm})(\text{Percent Solids})(24 \text{ hr/day})}$$

Substitute values and solve.

$$\text{Operating time, min/hr} = \frac{(2.16 \text{ mgd})(282 \text{ mg/L, SS} - 112 \text{ mg/L, SS})(100\%)}{(62 \text{ gpm})(4.45\%)(24 \text{ hr/day})}$$

$$\text{Operating time, min/hr} = \frac{(2.16 \text{ mgd})(170 \text{ mg/L, SS})(100\%)}{(62 \text{ gpm})(4.45\%)(24 \text{ hr/day})}$$

Operating time, min/hr = 5.545 min/hr, round to **5.5 min/hr**

7. **What is the biosolids retention time (BRT) in days for a digester that is 70.3 ft in diameter, has a working level of 20.75 ft, and receives an average flow of 14.5 gpm?**

First, convert gpm to gpd.

Digester influent, gpm = (14.5 gpm)(1,440 min/day) = 20,880 gpd

Next, determine the working volume in gallons for the digester.

Equation: **Digester volume, gal = (0.785)(Diameter, ft)^2(Height, ft)(7.48 gal/ft^3)**

Digester volume, gal = (0.785)(70.3 ft)(70.3 ft)(20.75 ft)(7.48 gal/ft^3) = 602,143.5 gal

Lastly, calculate the BRT.

Equation: **BRT, days** $= \dfrac{\text{Digester working volume, gal}}{\text{Influent flow, gpd}}$

BRT, days $= \dfrac{602,143.5 \text{ gal}}{20,880 \text{ gpd}} =$ **28.8 days**

8. **Determine the waste activated sludge (WAS) in mgd, given the following data.**

Influent flow = 2.59 mgd
Clarifier radius = 42.5 ft
Clarifier depth = 15.3 ft
Aerator = 0.840 mil gal
Mixed liquor suspended solids (MLSS) = 2,835 mg/L
Return activated sludge (RAS) SS = 6,970 mg/L
Secondary effluent SS = 17.2 mg/L

Target solids retention time (SRT) = 10 days exactly

First, determine the volume in mil gal for the clarifier and add to the aerator volume.

Equation: $\text{Clarifier, mil gal} = \dfrac{\pi\,(\text{radius})^2\,(\text{Depth, ft})\,(7.48\ \text{gal/ft}^3)}{1,000,000/\text{mil}}$

$\text{Clarifier, mil gal} = \dfrac{3.14\,(42.5\ \text{ft})\,(42.5\ \text{ft})\,(15.3\ \text{ft})\,(7.48\ \text{gal/ft}^3)}{1,000,000/\text{mil}} = 0.64908\ \text{mil gal}$

Total volume = 0.64908 mil gal + 0.840 mil gal = 1.48908 mil gal

Equation: **Target SRT =**

$$\dfrac{(\text{MLSS mg/L})\,(\text{Clarifier, Aerator Volume, mil gal})\,(8.34\ \text{lb/gal})}{(\text{RAS SS mg/L})\,(x\ \text{mgd})\,(8.34\ \text{lb/gal}) + (\text{Effluent SS, mg/L})\,(\text{Flow, mgd})\,(8.34\ \text{lb/gal})}$$

Substitute values and solve.

$10\ \text{days SRT} = \dfrac{(2,835\ \text{mg/L})\,(1.48908\ \text{mil gal})\,(8.34\ \text{lb/gal})}{(6,970\ \text{mg/L})\,(x\ \text{mgd})\,(8.34\ \text{lb/gal}) + (17.2\ \text{mg/L})\,(2.59\ \text{mgd})\,(8.34\ \text{lb/gal})}$

$10\ \text{days SRT} = \dfrac{35,207.66\ \text{lb MLSS}}{(6,970\ \text{mg/L})\,(x\ \text{mgd})\,(8.34\ \text{lb/gal}) + 371.53\ \text{lb/day}}$

Rearrange the equation so that x mgd is in the numerator.

$(6,970\ \text{mg/L})(\,x\ \text{mgd})(8.34\ \text{lb/gal}) + 371.53\ \text{lb/day} = \dfrac{35,207.66\ \text{lb MLSS}}{10\ \text{days SRT}}$

$(6{,}970 \text{ mg/L})(\,x \text{ mgd})(8.34 \text{ lb/gal}) + 371.53 \text{ lb/day} = 3{,}520.766 \text{ lb/day}$

Subtract 371.53 lb/day from both sides of the equation.

$(6{,}970 \text{ mg/L})(\,x \text{ mgd})(8.34 \text{ lb/gal}) = 3{,}149.236 \text{ lb/day}$

$$x \text{ mgd} = \frac{3{,}149.236 \text{ lb/day}}{(6{,}970 \text{ mg/L})(8.34 \text{ lb/gal})} = \mathbf{0.0542 \text{ mgd}}$$

9. **Given the following data, calculate the cost of running a pump in dollars and cents per day:**

Flow = 1.39 mgd
Total differential head (TDH) = 214 ft
Motor efficiency = 88.6%
Pump efficiency = 71.3%
Cost in kilowatt hours = $0.07568

First, convert mgd to gpm.

$$\text{Number of mgd} = \frac{(1.39 \text{ mgd})(1{,}000{,}000/\text{mil})}{(1{,}440 \text{ min/day})} = 965.28 \text{ gpm}$$

Next, determine the horsepower (hp).

Equation: $\mathbf{Motor\ hp} = \dfrac{\mathbf{(Flow,\ gpm)(TDH,\ ft)}}{\mathbf{(3{,}960)(Motor\ efficiency)(Pump\ efficiency)}}$

$$\text{Motor hp} = \frac{(965.28 \text{ gpm})(214 \text{ ft})}{(3{,}960)(88.6\%/100\%)(71.3\%/100\%)} = 82.575 \text{ hp}$$

Now, calculate the cost of running the pump in dollars and cents.

Equation: **Cost, $/day = (Motor hp)(24 hr/day)(0.746 kW/hp)(Cost/kW-hr)**

Cost, $/day = (82.575 hp)(24 hr/day)(0.746 kW/hp)($0.07568/kW-hr) = **$111.89/day**

10. **A wastewater treatment plant is treating 2,065,000 gpd with a 12.2% sodium hypochlorite solution. If the dosage required is 9.65 mg/L and the specific gravity of the hypochlorite is 1.03, how many gpd of sodium hypochlorite are required?**

First, convert gpd to mgd.

$$\text{Number of mgd} = \frac{2,065,000 \text{ gpd}}{1,000,000/\text{mil}} = 2.065 \text{ mgd}$$

Next, determine the lb/gal for the hypochlorite solution.

Hypochlorite, lb/gal = (8.34 lb/gal)(1.03 sp gr) = 8.5902 lb/gal

Next, using the "pounds equation," calculate the lb day of chlorine needed.

Equation: Chlorine, lb/day = (Dosage, mg/L)(mgd)(8.34 lb/gal)

Chlorine, lb/day = (9.65 mg/L)(2.065 mgd)(8.34 lb/gal) = 166.193 lb/day

Since the solution is not 100%, divide the percent hypochlorite into the lb/day of chlorine needed.

$$\text{Hypochlorite, lb/day} = \frac{166.193 \text{ lb/day}}{12.2\%/100\%} = 1,362.238 \text{ lb/day hypochlorite}$$

Lastly, determine the gpd of hypochlorite solution needed.

$$\text{Hypochlorite, gpd} = \frac{1,362.238 \text{ lb/day}}{8.5902 \text{ lb/gal}} = 158.58 \text{ gpd, round to } \textbf{159 gpd sodium hypochlorite}$$

11. **Determine the number of mgd that flows through a wastewater plant's effluent, given the following data:**

SO_2 dosage = **4.6 mg/L**
Pounds of chlorine used = 173 lb/day
Chlorine demand = 6.35 mg/L
Assume SO_2 is exactly 3.25 mg/L higher than the chlorine residual

First, based on the assumption, determine the chlorine residual in mg/L.

Chlorine residual, mg/L = SO_2 dosage − assumption that SO_2 is 3.25 mg/L higher

Chlorine residual, mg/L = 4.6 mg/L − 3.25 mg/L = 1.35 mg/L

Chlorine dosage, mg/L = Chlorine demand, mg/L + Chlorine residual, mg/L

Chlorine dosage, mg/L = 6.35 mg/L + 1.35 mg/L = 7.70 mg/L

Equation: **Number of lb/day Cl_2 = (Dosage, mg/L)(Number of mgd)(8.34 lb/gal)**

Rearrange the equation to solve for the number of mgd the plant is treating.

$$\text{Number of mgd} = \frac{\text{Number of lb/day } Cl_2}{(\text{Dosage, mg/L})(8.34 \text{ lb/gal})}$$

Substitute values and solve.

$$\text{Number of mgd} = \frac{173 \text{ lb/day}}{(7.70 \text{ mg/L})(8.34 \text{ lb/gal})} = 2.694 \text{ mgd, round to } \textbf{2.7 mgd}$$

12. **Given the following data, calculate the flocculant feed rate for a belt press in lb/hr and the flocculant dosage in lb/ton:**

Flocculant concentration = 1.25%
Flocculant feed rate = 1.20 gpm
Solids loading on belt filter = 2,830 lb/hr

Know: 1% = 10,000 mg/L, therefore 1.2% = 12,000 mg/L

Next, find the number of mgd of flocculant used.

$$\text{Flocculant, mgd} = \frac{(1.20\text{ gpm})(1,440\text{ min}/\text{day})}{1,000,000/\text{mil}} = 0.001728\text{ mgd of flocculant}$$

Next, determine the flocculant feed rate in lb/hr.

Equation: **Flocculant feed, lb/hr** $= \dfrac{(\text{Flocculant, mg/L})(\text{mgd flocculant})(8.34\text{ lb/gal})}{24\text{ hr/day}}$

Substitute values and solve.

$$\text{Flocculant feed, lb/hr} = \frac{(12,000\text{ mg/L})(0.001728\text{ mgd})(8.34\text{ lb/gal})}{24\text{ hr/day}}$$

Flocculant feed, lb/hr = 7.206 lb/hr, round to **7.21 lb/hr**

Next, determine flocculant dosage. Start with converting the solids loading from lb/hr to tons/hr.

$$\text{Solids loading, tons/hr} = \frac{2,830\text{ lb/hr}}{2,000\text{ lb/ton}} = 1.415\text{ tons/hr}$$

Lastly, calculate the flocculant dosage in lb/ton.

$$\text{Flocculant dosage, lb/ton} = \frac{7.206\text{ lb/hr}}{1.415\text{ tons/hr}} = 5.0926\text{ lb/ton, round to } \textbf{5.09 lb/ton}$$

13. **Given the following data, determine the sludge age in days for an oxidation ditch wastewater treatment plant:**

Mixed liquor suspended solids (MLSS) = 2,690 mg/L
Ditch volume = 0.634 mil gal
Solids added = 845 lb/day

First, calculate the amount of solids under aeration.

Know: **Solids under aeration, lb = (Ditch volume, mil gal)(MLSS, mg/L)(8.34 lb/gal)**

Solids under aeration, lb = (0.634 MD)(2,690 mg/L)(8.34 lb/gal) = 14,224 lb

Next, determine the sludge age in days.

Know: $\textbf{Sludge age, days} = \dfrac{\text{Solids under aeration, lb}}{\text{Solids added, lb/day}}$

Sludge age, days = $\dfrac{14,224\ \text{lb}}{845\ \text{lb/day}}$ = 16.83 days, round to **16.8 days**

14. **Calculate the mean cell residence time (MCRT), given the following data:**

Flow = 2.92 mgd
Aeration tank volume = 460,000 gallons
Clarifier tank volume = 308,000 gallons
MLSS = 2,885 mg/L
Waste rate = 25,600 gpd
Waste activated sludge (WAS) = 7,050 mg/L
Effluent TSS = 15.5 mg/L

First, convert the volumes for the tanks to mil gal.

Aeration tank, mgd = $\dfrac{460,000\ \text{gal}}{1,000,000/\text{mil}}$ = 0.46 mil gal

Clarifier tank, mgd = $\dfrac{308,000\ \text{gal}}{1,000,000/\text{mil}}$ = 0.308 mil gal

Next, convert the waste rate from gpd to mgd.

$$\text{Waste rate, mgd} = \frac{25{,}600 \text{ gal}}{1{,}000{,}000/\text{mil}} = 0.0256 \text{ mgd}$$

Next, calculate the MCRT.

Equation: **MCRT, days** $=$

$$\frac{(\text{MLSS, mg/L})(\text{Aeration tank mil gal} + \text{Clarifier tank mil gal})(8.34 \text{ lb/gal})}{(\text{WAS, mg/L})(\text{Waste rate, mgd})(8.34 \text{ lb/gal}) + (\text{TSS, mg/L})(\text{Flow, mgd})(8.34 \text{ lb/gal})}$$

$$\text{MCRT, days} = \frac{(2{,}885 \text{ mg/L MLSS})(0.46 \text{ mil gal} + 0.308 \text{ mil gal})(8.34 \text{ lb/gal})}{(7{,}050 \text{ mg/L})(0.0256 \text{ mgd})(8.34 \text{ lb/gal}) + (15.5 \text{ mg/L TSS})(2.92 \text{ mgd})(8.34 \text{ lb/gal})}$$

$$\text{MCRT, days} = \frac{(2{,}885 \text{ mg/L MLSS})(0.768 \text{ mil gal})(8.34 \text{ lb/gal})}{1{,}505.20 \text{ lb/day} + 377.47 \text{ lb/day}}$$

$$\text{MCRT, days} = \frac{18{,}478.771 \text{ lb}}{1{,}882.67 \text{ lb/day}} = \mathbf{9.8 \text{ days}}$$

15. **Given the following data, calculate whether a gravity thickener's sludge blanket will increase, remain the same, or decrease, and how much that change will be in lb/day solids, if it increases or decreases:**

Pumped sludge to thickener = 110 gpm
Pumped sludge from thickener = 53 gpm
Thickener's effluent suspended solids (SS) = 757 mg/L
Primary sludge solids = 3.45%
Thickened sludge solids = 7.18%

First, calculate the thickener's influent and effluent solids in lb/day.

Equation: **Influent solids, lb/day** $=$

(Influent flow gpm)(1,440 min/day)(Percent sludge solids)(8.34 lb/gal)

Influent solids, lb/day = (110 gpm)(1,440 min/day)(3.45%/100%)(8.34 lb/gal)

Influent solids, lb/day = 45,576.432 lb/day

Effluent solids, lb/day = (53 gpm)(1,440 min/day)(7.18%/100%)(8.34 lb/gal)

Effluent solids, lb/day = 45,701.332 lb/day

Next, calculate the flow leaving the thickener.

Effluent flow = 110 gpm − 53 gpm = 57 gpm

Next, convert thickened sludge solids to percent.

Know: 1% = 10,000 mg/L

Percent sludge solids = (757 mg/L)(1%/10,000 mg/L) = 0.0757%

Now, calculate the sludge solids that are leaving the thickener.

Sludge solids, lb/day = (57 gpm)(1,440 min/day)(0.0757%/100%)(8.34 lb/gal)

Sludge solids, lb/day = 518.202 lb/day

Now, calculate whether the sludge blanket will increase, remain the same, or decrease.

Total solids, lb/day = 45,576.432 lb/day − 45,701.332 lb/day − 518.202 lb/day

Total solids, lb/day = − 643.102 lb/day, round to − **640 lb/day**

The sludge blanket will decrease because the total solids are negative.

16. **What must have been the concentration of volatile acids in mg/L for a sour digester with a radius of 29.7 ft and a sludge level of 16.3 ft, if the number of pounds of lime to neutralize the volatile acids was 2,930 lb?**

Know: 1 mg/L of lime will neutralize 1 mg/L of volatile acids

First, determine the sludge volume in gallons that is in the digester.

Number of gallons $= \pi(r^2)$(Height)(7.48 gal/ft^3)

Number of gallons $= 3.14(29.7 \text{ ft})(29.7 \text{ ft})(16.3 \text{ ft})(7.48 \text{ gal/ft}^3) = 337,701$ gal

First, convert the digester's volume from gallons to mil gal.

Number of mil gal $= \dfrac{337,701 \text{ gal}}{1,000,000/\text{mil}} = 0.3377$ mil gal

Equation: **Number of lb Lime = (Volatile acids, mg/L)(mil gal)(8.34 lb/gal)**

Rearrange the equation to solve for the concentration of volatile acids.

Volatile acids, mg/L $= \dfrac{\text{Number of lb, lim e}}{(\text{mil gal})(8.34 \text{ lb/gal})}$

Substitute and solve.

Volatile acids, mg/L $= \dfrac{2,930 \text{ lb, lim e}}{(0.3377 \text{ mil gal})(8.34 \text{ lb/gal})}$

Volatile acids, mg/L $= 1,040.33$ mg/L, round to **1,040 mg/L Volatile acids**

17. Calculate the required waste rate from an aeration tank in mgd and gpm, given the following data:

Aeration tank volume = 0.698 mil gal
Desired COD lb/MLVSS lb = 0.18
Primary effluent flow = 3.10 mgd
Primary effluent COD = 136 mg/L
Mixed liquor volatile suspended solids (MLVSS) = 3,755 mg/L
Waste volatile solids (WVS) concentration = 4,075 mg/L

First, find the existing MLVSS in pounds.

Equation: **Existing MLVSS, lb = (MLVSS, mg/L)(Aeration tank, mil gal)(8.34 lb/gal)**

Existing MLVSS, lb = (3,755 mg/L)(0.698 mil gal)(8.34 lb/gal) = 21,859 lb MLVSS

Next, determine the desired MLVSS in pounds.

Equation: $$\textbf{Desired MLVSS, lb} = \frac{\textbf{(Primary effluent COD, mg/L) (mgd) (8.34 lb/gal)}}{\textbf{Desired COD lb/MLVSS lb}}$$

$$\text{Desired MLVSS, lb} = \frac{(136 \text{ mg/L}) (3.10 \text{ mgd}) (8.34 \text{ lb/gal})}{0.18 \text{ COD lb/MLVSS lb}} = 19{,}534 \text{ lb MLVSS}$$

Next, subtract the existing MLVSS from the desired MLVSS to find the waste in pounds.

Waste, lb = 21,859 lb – 19,534 lb = 2,325 lb

Next, calculate the waste rate in mgd.

Equation: $$\textbf{Waste rate, mgd} = \frac{\text{Waste, lb}}{\text{(WVS concentration, mg/L) (8.34 lb/gal)}}$$

$$\text{Waste rate, mgd} = \frac{2{,}325 \text{ lb}}{(4{,}075 \text{ mg/L}) (8.34 \text{ lb/gal})} = 0.06841 \text{ mgd, round to } \textbf{0.68 mgd}$$

Lastly, calculate the waste rate in gpm.

$$\text{Waste rate, gpm} = \frac{(0.06841 \text{ mgd})(1{,}000{,}000 \text{ gpd/mgd})}{1{,}440 \text{ min/day}} = 47.51 \text{ gpm, round to } \textbf{48 gpm}$$

18. Given the following data, determine the solids loading in lb/d/ft² on a dissolved air flotation (DAF) thickener unit:

DAF unit length = 79.9 ft
DAF unit width = 32.1 ft
Waste-activated sludge (WAS) = 7,828 mg/L
Sludge flow = 113 gpm
Sludge specific gravity = 1.04 lb/gal

First, determine the area of the DAF.

DAF area, ft² = (Length, ft)(Width, ft)

DAF area, ft² = (79.9 ft)(32.1 ft) = 2,564.79 ft²

Next, convert gpm to mgd.

$$\text{Number of gpd} = \frac{(113 \text{ gpm})(1,440 \text{ min/day})}{1,000,000/\text{mil}} = 0.16272 \text{ mgd}$$

Next, determine the lb/gal for the sludge using the specific gravity.

Sludge, lb/gal = (8.34 lb/gal)(1.04 sp gr) = 8.6736 lb/gal

Equation is: **Solids loading, lb/d/ft²** $= \dfrac{\textbf{(WAS, mg/L)(mgd)(lb/gal, Sludge)}}{\textbf{DAF area, ft}^2}$

$$\text{Solids loading, lb/d/ft}^2 = \frac{(7,828 \text{ mg/L, WAS})(0.16272 \text{ mgd})(8.6736 \text{ lb/gal, Sludge})}{2,564.79 \text{ ft}^2 \text{ DAF}}$$

Solids loading, lb/d/ft² = **4.31 lb/d/ft²**

19. **What is the air-to-solids ratio for a dissolved air flotation (DAF) unit that has an air flow rate equal to 10.3 ft³/min, a solids concentration of 0.78%, and a flow of 144,000 gpd?**

Know: Air = 0.0807 lb/ft³ at standard temperature, pressure, and average composition

Equation is: **Air-to-solids ratio** $= \dfrac{(\text{Air flow, ft}^3/\text{min})(\text{Air, lb/ft}^3)}{(\text{gpm})(\text{Percent solids}/100\%)(8.34\ \text{lb/gal})}$

Because equation requires gpm, first convert gpd to gpm.

Number of gpm $= \dfrac{144,000\ \text{gpd}}{1,440\ \text{min}/\text{day}} = 100\ \text{gpm}$

Now, using the air-to-solids equation, substitute and solve.

Air-to-solids ratio $= \dfrac{(10.3\ \text{ft}^3/\text{min})(0.0807\ \text{lb/ft}^3)}{(100\ \text{gpm})(0.78\%/100\%)(8.34\ \text{lb/gal})} = 0.12778$ round to **0.13**

20. **A drying bed is 289 feet long and 50.0 feet wide. If on average 2.87 inches of sludge that has 5.36% solids were applied to the drying bed and the drying and removal cycle on average takes 23.3 days, how many gallons of sludge were applied on average for each cycle and what is the solids loading rate in lb/yr/ft²?**

First, convert 2.87 inches to feet.

Number of feet = 2.87 in./12 in./ft = 0.2392 ft

Next, determine the volume in ft³ sent to the drying bed.

Volume, ft³ = (289 ft)(50.0 ft)(0.2392 ft) = 3,456.44 ft³

Next, calculate the volume in gallons sent to the sand drying beds.

Number of gal = (3,456.44 ft³)(7.48 gal/ft³) = 25,854.17 gal, round to **25,900 gal**

Next, calculate the number of lb.

Number of lb = (3,456.44 ft³)(7.48 gal/ft³)(8.34 lb/gal) = 215,624 lb

Lastly, calculate the solids loading rate.

Equation: **Solids loading rate, lb/yr/ft²** $= \dfrac{\dfrac{(\text{lb})\,(365\ \text{days})\,(\text{Percent solids})}{(\text{Drying cycle})\,(\text{yr})\,(100\%)}}{\text{Drying bed area, ft}^2}$

Substitute values and solve.

Solids loading rate, lb/yr/ft² $= \dfrac{\dfrac{(215,624\ \text{lb})\,(365\ \text{days})\,(5.36\%\ \text{solids})}{(23.3\ \text{days})\,(\text{yr})\,(100\%)}}{(289\ \text{ft})\,(50.0\ \text{ft})}$

Solids loading rate, lb/yr/ft² $= \dfrac{(9,254.25\ \text{lb/day})\,(365\ \text{days/yr})\,(0.0536)}{14,450\ \text{ft}^2}$

Solids loading rate, lb/yr/ft² $=$ **12.5 lb/yr/ft²**

21. **Given the following data, calculate the amount of compost in lb/day that needs to be blended with a dewatered sludge to make a mixture that has a moisture content of 42.0%.**

Dewatered digester primary sludge = 9,150 lb/day
Dewatered sludge solids = 38.4%
Compost solids = 70.6%

First, determine the moisture content of the dewatered sludge and the compost.

Dewatered sludge moisture = 100% − 38.4% = 61.6%

Compost percent moisture = 100% − 70.6% = 29.4%

Equation: **Mixture's % moisture =**

$$\frac{[(\text{Sludge, lb})\,(\%\ \text{moisture}) + (\text{Compost, lb})\,(\%\ \text{moisture})]\,100\%}{\text{Sludge, lb} + \text{Compost, lb}}$$

Substitute values and solve.

$$42.0\% \text{ moisture content} = \frac{[(9{,}150\text{ lb})(61.6\%/100\%) + (x\text{ lb})(29.4\%/100\%)]\,100\%}{9{,}150\text{ lb} + x\text{ lb}}$$

Solve for x.

Divide both sides of the equation by 100%.

$$0.420 = \frac{(9{,}150\text{ lb})(61.6\%/100\%) + (x\text{ lb})(29.4\%/100\%)}{9{,}150\text{ lb} + x\text{ lb}}$$

Simplify terms in the numerator.

$$0.420 = \frac{5{,}636.4\text{ lb} + 0.294x}{9{,}150\text{ lb} + x\text{ lb}}$$

Multiply both sides of the equation by $(9{,}150\text{ lb} + x\text{ lb})$

$$0.420(9{,}150\text{ lb} + x\text{ lb}) = 5{,}636.4\text{ lb} + 0.294x$$

Multiply terms on the left side of the equation.

$$4{,}392\text{ lb} + 0.420x = 5{,}636.4\text{ lb} + 0.294x$$

Subtract $0.294x$ and 4,392 lb from both sides of the equation.

$$0.420x - 0.294x = 5{,}636.4\text{ lb} - 4{,}392\text{ lb}$$

Simplify both sides of the equation.

$$0.126x = 1{,}244.4\text{ lb}$$

$x = 9{,}876.19$ lb, round to **9,880 lb of Compost required**

22. **Determine the number of acres used in the following land application of biosolids, given the following data:**

Hydraulic loading rate = 0.52 in./day
Flow = 1,620,000 gpd

First, determine the number of gallons per acre-inch.

Number of gals/acre-in. = (43,560 ft^2/acre)(1 ft/12 in.)(7.48 gal/ft^3) = 27,152 gal/acre-in.

Equation: **Hydraulic loading rate, in./day** $= \dfrac{\text{Flow, gpd}}{(27,152 \text{ gal/acre-in.})(\text{Area, acres})}$

Rearrange the equation and solve for area in acres.

Area, acres $= \dfrac{\text{Flow, gpd}}{(27,152 \text{ gal/acre-in.})(\text{Hydraulic loading rate, in/day})}$

Substitute values and solve.

Area, acres $= \dfrac{1,620,000 \text{ gpd}}{(27,152 \text{ gal/acre-in.})(0.52 \text{ in./day})}$

Area, acres = 114.74 acres, round to **110 acres**

23. **A 0.1% stock solution (1,000 ppm or 1,000 mg/L) is required for doing jar tests. If the alum has a specific gravity of 1.293 and is 48.2% aluminum sulfate, how many milliliters of alum are required to make exactly 1,000 mL of stock solution?**

First, find the number of lb/gal of alum.

Alum, lb/gal = (sp gr)(8.34 lb/gal) = (1.293)(8.34 lb/gal) = 10.7836 lb/gal

Next, determine the number of grams/mL.

$$\text{Number grams/mL, Alum} = \frac{(10.7836 \text{ lb/gal})(48.2\% \text{ Al}_2\text{SO}_4, \text{Purity})(454 \text{ grams/lb})}{(3,785 \text{ mL/gal})(100\%)}$$

Number grams/mL, Alum $= 0.62345$ grams/mL

Convert grams/mL to milligrams/mL

Number mg/mL $= (0.62345 \text{ grams/mL})(1,000 \text{ mg/g}) = 623.45$ mg/mL

Next, convert mL to liters by multiplying by 1,000.

Number mg/L $= (623.45 \text{ mg/mL})(1,000 \text{ mL/liter}) = 623,450$ mg/Liter

Next, determine the number of mL required.

Equation: $\mathbf{C_1V_1 = C_2V_2}$

$(623,450 \text{ mg/mL})(x, \text{mL}) = (1,000 \text{ mg/liter})(1,000 \text{ mL})$

$$x, \text{mL} = \frac{(1,000 \text{ mg/Liter})(1,000 \text{ mL})}{623,450 \text{ mg/liter}} = \textbf{1.60 mL Alum}$$

Now, using a pipette, add 1.60 mL of the 48.2% alum solution to a clean, dry 1,000-mL flask. Dilute the alum to the 1,000-mL mark with deiorized water. Add a magnetic stir bar and place the flask on a magnetic stirrer. Turn the magnetic stirrer on and mix this solution with the bar as vigorously as possible for at least 10 minutes.

Thus, every 1 mL of alum solution that is added to 1,000-mL sample of raw water will add a 1 mg/L dose (because of second dilution with raw water; 1,000 mg/L / 1,000 mg/L raw water sample = 1 mg/L). If 10 mL of this stock solution were added to the 1,000-mL raw water sample, it would be a dose of 10 mg/L. If you are using the 2-liter square jars, simply double the mL added for each mg/L dosage increase desired. Another way is to feed the alum neat by using a micropipette; pipette 0.00160 mL of alum into a 1,000-mL raw water sample.

24. **Calculate the seeded BOD$_5$ in mg/L, given the following data:**

Sample size = 150 mL
Initial dissolved oxygen (DO) = 8.8 mg/L
Final DO = 3.1 mg/L
BOD$_5$ of seed stock = 75 mg/L
Seed stock = 4.0 mL
Total volume = 300 mL

First, calculate the seed correction in mg/L.

Equation: **Seed correction, mg/L** $= \dfrac{(\text{BOD}_5 \text{ of seed stock, mg/L})(\text{Seed stock, mg/L})}{\text{Total volume, mL}}$

Seed correction, mg/L $= \dfrac{(75 \text{ mg/L})(4.0 \text{ mg/L})}{300 \text{ mL}} = 1.0 \text{ mg/L}$

Next, calculate the BOD$_5$ seeded in mg/L.

Equation: **BOD$_5$ seeded, mg/L =**

$\dfrac{(\text{Initial DO, mg/L} - \text{Final DO, mg/L} - \text{Seed correction, mg/L})(\text{Total volume, mL})}{\text{Sample volume, mL}}$

BOD$_5$ seeded, mg/L $= \dfrac{(8.8 \text{ mg/L} - 3.1 \text{ mg/L} - 1.0 \text{ mg/L})(300 \text{ mL})}{150 \text{ mL}}$

BOD$_5$ seeded, mg/L $= \dfrac{(4.7 \text{ mg/L})(300 \text{ mL})}{150 \text{ mL}} = \mathbf{9.4 \text{ mg/L}}$

25. **The valve on a secondary clarifier is controlled by a 4- to 20-mA signal from the SCADA system. If the valve is 43.2% open and has a 0% to 100% range, what must the signal be in mA from the SCADA system?**

Use the following equation:

$$\text{Current process reading} = \frac{(\text{Live signal, mA} - 4\,\text{mA offset})(\text{Maximum capacity})}{16\,\text{milliamp span}}$$

Rearrange the formula to solve for the mA signal from the SCADA system.

$$\text{Live signal, mA} = \frac{(\text{Current process reading})(16\,\text{mA span})}{\text{maximum capacity}} + 4\,\text{mA offset}$$

Substitute values and solve.

$$\text{Live signal, mA} = \frac{(43.2\%)(16\,\text{mA span})}{100\%} + 4\,\text{mA offset}$$

Live signal, mA = **10.912 mA, round to 10.9 mA**

WASTEWATER TREATMENT
Grade 4

Students preparing for the Grade 4 certification test should understand these problems and all the problems presented in Grade 3, Chapter 1.

PERCENT CALCULATIONS

Percent calculations are used throughout this book and are thus essential to understand. They are also a good refresher for the student or operator.

1. **What is the percent removal across a primary clarifier, if the influent biochemical oxygen demand (BOD$_5$) is 316 and the effluent BOD$_5$ is 186?**

Equation: $\textbf{Percent BOD}_5 \textbf{ removal} = \dfrac{(\textbf{BOD}_5 \textbf{ in} - \textbf{BOD}_5 \textbf{ out})(\textbf{100\%})}{\textbf{BOD}_5 \textbf{ in}}$

$\text{Percent BOD}_5 \text{ removal} = \dfrac{(316\,\text{BOD}_5 - 186\,\text{BOD}_5)(100\%)}{316\,\text{BOD}_5} = \dfrac{(130\,\text{BOD}_5)(100\%)}{316\,\text{BOD}_5} = \textbf{41.1\%}$

2. **What is the percent recovery of solids for a belt filter press, if the feed sludge total suspended solids (TSS) is 4.18% by weight, the return flow TSS is 0.037% by weight, and the cake total solids (TS) are 18.4% by weight?**

Equation: **Percent recovery** $= \dfrac{\text{Cake TS, \% (Feed sludge TSS, \% } - \text{ Return flow TSS, \%) (100\%)}}{\text{Feed sludge TSS, \% (Cake TS, \% } - \text{ Return flow TSS, \%)}}$

Substitute values and solve.

Percent recovery $= \dfrac{18.4\% (4.18\% - 0.037\%) (100\%)}{4.18\% (18.4\% - 0.037\%)} = \dfrac{(76.2312)(100\%)}{76.75734}$

Percent recovery $= 99.31\%$, round to **99% Recovery by weight**

PERCENT STRENGTH BY WEIGHT SOLUTION PROBLEMS

The strength of solution calculations are important to determine so that operators can properly mix chemicals in the percentages they need for dosing a particular wastewater process or other application.

3. **If 215 kilograms of magnesium hydroxide [$Mg(OH)_2$] are dissolved 215 gallons of water, what is the percent strength by weight of the magnesium hydroxide solution?**

First, convert kilograms (kg) to lb.

Number of lb $= (215 \text{ kg})(2.205 \text{ lb/kg}) = 474.075 \text{ lb}$

Equation: **Percent strength** $= \dfrac{(\text{Number of lb, chemical})(100\%)}{(\text{Number of gal})(8.34 \text{ lb/gal}) + \text{Number of lb, chemical}}$

Substitute values and solve.

Percent strength $= \dfrac{(474.075 \text{ lb})(100\%)}{(215 \text{ gal})(8.34 \text{ lb/gal}) + 474.075 \text{ lb}} = \dfrac{47{,}407.5 \text{ lb \%}}{1{,}793.1 \text{ lb} + 474.075 \text{ lb}}$

Percent strength $= \dfrac{47{,}407.5 \text{ lb \%}}{2{,}267.175 \text{ lb}} = 20.91\%$, round to **20.9% $Mg(OH)_2$ Solution by weight**

4. How much of a 30.0% polymer solution should be added to 450 gallons of water to make a 0.500 percent by weight polymer solution? The 30.0% polymer solution has a specific gravity of 1.12 and the 0.500 polymer solution will have a specific gravity of 1.02.

First, convert the specific gravity of each polymer solution to lb/gal.

Polymer solution, 30.0% = (8.34 lb/gal)(1.12 sp gr) = 9.3408 lb/gal

Polymer solution, 0.500% = (8.34 lb/gal)(1.02 sp gr) = 8.5068 lb/gal

Next, write the equation.

(Solution$_1$ percent)(x gal, Solution$_1$)(Solution$_1$, lb/gal) =

(Solution$_2$ percent)(Solution$_2$, gal)(Solution$_2$, lb/gal)

Where Solution$_1$ is the 30.0% polymer solution and Solution$_2$ is the final 0.500% polymer solution.

Rearrange the equation to solve for the number of gallons of Solution$_1$ needed.

$$x \text{ gal, Solution}_1 = \frac{(\text{Solution}_2 \text{ percent})(\text{Solution}_2, \text{gal})(\text{Solution}_2, \text{lb/gal})}{(\text{Solution}_1 \text{ percent})(\text{Solution}_1, \text{lb/gal})}$$

Substitute values and solve.

$$x \text{ gal, Solution}_1 = \frac{(0.500\%)(450 \text{ gal})(8.5068 \text{ lb/gal})}{(30.0\%)(9.3408 \text{ lb/gal})}$$

x gal, Solution$_1$ = **6.83 gal of 30.0% Polymer solution is needed**

5. How much of a 48.5% alum solution by weight should be added to 225 gallons of water to make a 1.00 percent by weight alum solution? The alum solution has a specific gravity of 1.325 and the 1.00% alum solution will have a specific gravity of 1.01.

First, convert the specific gravity of each polymer solution to lb/gal.

Alum solution, 48.5% = (8.34 lb/gal)(1.325 sp gr) = 11.0505 lb/gal

Alum solution, 1.00% = (8.34 lb/gal)(1.01 sp gr) = 8.4234 lb/gal

Next, write the equation.

(Percent solution$_1$)(x gal, Solution$_1$)(Solution$_1$, lb/gal) =

(Percent solution$_2$)(Solution$_2$, gal)(Solution$_2$, lb/gal)

Where Solution$_1$ is the 48.5% alum solution and Solution$_2$ is the final 1.00% alum solution.

Rearrange the equation to solve for the number of gallons of Solution$_1$ needed.

$$x \text{ gal, Solution}_1 = \frac{(\text{Percent Solution}_2)(\text{Solution}_2, \text{gal})(\text{Solution}_2, \text{lb/gal})}{(\text{Percent Solution}_1)(\text{Solution}_1, \text{lb/gal})}$$

Substitute values and solve.

$$x \text{ gal, Solution}_1 = \frac{(1.00\%)(225 \text{ gal})(8.4234 \text{ lb/gal})}{(48.5\%)(11.0505 \text{ lb/gal})}$$

x gal, Solution$_1$ = **3.54 gal of 48.5% Alum solution by weight is needed**

PERCENT SOLIDS BY WEIGHT CALCULATIONS

Operators use percent solids calculations to determine efficiency of different unit processes, as well as to determine how much waste will require disposal.

6.   What is the percent by weight of total inorganic solids in a sludge sample, given the following data?

Sludge sample wet weight = 1,151 grams
Total solids dry weight = 92.45 grams
Inorganic dry weight = 11.07 grams

Equation: **Percent inorganic solids** $= \dfrac{\text{(Dry sample in grams)}\,(100\%)}{\text{Sludge sample in grams}}$

Percent total solids $= \dfrac{(11.07\ \text{grams})\,(100\%)}{1,151\ \text{grams}} =$ **0.962% Total inorganic solids by weight**

7.   Calculate the percent by weight of solids in thickened sludge, given the following data:

Primary sludge flow = 2,815 gpd
Thickened sludge flow = 2,092 gpd
Primary sludge percent solids = 5.22%
Primary sludge lb/gal = 8.39 lb/gal
Thickened sludge lb/gal = 8.59 lb/gal

Equation: **(2° gpd)(2° sludge lb/gal)(x% 2° sludge) =**

(1° sludge, gpd)(1° sludge lb/gal)(% 1° sludge)

Where the primary sludge = 1° and the thickened sludge = 2°

Rearrange the equation to solve for the thickened sludge percent.

x% 2° sludge $= \dfrac{(1°\ \text{sludge, gpd})(1°\ \text{sludge lb/gal})(\text{Percent }1°\ \text{sludge})}{(2°\ \text{gpd})(2°\ \text{sludge lb/gal})}$

x% 2° sludge $= \dfrac{(2,815\ \text{gpd})\,(8.39\ \text{lb/gal})\,(5.22\%)}{(2,092\ \text{gpd})\,(8.59\ \text{lb/gal})} =$ **6.86% Solids by weight from thickener**

8.   A total of 3,890 gpd of sludge (primary sludge) with 5.05% by weight solids content and weighing 8.38 lb/gal is pumped to a wastewater thickener tank. Calculate the final concentration of the sludge, if 2,465 gpd of sludge flows from the thickener tank and the thickened sludge weighs 8.50 lb/gal. *Note:* For convenience, primary sludge is abbreviated to 1° sludge and secondary sludge to 2° sludge.

Equation: **(2° sludge, gpd)(2° sludge, lb/gal)(*x* Percent 2° sludge)** =

(1° sludge, gpd)(1° sludge, lb/gal)(Percent 1° sludge)

Substitute values and solve.

(2,365 gpd)(8.50 lb/gal)(*x*% 2°/100%) = (3,890 gpd)(8.38 lb/gal)(5.05%/100%)

$$x\% \, 2° = \frac{[(3,890 \text{ gpd})(8.38 \text{ lb/gal})(5.05\%/100\%)]100\%}{(2,465 \text{ gpd})(8.50 \text{ lb/gal})} = \textbf{7.86\% By weight}$$

PERCENT VOLATILE SOLIDS REDUCTION AND VOLATILE SOLIDS DESTROYED

The percent volatile solids reduction calculations indicate the effectiveness of the digested sludge process when compared to the volatile solids in the influent. The higher the percent volatile solids reduced or destroyed, the more stable the organic matter in the digester becomes and the more gas that is produced. Volatile solids destroyed are a measure of the effectiveness of the digester process. It tells the operator the number of pounds of volatile solids destroyed per cubic foot of digester volume. See Figures E-2, E-4, E-5, and E-6 in Appendix E for four types of wastewater plants using a digester.

9.   A digester has a radius of 24.0 ft and an average sludge level of 16.35 ft throughout the day. If the digester influent sludge has a VS content of 54.9% by weight and the digester effluent sludge has a VS content of 35.8% by weight, what is the volatile solids reduction?

First, convert percentage to decimal form by dividing by 100%.

54.9%/100% = 0.549 and 35.8%/100% = 0.358

Equation: **Percent VS content** $= \dfrac{(\text{Influent} - \text{Effluent})(100\%)}{\text{Effluent} - (\text{Influent})(\text{Effluent})}$

$$\text{Percent VS content} = \frac{(0.549 - 0.358)(100\%)}{0.549 - (0.549)(0.358)} = \frac{0.191(100\%)}{0.549 - 0.196542}$$

$$\text{Percent VS content} = \frac{19.1\%}{0.352458} = 54.19\%, \text{ round to } \mathbf{54.2\%} \textbf{ VS reduction by weight}$$

10. **Given the following data, calculate the amount of volatile solids (VS) destroyed in lb/day/ft³ of digester capacity:**

Digester radius = 30.1 ft
Average sludge height = 18.35 ft
Sludge flow (Flow) = 4,420 gpd
Sludge solids concentration (SSC) = 5.76% by weight
Volatile solids content (VSC) = 63.8% by weight
Volatile solids reduction (VSR) = 52.9% by weight
Specific gravity of sludge = 1.03

First, determine the number of lb/gal for the sludge.

Sludge, lb/gal = (8.34 lb/gal)(1.03 sp gr) = 8.59 lb/gal

Next, calculate the digester capacity in ft³.

Digester capacity, ft³ = π(radius)²(Height, ft)

Digester capacity, ft³ = 3.14(30.1 ft)(30.1 ft)(18.35 ft) = 52,203.39 ft³

Next, write the equation.

$$\textbf{VS destroyed, lb/day/ft}^3 = \frac{(\text{Flow, gpd})(\text{Sludge, lb/gal})(\text{SSC, \%})(\text{VSC, \%})(\text{VSR, \%})}{\text{Digester capacity, ft}^3}$$

Substitute values and solve.

$$\text{VS destroyed, lb/day/ft}^3 = \frac{(4,420 \text{ gpd})(8.59 \text{ lb/gal})(5.76\%/100\%)(63.8\%/100\%)(52.9\%/100\%)}{52,203.39 \text{ ft}^3}$$

VS destroyed, lb/day/ft³ = **0.014 lb/day/ft³ VS destroyed**

11. Given the following data, calculate the amount of volatile solids (VS) destroyed in lb/day/1,000 ft³ of digester capacity:

Digester radius = 27.5 ft
Average sludge height = 16.5 ft
Sludge flow (Flow) = 3,420 gpd
Sludge solids concentration (SSC) = 6.18%
Volatile solids content (VSC) = 66.7%
Volatile solids reduction (VSR) = 54.1%
Specific gravity of sludge = 1.04

First, determine the number of lb/gal for the sludge.

Sludge, lb/gal = (8.34 lb/gal)(1.04 sp gr) = 8.674 lb/gal

Next, calculate the digester capacity in ft³.

Digester capacity, ft³ = π(radius)²(Height, ft)

Digester capacity, ft³ = 3.14(27.5 ft)(27.5 ft)(16.5 ft) = 39,181.31 ft³

Factor out 1,000 ft³ from the digester capacity.

Digester capacity, 1,000 ft³ = 39,181.31 ft³ = (39.18131)(1,000 ft³)

Next, write the equation.

$$\textbf{VS destroyed, lb/day/ft}^3 = \frac{(\text{Flow, gpd})\,(\text{Sludge, lb/gal})\,(\text{SSC, \%})\,(\text{VSC, \%})\,(\text{VSR, \%})}{\text{Digester capacity, 1,000 ft}^3}$$

Substitute values and solve.

$$\text{VS destroyed, lb/day/ft}^3 = \frac{(3,420 \text{ gpd})\,(8.674 \text{ lb/gal})\,(6.18\%)\,(66.7\%)\,(54.1\%)}{(39.18131)\,(1,000 \text{ ft}^3)\,(100\%)\,(100\%)\,(100\%)}$$

VS destroyed, lb/day/ft³ = **16.9 lb/day/1,000 ft³ VS destroyed**

PERCENT SEED SLUDGE PROBLEMS

These calculations are used by operators when starting up a new digester or one that has been cleaned, and are based on volume. The other method presented later in this chapter is based on volatile solids added per pound of volatile solids contained in the digester.

12. **How many gallons of seed sludge are needed, if the digester has a radius of 20.5 ft, a maximum side depth of 21.5 ft, a wastewater depth of 15.4 ft, and requires the seed sludge to be 22.0% of the digester volume?**

First, determine the volume of the digester in gallons.

Number of gallons $= \pi(\text{radius})^2(\text{Depth, ft})(7.48 \text{ gal/ft}^3)$

Number of gallons $= (3.14)(20.5 \text{ ft})(20.5 \text{ ft})(15.4 \text{ ft})(7.48 \text{ gal/ft}^3) = 152{,}006 \text{ gal}$

Next, calculate the seed sludge required in gallons.

Equation: **Percent seed sludge** $= \dfrac{(\text{Seed sludge, gal})(100\%)}{\text{Total digester volume, gal}}$

Rearrange the equation to solve for the number of gallons.

Seed sludge, gal $= \dfrac{(\text{Total digester volume, gal})(\text{Percent seed sludge})}{100\%}$

Seed sludge, gal $= \dfrac{(152{,}006 \text{ gal})(22.0\%)}{100\%} = 33{,}441 \text{ gal, round to } \textbf{33,400 gal}$

13. **What must have been the percent seed sludge added to a digester, given the following data?**

Diameter of digester = 49.85 ft
Maximum side height = 24.8 ft
Seed sludge added to digester = 72,350 gal

First, determine the volume of the digester in gallons.

Number of gallons = $(0.785)(\text{Diameter})^2(\text{Depth, ft})(7.48 \text{ gal/ft}^3)$

Number of gallons = $(0.785)(49.85 \text{ ft})(49.85 \text{ ft})(24.8 \text{ ft})(7.48 \text{ gal/ft}^3) = 361{,}871 \text{ gal}$

Next, calculate the seed sludge required in gallons.

Equation: **Percent seed sludge** $= \dfrac{(\text{Seed sludge, gal})(100\%)}{\text{Total digester volume, gal}}$

Percent seed sludge $= \dfrac{(72{,}350 \text{ gal})(100\%)}{361{,}871 \text{ gal}} = 19.99\%$, round to **20.0% Seed sludge**

VOLATILE SOLIDS PUMPING CALCULATIONS

These calculations are used as a planning tool by the operator. By knowing the pumping rate of volatile solids into a digester, an operator can make sure it is not overloaded, which would adversely affect the digester's operation and performance.

14. **How many lb/day of volatile solids (VS) are pumped to a digester, given the following data?**

Influent pumping rate = 3,450 gpd
Solids content = 5.33%
Volatile solids = 64.7%
Specific gravity of sludge = 1.03

First, determine the lb/gal for the sludge.

Sludge, lb/gal = (8.34 lb/gal)(1.03 sp gr) = 8.59 lb/gal

Equation: **VS, lb/day =**

(Number of gpd to digester) $\dfrac{\text{(Percent solids)}}{100\%}$ $\dfrac{\text{(Percent VS)}}{100\%}$ **(Sludge, lb/gal)**

VS, lb/day = (3,450 gpd Solids) $\dfrac{(5.33\%)}{100\%}$ $\dfrac{(64.7\% \text{ VS})}{100\%}$ (8.59 lb/gal)

VS, lb/day = 1,021.98 lb/day, round to **1,020 lb/day VS**

15. Given the following data, how much sludge in gpd is pumped to a digester?

Volatile solids (VS) pumped to digester = 1,645 lb/day
Solids content = 5.14%
Volatile solids = 60.8%
Specific gravity of sludge = 1.03

First, determine the lb/gal for the sludge.

Sludge, lb/gal = (8.34 lb/gal)(1.03 sp gr) = 8.59 lb/gal

Equation: **VS, lb/day =**

(Number of gpd to digester) $\dfrac{\text{(Percent solids)}}{100\%}$ $\dfrac{\text{(Percent VS)}}{100\%}$ **(Sludge, lb/gal)**

Substitute values and solve.

1,645 lb/day, VS = (x gpd Solids) $\dfrac{(5.14\%)}{100\%}$ $\dfrac{(60.8\% \text{ VS})}{100\%}$ (8.59 lb/gal)

1,645 lb/day, VS = (x gpd Solids)(0.26845 lb/gal)

x gpd Solids = $\dfrac{1,645 \text{ lb/day, VS}}{0.26845 \text{ lb/gal}}$ = **6,128 gpd Sludge**

16. **Given the following data, how much sludge in gal/day is pumped to a digester?**

Volatile solids (VS) pumped to digester = 1,475 lb/day
Solids content = 5.21%
Volatile solids = 61.3%
Specific gravity of sludge = 1.03

First, determine the lb/gal for the sludge.

Sludge, lb/gal = (8.34 lb/gal)(1.03 sp gr) = 8.59 lb/gal

Equation: **VS, lb/day = (Sludge, lb/day)** $\dfrac{\text{(Percent solids)}}{100\%} \dfrac{\text{(Percent VS)}}{100\%}$

Substitute values and solve.

1,475 lb/day, VS = (x Sludge, lb/day) $\dfrac{(5.21\%)}{100\%} \dfrac{(61.3\% \text{ VS})}{100\%}$

1,475 lb/day, VS = (x Sludge, lb/day)(0.03194 lb/day)

x lb/day = $\dfrac{1,475 \text{ lb/day, VS}}{0.03194 \text{ lb/day}}$ = 46,180 lb/day Sludge

Lastly, convert lb/day sludge to gal/day.

Sludge, gal/day = $\dfrac{46,180 \text{ lb/day}}{8.59 \text{ lb/gal}}$ = 5,376 gal/day, round to **5,380 gal/day of Sludge**

SOLUTION MIXTURE CALCULATIONS

These calculations are used when mixing two of the same solutions that have different strengths given a volume target. They are important to the operator to understand because there most probably will be times when solutions will require mixing. Three ways to solve these problems follow. The method scientists and chemists use is $C_1V_1 + C_2V_2 = C_3V_3$.

17. **What percent sodium hypochlorite solution would result, if 675 gallons of a 12.5% solution were mixed with 315 gallons of a 7.45% solution? Assume both solutions have the same density.**

First, find the total volume that would result from mixing these two solutions.

Total Volume = 675 gal + 315 gal = 990 gal

Then, write the equation.

(Concentration$_1$)(Volume$_1$) + (Concentration$_2$)(Volume$_2$) = (Concentration$_3$)(Volume$_3$)

Condensed as $C_1V_1 + C_2V_2 = C_3V_3$, where C_1 and C_2 = % Concentration of the two solutions before being mixed, V_1 and V_2 = Volume of the two solutions before being mixed, and C_3 and V_3 = the resulting % Concentration and Volume, respectively.

Substitute values and solve.

$$\frac{(12.5\%)\,(675\ \text{gal})}{100\%} + \frac{(7.45\%)\,(315\ \text{gal})}{100\%} = \frac{C_3\,(990\ \text{gal})}{100\%}$$

$$84.375\ \text{gal} + 23.4675\ \text{gal} = \frac{C_3\,(990\ \text{gal})}{100\%}$$

Solve for C_3.

$$C_3 = \frac{(84.375\ \text{gal} + 23.4675\ \text{gal})\,(100\%)}{990\ \text{gal}} = \frac{(107.8425\ \text{gal})\,(100\%)}{990\ \text{gal}}$$

C_3 = 10.89, round to **10.9% Final sodium hypochlorite solution**

18. **How many gallons of a 48.5% solution must be mixed with a 17.8% solution to make exactly 1,000 gallons of a 25.0% solution?**

There are two ways to solve dilution problems. The dilution triangle is perhaps the easiest, and is shown below for the next two problems.

How to solve the problem using the dilution triangle: The two numbers on the left are the existing concentrations of 48.5% and 17.8%. The number in the center, 25.0%, is the desired concentration. The numbers on the right are determined by subtracting diagonally the existing concentrations from the desired concentration.

48.5%   7.2 *[1]   7.2 parts of the 48.5% solution are required for every 30.7 parts.

    25.0%

17.8% $\dfrac{23.5 *[2]}{30.7 \text{ total parts}}$ 23.5 parts of the 17.8% solution are required for every 30.7 parts.

*[1] **7.2 is determined by subtracting diagonally 17.8% from 25.0%.**
*[2] **23.5 is determined by subtracting diagonally 25.0% from 48.5%.**

$$\dfrac{7.2 \text{ parts} (1,000 \text{ gal})}{30.7 \text{ parts}} = 234.53 \text{ gallons, round to} \quad \textbf{235 gallons of the 48.5\% solution}$$

$$\dfrac{23.5 \text{ parts} (1,000 \text{ gal})}{30.7 \text{ parts}} = 765.47 \text{ gallons, round to} \quad \underline{765} \text{ gallons of the 17.8\% solution}$$
$$\phantom{\dfrac{23.5 \text{ parts} (1,000 \text{ gal})}{30.7 \text{ parts}} = 765.47 \text{ gallons, round to} \quad} \overline{1,000} \text{ gallons—added here to cross check math}$$

To make the 1,000 gallons of the 25.0% solution, mix 235 gallons of the 48.5% solution with 765 gallons of the 17.8% solution.

19. **A 9.70% hypochlorite solution is required. If exactly 225 gallons are needed, how many gallons of a 12.2% solution must be mixed with a 3.75% solution to make the required solution? Solve the problem using the dilution triangle.**

12.2%   5.95   5.95 parts of the 12.2% solution are required for every 8.45 parts.

    9.70%

3.75% $\dfrac{2.50}{8.45 \text{ total parts}}$ 2.50 parts of the 3.75% solution are required for every 8.45 parts.

$$\frac{5.95 \text{ parts} (225 \text{ gal})}{8.45 \text{ parts}} = \textbf{158 gallons of the 12.2\% solution}$$

$$\frac{2.50 \text{ parts} (225 \text{ gal})}{8.45 \text{ parts}} = \frac{67 \text{ gallons of the 3.75\% solution}}{225 \text{ gallons}}$$

Mix 158 gal of the 12.2% solution with 67 gal of the 3.75% solution to get the final solution of 9.70% hypochlorite.

20. **What percent polymer solution would result, if 245 gallons of a 40.5% solution were mixed with 795 gallons of a 3.75% solution? Assume both solutions have the same density.**

First, find the total volume that would result from mixing these two solutions.

Total Volume = 245 gal + 795 gal = 1,040 gal

Write the equation.

$\mathbf{C_1 V_1 + C_2 V_2 = C_3 V_3}$, where C_1 and C_2 = % Concentration of the two solutions before being mixed,
V_1 and V_2 = Volume of the two solutions before being mixed, and
C_3 and V_3 = the resulting % Concentration and Volume, respectively.

Substitute values and solve.

$$\frac{(40.5\%)(245 \text{ gal})}{100\%} + \frac{(3.75\%)(795 \text{ gal})}{100\%} = \frac{C_3 (1,040 \text{ gal})}{100\%}$$

Solving for C_3 and reducing the left side of the equation.

$$C_3 = \frac{(99.225 \text{ gal} + 29.8125 \text{ gal})(100\%)}{1,040 \text{ gal}} = \frac{(129.0375 \text{ gal})(100\%)}{1,040 \text{ gal}}$$

$C_3 = 12.407$, round to **12.4% Final polymer solution**

21. Calculate the percent strength of a solution mixture, when 50.2 lb of a 98.1% solution are mixed with 225 lb of a 3.50% solution.

Equation: **% Mixture strength =**

$$\frac{(\text{Solution}_1 \text{ lb})(\text{Available \%}/100\%) + (\text{Solution}_2, \text{ lb})(\text{Available \%}/100\%)(100\%)}{\text{Solution}_1, \text{ lb} + \text{Solution}_2, \text{ lb}}$$

Substitute values and solve.

$$\text{\% Mixture strength} = \frac{[(50.2 \text{ lb})(98.1\%/100\%) + (225 \text{ lb})(3.50 \%/100\%)](100\%)}{50.2 \text{ lb} + 225 \text{ lb}}$$

$$\text{\% Mixture strength} = \frac{[49.2462 \text{ lb} + 7.875 \text{ lb}](100\%)}{275.2 \text{ lb}}$$

$$\text{\% Mixture strength} = \frac{[57.1212 \text{ lb}](100\%)}{275.2 \text{ lb}} = 20.756\%, \text{ round to } \textbf{20.8\% Mixture strength}$$

22. What percent strength of a solution mixture results, when exactly 210 gallons of a 26.5% solution that weighs 9.48 lb/gal are mixed with 375 gallons of a 5.65% solution that weighs 9.51 lb/gal?

Equation: **Percent mixture strength =**

$$\frac{(\text{Solution}_1 \text{ lb})(\text{Available \%}/100\%) + (\text{Solution}_2, \text{ lb})(\text{Available \%}/100\%)(100\%)}{\text{Solution}_1, \text{ lb} + \text{Solution}_2, \text{ lb}}$$

% Mixture strength =

$$\frac{[(210 \text{ gal})(9.48 \text{ lb/gal})(26.5\%/100\%) + (375 \text{ gal})(9.51 \text{ lb/gal})(5.65\%/100\%)]100\%}{(210 \text{ gal})(9.48 \text{ lb/gal}) + (375 \text{ gal})(9.51 \text{ lb/gal})}$$

$$\text{\% Mixture strength} = \frac{(527.562 \text{ lb} + 201.493 \text{ lb})(100\%)}{1,990.8 \text{ lb} + 3,566.25 \text{ lb}} = \frac{(729.055 \text{ lb})(100\%)}{5,557.05 \text{ lb}}$$

% Mixture strength = 13.119%, round to **13.1% Mixture strength**

23. **What percent sodium hypochlorite solution would result, if 225 gallons of a 15% solution were mixed with exactly 130 gallons of a 4.75% solution? Assume both solutions have the same density.**

First, find the total volume that would result from mixing these two solutions.

Total Volume = 225 gal + 130 gal = 355 gal

Then, write the equation.

(Concentration$_1$)(Volume$_1$) + (Concentration$_2$)(Volume$_2$) = (Concentration$_3$)(Volume$_3$)

Condensed as $C_1V_1 + C_2V_2 = C_3V_3$, where
C_1 and C_2 = % Concentration of the two solutions before being mixed,
V_1 and V_2 = Volume of the two solutions before being mixed, and
C_3 and V_3 = the resulting % Concentration and Volume, respectively.

Substitute values and solve.

$$\frac{(15\%)(225\,\text{gal})}{100\%} + \frac{(4.75\%)(130\,\text{gal})}{100\%} = \frac{C_3(355\,\text{gal})}{100\%}$$

$$33.75\,\text{gal} + 6.175\,\text{gal} = \frac{C_3(355\,\text{gal})}{100\%}$$

Solve for C_3.

$$C_3 = \frac{(33.75\,\text{gal} + 6.175\,\text{gal})(100\%)}{355\,\text{gal}} = \frac{(39.925\,\text{gal})(100\%)}{355\,\text{gal}}$$

C_3 = 11.25, round to **11% Final sodium hypochlorite solution**

24. **What percent polymer solution would result, if 730 gallons of a 30.5% solution were mixed with 265 gallons of a 15.8% solution? Assume both solutions have the same density.**

First, find the total volume that would result from mixing these two solutions.

Total Volume = 730 gal + 265 gal = 995 gal

Write the equation.

$C_1V_1 + C_2V_2 = C_3V_3$, where
C_1 and C_2 = % Concentration of the two solutions before being mixed,
V_1 and V_2 = Volume of the two solutions before being mixed, and
C_3 and V_3 = the resulting % Concentration and Volume, respectively.

Substitute values and solve.

$$\frac{(30.5\%)\,(730\text{ gal})}{100\%} + \frac{(15.8\%)\,(265\text{ gal})}{100\%} = \frac{C_3\,(995\text{ gal})}{100\%}$$

Solving for C_3

$$C_3 = \frac{(222.65\text{ gal} + 41.87\text{ gal})\,(100\%)}{995\text{ gal}} = \frac{(264.52\text{ gal})\,(100\%)}{995\text{ gal}}$$

C_3 = 26.58, round to **27% Final polymer solution**

25. **What percent strength of a solution mixture results, when 85 gallons of a 26% solution that weighs 10.86 lb/gal are mixed with 310 gallons of an 12% solution that weighs 10.82 lb/gal?**

Equation: **Percent mixture strength =**

$$\frac{[(\text{Solution}_1\text{ gal})(\text{lb/gal})(\text{Avail }\%/100\%) + (\text{Solution}_2\text{ gal})(\text{lb/gal})(\text{Avail }\%/100\%)]100\%}{(\text{Solution}_1,\text{ gal})(\text{lb/gal}) + (\text{Solution}_2,\text{ gal})(\text{lb/gal})}$$

Where Avail = Available

% Mixture strength =

$$\frac{[(85 \text{ gal})(10.86 \text{ lb/gal})(26\%/100\%) + (310 \text{ gal})(10.82 \text{ lb/gal})(12\%/100\%)]100\%}{(85 \text{ gal})(10.86 \text{ lb/gal}) - (310 \text{ gal})(10.82 \text{ lb/gal})}$$

% Mixture strength $= \dfrac{(240.006 \text{ lb} + 402.504 \text{ lb})(100\%)}{923.1 \text{ lb} + 3,354.2 \text{ lb}} = \dfrac{(642.51 \text{ lb})(100\%)}{4,277.3 \text{ lb}}$

% Mixture strength $= 15.02\%$, round to **15% Mixture strength**

VOLUME PROBLEMS

Volumes are very important to determine because many problems in the wastewater field require the volume to be known first before the rest of the calculations can be made. Knowing the volume of a particular process can also help the operator plan and make proper decisions in the treatment of wastewater.

26. **Calculate the volume in gallons for a pipeline that is 20.0 inches in diameter and 2,348 ft long.**

First, convert the diameter to feet.

$(20.0 \text{ in.})\dfrac{(1 \text{ ft})}{12 \text{ in.}} = 1.6667 \text{ ft (Diameter)}$

Then, convert the diameter to the radius.

Radius = Diameter /2 = 1.6667 ft/2 = 0.833 ft in radius

Formula for the volume of a pipe in gallons is:

Volume, gal $= \pi r^2$**(Length) or (0.785)(Diameter)^2 (Length)(7.48 gal/ft^3)**

Volume, gal $= (3.14)(0.833 \text{ ft})(0.833 \text{ ft})(2,348 \text{ ft})(7.48 \text{ gal/ft}^3)$

Volume, gal $= 38,266.55$ gal, round to **38,300 gal**

27. **What is the volume of a conical tank in cubic feet that has a radius of 18.4 ft and a height of 25.4 ft?**

Equation: **Volume, ft³ = 1/3πr²(Height or Depth)**

Volume, ft³ = 1/3(3.14)(18.4 ft)(18.4 ft)(25.4 ft) = 9,000.73 ft³, round to **9,000 ft³**

28. **A chemical storage tank is conical at the bottom and cylindrical at the top. If the diameter of the cylinder is 19.8 ft with a depth of 34.9 ft, a working height of 31.5 ft and the cone depth is 13.25 ft, what is the working volume of the tank in gallons?**

First, find the volume of the cone in gallons.

Volume, gal = 1/3(0.785)(Diameter)²(Depth)(7.48 gal/ft³)

Volume, gal = 1/3(0.785)(19.8 ft)(19.8 ft)(13.25 ft)(7.48 gal/ft³) = 10,167 gal

Next, find the volume of the cylindrical part of the tank in gallons.

Volume of cylinder, gal = (0.785)(Diameter)²(Depth)(7.48 gal/ft³)

Volume of cylinder, gal = (0.785)(19.8 ft)(19.8 ft)(31.5 ft)(7.48 gal/ft³)

Volume of cylinder, gal = 72,512 gal

Lastly, add the two volumes for the answer.

Total volume, gal = 10,167 gal + 72,512 gal = 82,679 gal, round to **82,700 gal**

29. **What is the volume in gallons for an oxidation ditch, given the following data:**

Average top width of ditch at water surface = 8.55 ft
Average depth = 4.82 ft
Average bottom width = 5.48 ft
Length of ditch = 157 ft

Diameter of half circles = 103 ft

Equation: $\dfrac{[(b_1 + b_2)\,\text{Depth}]}{2}$ (**Length of 2 sides + Length of 2 half circles**)(**7.48 gal/ft³**)

Where b_1 = bottom width of ditch and b_2 = top width of ditch at water surface, and the length of the two half circles is π(Diameter)

Substitute values and solve.

$$\text{Volume, gal} = \frac{[(5.48 \text{ ft} + 8.55 \text{ ft})(4.82 \text{ ft})]}{2}[(2)(157 \text{ ft}) + (3.14)(103 \text{ ft})](7.48 \text{ gal/ft}^3)$$

$$\text{Volume, gal} = (33.81 \text{ ft}^2)(314 \text{ ft} + 323.42 \text{ ft})(7.48 \text{ gal/ft}^3)$$

$$\text{Volume, gal} = (33.81 \text{ ft}^2)(637.42 \text{ ft})(7.48 \text{ gal/ft}^3)$$

$$\text{Volume, gal} = 161{,}203 \text{ gal, round to } \mathbf{161{,}000 \text{ gal}}$$

30. **Calculate the volume of a rectangular pond, given the following data:**

Average pond depth = 5.37 ft
Surface water dimensions = 234 ft by 418 ft
Bottom pond dimensions = 214 ft by 398 ft
Slope of sides average = 2 ft horizontal to 1 ft vertical (2:1)

Note: Since the slope is 2:1, the following equation applies.

Equation: **Volume, gal** $= \dfrac{(\text{Length}_1 + \text{Length}_2)}{2} \dfrac{(\text{Width}_1 + \text{Width}_2)}{2}$ (**Depth, ft**)(**7.48 gal/ft³**)

Where Length_1 and Width_1 are the bottom dimensions of the pond and Length_2 and Width_2 are the surface water dimensions of the pond.

Substitute values and solve.

$$\text{Volume, gal} = \frac{(398 \text{ ft} + 418 \text{ ft})}{2} \frac{(214 \text{ ft} + 234 \text{ ft})}{2}(5.37 \text{ ft})(7.48 \text{ gal/ft}^3)$$

$$\text{Volume, gal} = \frac{(816 \text{ ft})}{2} \frac{(448 \text{ ft})}{2}(5.37 \text{ ft})(7.48 \text{ gal/ft}^3) = 3{,}670{,}997 \text{ gal, round to } \mathbf{3{,}670{,}000 \text{ gal}}$$

31. **Digester gas is stored in a spherical tank that is 15.24 ft in diameter. What is the capacity of the sphere in ft³?**

First, convert the diameter in feet to the radius in feet.

Radius, ft = Diameter/2 = 15.24 ft/2 = 7.62 ft

Next, calculate the sphere's capacity.

Equation: **Sphere volume ft³** $= \dfrac{4\pi r^3}{3}$

Sphere volume ft³ $= \dfrac{4\,(3.14)\,(7.62\text{ ft})\,(7.62\text{ ft})\,(7.62\text{ ft})}{3} = 1{,}852.39 \text{ ft}^3 = \textbf{1,850 ft}^3$

DENSITY AND SPECIFIC GRAVITY OF LIQUIDS AND SOLIDS

The density of a substance is the amount of mass for a given volume. It is usually expressed as lb/gal or lb/ft³ in the English system or as g/cm³, kg/L, or kg/m³ in the metric system. Specific gravity compares the density of one substance to another. Water is the standard for liquids and solids and is equal to 1.

32. **What is the specific gravity for a solution that weighs 1,324 grams/liter?**

Know: Density of water equals 1 gram per milliliter

sp gr $= \dfrac{(1{,}324 \text{ g/L})\,(1\text{ L})}{(1 \text{ g/mL})\,(1{,}000 \text{ mL})} = \textbf{1.324 sp gr}$

33. **Determine how much space a liquid substance would occupy in liters, if it weighed 438 lb and its specific gravity was 1.45.**

First, convert the number of lb to grams (g).

Number of g = (Number of lb)(454 g/1 lb)

Substitution: Number of g = (438 lb)(454 g/1 lb) = 198,852 g

Because the specific gravity is given as 1.45, we know that 1 cubic centimeter of the substance equals 1.45 g/cm³.

Know: 1 milliliter = 1 cubic centimeter (cm³)

Therefore, it follows that 3,785 mL = 3,785 cm³/gal

$$\text{Number of g/cm}^3 = \frac{(8.34 \text{ lb/gal})(454 \text{ g/lb})(1.45)}{3,785 \text{ cm}^3/\text{gal}} = 1.4505 \text{ g/cm}^3, \text{ round to } 1.45 \text{ g/cm}^3$$

We know that 1.45 grams of the substance occupies 1 cm³ by knowing its specific gravity, which was converted to density. To get the space 198,852 grams occupies, we only need to divide by the density.

$$\text{Space occupied in liters} = \frac{198,852 \text{ g}}{(1.45 \text{ g/cm}^3)(1,000 \text{ cm}^3/\text{L})} = 137.14 \text{ liters, round to } \mathbf{137 \text{ liters}}$$

34. What is the specific gravity of a rock, if it weighs 748 grams in air and weighs 462 grams in water?

First, subtract the weight in air from the weight in water to determine the loss of weight in water.

Number of kilograms = 748 g − 462 g = 286 g is weight loss in water

Next, find the specific gravity by dividing the weight of the rock in air by the weight loss in water.

Sp gr = 748 g/286 g = 2.615, round to **2.62 sp gr**

PRESSURE PROBLEMS

Pressure is the measure of force against a surface and is usually expressed as force per unit area. In the English system the units are usually in lb/in.² or lb/ft². Scientists and engineers usually use the metric system, where pressure is measured in Pascals (Pa). One Pascal is equal to a force of 1 Newton per square meter. A Newton is equal to the force required to accelerate 1 kilogram 1 meter per second per second (1 kg·m/s²). You can also have kilopascals (kPa), megapascals (mPa), and giga-pascals (gPa). Also: 1 Pascal = 10 dyne/cm² = 0.01 mbar. 1 atm = 101,325 Pascals = 760 mm Hg = 760 torr (Torricelli barometer) = 14.7 psi. *Note:* psi = pounds per square inch.

35. **A 16-inch diameter pipe is flowing wastewater at a flow rate of 1.5 ft³/s. The pressure in this pipe is 65 psi. If the pipe size decreases to 8 inches, what is the pressure in the 8-inch pipe?**

First, convert each pipe from inches to feet and from the diameter to the radius.

$$\text{Radius of 16-inch pipe, ft} = \frac{(16\,\text{in.})(1\,\text{ft})}{(2)(12\,\text{in.})} = 0.6667 \text{ ft in radius}$$

$$\text{Radius of 8-inch pipe, ft} = \frac{(8\,\text{in.})(1\,\text{ft})}{(2)(12\,\text{in.})} = 0.3333 \text{ ft in radius}$$

Now, determine the area of each pipe. Use the equation: Area $= \pi r^2$ or $\frac{(\pi)(\text{Diameter})^2}{4}$

Area of 16-inch pipe $= (3.14)(0.6667\,\text{ft})(0.6667\,\text{ft}) = 1.40\,\text{ft}^2$

Area of 8-inch pipe $= (3.14)(0.3333\,\text{ft})(0.3333\,\text{ft}) = 0.349\,\text{ft}^2$

Next, determine the flow in feet per second (ft/s) for each pipe.

Equation: Q (flow) = (Velocity)(Area) or for our purposes V = Q/A

Know: Velocity in 16-inch pipe = Velocity in 8-inch pipe

Velocity, 16-inch pipe $= (1.5\,\text{ft}^3/\text{s})/(1.40\,\text{ft}^2) = 1.07\,\text{ft/s}$

Velocity, 8-inch pipe $= (1.5\,\text{ft}^3/\text{s})/(0.349\,\text{ft}^2) = 4.30\,\text{ft/s}$

Lastly, solve for the pressure in 8-inch pipe using the Bernoulli's equation.

$$\frac{\text{Pressure}_A}{w} + \frac{(\text{Velocity}_A)^2}{2g} = \frac{\text{Pressure}_B}{w} + \frac{(\text{Velocity}_B)^2}{2g}$$

Where in this case, A is the 16-inch pipe and B is the 8-inch pipe.

Substitute values and convert psi to lb/ft², then solve.

$$\frac{(65\,\text{psi})(144\,\text{in}^2/\text{ft}^2)}{62.4\,\text{lb/ft}^2} + \frac{(1.07\,\text{ft/s})^2}{2(32.2\,\text{ft/s}^2)} = \frac{(\text{Pressure}_B)(144\,\text{in}^2/\text{ft}^2)}{62.4\,\text{lb/ft}^2} + \frac{(4.30\,\text{ft/s})^2}{2(32.2\,\text{ft/s}^2)}$$

Reduce equation and solve for pressure in 8-inch pipe.

$(2.31)(\text{Pressure}_B) = 150 + 0.01778 - 0.2871$

$\text{Pressure}_B = 149.731/2.31 = \textbf{64.8 psi}$

36. **What is the pressure 6.25 ft from the bottom of a tank at exactly 36 hours after the pumping started, given the following data?**

Tank radius = **40.25 ft**
Sludge depth at exactly 12 noon = **3.42 ft**
Sludge flow into tank = **64.5 gpm**

First, calculate the amount the pump added to the tank in 36 hours.

Volume, gal = (64.5 gpm)(60 min/hr)(36 hr) = 139,320 gal

Next, determine the depth of sludge this would add to the tank.

Equation: **Volume, gal** = π**(radius)²(Depth, ft)(7.48 gal/ft³)**

Rearrange the equation to solve for depth.

$$\text{Depth, ft} = \frac{\text{Volume, gal}}{\pi\,(\text{radius})^2\,(7.48\text{ gal/ft}^3)}$$

$$\text{Depth, ft} = \frac{139,320\text{ gal}}{3.14\,(40.5\text{ ft})\,(40.5\text{ ft})\,(7.48\text{ gal/ft}^3)} = 3.616\text{ ft}$$

Next, add the level in feet at the beginning to the level in feet that was added.

Total level, ft = 3.42 ft + 3.616 ft = 7.036 ft

Next, subtract the final level in feet from the level in feet that is in question.

Level, ft = 7.036 ft − 6.25 ft = 0.786 ft

Finally, determine the pressure 6.25 ft above the tank bottom.

Pressure = (0.786 ft)(0.4335 psi/ft) = **0.341 psi**

SCREENING MATERIAL REMOVAL CALCULATIONS

The amount of screening debris should be calculated by operators so that they can plan and properly dispose of the material. A record should be kept each time for the amount of material removed from the screening pits. Screenings are usually disposed of by landfill, incinerated, or ground and returned to the wastewater process. They are very odorous and will attract flies. See the figures in Appendix E for placement of wastewater screens.

37. **How many gallons of screenings were removed from a wastewater plant, if the plant processed 5,142,000 gallons and the screenings removed per million gallons was 3.19 ft³/mil gal?**

First, convert gallons to mil gal.

$$\text{Number of mil gal} = \frac{5,142,000 \text{ gal}}{1,000,000/\text{mil}} = 5.142 \text{ mil gal}$$

Next, calculate the number of ft³ of screenings removed by rearranging the following equation.

Equation: **Screenings, ft³/mil gal** $= \dfrac{\text{Number of ft}^3}{\text{Number of mil gal}}$

Rearrange the equation.

Number of ft³ = (Screenings, ft³/mil gal)(Number of mil gal)

Substitute values and solve.

Number of ft³ = (3.19 ft³/mil gal)(5.142 mil gal) = 16.403 ft³

Lastly, convert the ft³ of screenings removed to gallons of screenings removed.

Screenings removed, gal = (16.403 ft³)(7.48 gal/ft³) = 122.69 gal, round to **123 gal**

38. **A wastewater treatment plant processes 2,150 gpm. If the total screenings for one day were 98 gallons, what were the screenings in cubic feet per mil gal?**

First, determine the amount of cubic feet in 98 gallons.

Number of ft^3 = 98 gal/7.48 gal/ft^3 = 13.1 ft^3

Next, convert gpm to mgd.

$$\text{Number of mgd} = \frac{(2,150 \text{ gpm})(1,440 \text{ min/day})}{1,000,000/\text{mil}} = 3.096 \text{ mgd}$$

The day (D) can be dropped in this problem. Next, determine the screenings removed in ft^3/mil gal.

Equation: **Screenings, ft^3/mil gal** $= \dfrac{\text{Number of ft}^3}{\text{Number of mil gal}}$

Screenings, ft^3/mil gal $= \dfrac{13.1 \text{ ft}^3}{3.096 \text{ mgd}} = $ **4.2 ft^3/mil gal**

SCREENING PIT CAPACITY CALCULATIONS

The operator needs to know the capacity of a screening pit so he or she knows when it should be cleaned based on past records of material removed (above calculations).

39. **How many days will it take to fill a screening pit, if the pit is 5.45 ft by 15.1 ft and 5.12 ft deep, and the average screenings are 2.88 ft^3/day?**

First, determine the volume of the pit in ft^3.

Pit volume, ft^3 = (5.45 ft)(15.1 ft)(5.12 ft) = 421.35 ft^3

Equation: **Number of days to fill** $= \dfrac{\text{Pit volume, ft}^3}{\text{Screenings removed, ft}^3/\text{day}}$

Substitute values and solve.

Number of days to fill $= \dfrac{421.35 \text{ ft}^3}{2.88 \text{ ft}^3/\text{days}} = 146.3 \text{ days, round to } \textbf{146 days}$

40. **What is the capacity of a screening pit in ft³, if it would fill in 89.5 days and the average screenings each day are 2.75 ft³?**

Equation: **Number of days to fill** $= \dfrac{\text{Pit volume, ft}^3}{\text{Screenings removed, ft}^3/\text{day}}$

Rearrange the equation to solve for pit volume.

Pit volume, ft³ = (Number of days to fill)(Screenings removed, ft³/day)

Substitute values and solve.

Pit volume, ft³ = (89.5 days)(2.75 ft³/day) = 246.125 ft³, round to **246 ft³**

GRIT REMOVAL CALCULATIONS

Grit removal is important for the same reason as screening removal—planning for proper disposal. Grit channels are important in wastewater treatment because by removing the grit from the waste they prevent wear on pumps and deposition in pipelines or channels. They also prevent grit from accumulating in other processes such as digesters or biological contactors. Not all wastewater treatment plants have grit channels and they are not always placed after screens or comminutors. See the figures in Appendix E for where grit channels are commonly placed in different treatment plants.

41. **Given the following data, determine the number of cubic yards of grit generated at a wastewater plant in a 30-day month:**

Grit = 2.58 ft³/mil gal
Average flow = 1,850 gpm

First, convert gpm to mgd.

Number of mgd $= \dfrac{(1,850 \text{ gpm})(1,440 \text{ min}/\text{day})}{1,000,000/\text{mil}} = 2.664 \text{ mgd}$

Next, calculate the number of cubic feet removed during the month.

Number of ft³ = (2.58 ft³/mil gal)(2.664 mgd)(30 days/month) = 206.194 ft³/month

Know: 1 yard^3 (yd^3) = 27 ft^3

Number of yd^3 = $\dfrac{206.194 \text{ ft}^3/\text{month}}{27 \text{ ft}^3/\text{yd}^3}$ = 7.6368 yd^3/month, round to **7.64 yd^3/month**

42. **A wastewater plant is building a new grit channel because the old one is too small. What should the capacity for a new grit channel be in ft^3, if the daily average grit removed from the existing pit is 32.5 gal/day and the desired capacity for the new pit is exactly 60 days?**

First, determine the number of gallons of grit that would be removed in 90 days.

Number of gal = (32.5 gal/day)(60 days) = 1,950 gal

Next, determine the number of ft^3 in 1,950 gallons.

Number of ft^3 = $\dfrac{1,950 \text{ gal}}{7.48 \text{ gal/ft}^3}$ = 260.695 ft^3, round to **261 ft^3**

43. **How many mil gal of waste was treated by a plant, if the number of gallons of grit removed were 231 liters and the grit removal rate were 3.72 ft^3/mil gal?**

First, convert liters to gallons.

Number of gallons = (231 liters)(1 gal/3.785 liters) = 61.03 gal

Equation: **Grit removal, ft^3/mil gal** = $\dfrac{\text{Number of gallons removed}}{(7.48 \text{ gal/ft}^3)(\text{mil gal treated})}$

Rearrange to solve for mil gal treated.

mil gal treated = $\dfrac{\text{Number of gallons removed}}{(\text{Grit removal, ft}^3/\text{mil gal})(7.48 \text{ gal/ft}^3)}$

Substitute values and solve.

mil gal treated = $\dfrac{61.03 \text{ gal}}{(3.72 \text{ ft}^3/\text{mil gal})(7.48 \text{ gal/ft}^3)}$ = **2.19 mil gal**

DETENTION TIME CALCULATIONS

Detention time is simply the time period that starts when wastewater flows into a basin or tank and that ends when it flows out of the basin or tank. Detention time is usually calculated for wastewater ponds, oxidation (aerobic) ditches, and clarifiers. Detention times are theoretical, because basins begin to fill with settled sludge and other debris. This causes the true detention time to constantly change (decrease). While it is true that sludge removals will cause the detention time to increase, the true detention time will always be less than theoretical. Also, flows through a basin are never perfectly laminar and thus cause a further decrease in the true detention time. See Figures E-5 and E-6 in Appendix E for two types of wastewater plants using ponds or Figure E-8 in Appendix E for a wastewater plant using an oxidation ditch.

44. **What is the detention time in days for a series of two wastewater treatment ponds, given the following data?**

Pond 1 = averages 286 ft by 142 ft with a depth of 5.09 ft
Pond 2 = averages 233 ft by 106 ft with a depth of 5.27 ft
Flow = 68,350 gpd

First, calculate the volume of both waste ponds in gallons.

Equation: **Volume, gal = (Length, ft)(Width, ft)(Depth, ft)(7.48 gal/ft³)**

Pond 1 volume, gal = (286 ft)(142 ft)(5.09 ft)(7.48 gal/ft³) = 1,546,229 gal

Pond 2 volume, gal = (233 ft)(106 ft)(5.27 ft)(7.48 gal/ft³) = $\underline{973,585 \text{ gal}}$
Add the volume of both ponds = 2,519,814 gal

Equation: **Detention time, days** $= \dfrac{\text{Volume, gal}}{\text{Flow, gpd}}$

Detention time, days $= \dfrac{2,519,814 \text{ gal}}{68,350 \text{ gpd}}$ = **36.9 days**

45. What is the detention time in days for a wastewater treatment system, given the following data?

Primary clarifier = 51.2 ft in diameter
Primary clarifier's working depth = 15.75 ft
Aeration tank = 298 ft by 43.5 ft with a working depth of 19.5 ft
Secondary clarifier = 50.3 ft in diameter
Secondary clarifier's working depth = 16.0 ft
Flow = 480,000 gpd

First, calculate the volume of the clarifiers in gallons.

Equation: **Volume, gal = (0.785)(Diameter)^2(Depth, ft)(7.48 gal/ft^3)**

Primary Clarifier, gal = (0.785 ft)(51.2 ft)(51.2 ft)(15.75 ft)(7.48 gal/ft^3) = 242,433 gal

Primary Clarifier, gal = (0.785 ft)(50.3 ft)(50.3 ft)(16.0 ft)(7.48 gal/ft^3) = 237,699 gal

Next, calculate the volume of the aeration tank.

Equation: **Volume, gal = (Length, ft)(Width, ft)(Depth, ft)(7.48 gal/ft^3)**

Aeration tank, gal = (298 ft)(43.5 ft)(19.5 ft)(7.48 gal/ft^3)   = 1,890,783 gal
                         Add the volume of the system   = 2,370,915 gal

Equation: **Detention time, days** $= \dfrac{\text{Volume, gal}}{\text{Flow, gpd}}$

Detention time, days $= \dfrac{2,370,915 \text{ gal}}{480,000 \text{ gpd}}$ = **4.94 days**

46. **What is the detention time in hours, if an oxidation ditch has an influent flow of 0.209 mgd, given the following data?**

Average top width of ditch at water surface = 14.8 ft
Average depth = 4.95 ft
Average bottom width = 10.2 ft
Length of ditch = 168 ft
Diameter of half circles = 114 ft

Equation: $\dfrac{[(b_1 + b_2)\,\text{Depth}]}{2}$ **(Length of 2 sides + Length of 2 half circles)(7.48 gal/ft³)**

Where b_1 = bottom width of ditch and b_2 = top width of ditch at water surface, and the length of the two half circles is π(Diameter)

Substitute values and solve.

$$\text{Volume, gal} = \frac{[(10.2\ \text{ft} + 14.8\ \text{ft})\,(4.95\ \text{ft})]}{2}[(2)(168\ \text{ft}) + (3.14)(114\ \text{ft})](7.48\ \text{gal/ft}^3)$$

Volume, gal = (61.875)(336 ft + 357.96 ft)(7.48 gal/ft³)

Volume, gal = (61.875)(693.96 ft)(7.48 gal/ft³)

Volume, gal = 321,182 gal

Next, convert mgd to gpd.

Number of gpd = (0.209 mgd)(1,000,000 gal/mil) = 209,000 gpd

Lastly, calculate the detention time in hours.

Equation: **Detention time, hr** $= \dfrac{(\text{Volume, gal})\,(24\ \text{hr/day})}{\text{Flow, gpd}}$

Detention time, hr $= \dfrac{(321,182\ \text{gal})\,(24\ \text{hr/day})}{209,000\ \text{gpd}}$ = **36.9 hr**

WEIR AND SURFACE OVERFLOW RATE PROBLEMS

A weir is like a small dam, gate, notch, or other barrier placed across a basin to help regulate water out of the basin. The weir overflow rate is used to determine the velocity of wastewater over the weir. The velocity informs the operator about the efficiency of the sedimentation process. At constant wastewater flow, the shorter the length of the weir, the faster the water velocity will be out of the basin. Conversely, the longer the weir length, the slower the velocity will be out of the basin. See Figures E-1, E-2, E-3, E-7, and E-8 in Appendix E for five types of wastewater plants using a clarifier.

47. What is the weir overflow rate in gpm/ft, if the flow is 0.442 mgd and the radius of the clarifier is 44.5 ft?

First, convert mgd to gpm.

$$\text{Number of gpm} = \frac{(0.442 \text{ mgd})(1,000,000 \text{ mil gal})}{1,440 \text{ min/day}} = 306.94 \text{ gpm}$$

Next, calculate the length of the weir.

Weir length, ft = 2π(radius, ft)

Weir length, ft = 2(3.14)(44.5 ft) = 279.46 ft

Next, solve for the weir overflow rate.

Equation: **Weir overflow rate, gpm/ft** $= \dfrac{\text{Flow, gpm}}{\text{Weir length, ft}}$

Weir overflow rate, gpm/ft $= \dfrac{306.94 \text{ gpm}}{279.46 \text{ ft}} = 1.098$ gpm/ft, round to **1.10 gpm/ft**

48. **What is the surface overflow rate in gpd/ft^2, if the clarifier is 40.1 ft in radius and the flow into the basin is 1,220 gpm?**

First, determine the area of the clarifier.

Area $= \pi r^2$ where $\pi = 3.14$

Area $= (3.14)(40.1 \text{ ft})(40.1 \text{ ft}) = 5{,}049.15 \text{ ft}^2$

Next, convert gpm to gpd.

Number of gpd $= (1{,}220 \text{ gpm})(1{,}440 \text{ min/day}) = 1{,}756{,}800 \text{ gpd}$

Lastly, calculate the surface overflow rate.

Equation: **Surface overflow rate** $= \dfrac{\text{Flow, gpd}}{\text{Area, ft}^2}$

Surface overflow rate $= \dfrac{1{,}756{,}800 \text{ gpd}}{5{,}049.15 \text{ ft}^2} = 347.94 \text{ gpd/ft}^2$, round to **348 gpd/ft^2**

FLOW AND VELOCITY CALCULATIONS

Operators need to know the flow and velocity of the wastewater throughout the different plant processes, for example to feed proper dosages of chemicals to treat wastewaters, to know how many clarifiers or ponds to use or how much supernatant to recirculate, or for settling purposes, among other uses.

49. **If an 8.50-ft diameter chemical tank drops 3.45 inches in exactly 6.25 hours, what is the pumping rate for the chemical in gpm?**

First, convert 6.25 hours to minutes.

Number of minutes = (6.25 hr)(60 min/hr) = 375 min

Next, determine the amount in feet the tank level dropped.

Drop, ft = (3.45 in.)(1 ft/12 in.) = 0.2875 ft

Then, determine the volume in gallons for the drop in level of the tank.

Equation: **Volume, gal = (0.785)(Diameter)^2(Drop, ft)(7.48 gal/ft^3)**

Substitute values and solve.

Volume, gal = (0.785)(8.50 ft)(8.50 ft)(0.2875 ft)(7.48 gal/ft^3) = 121.968 gal

Now, calculate the pumping rate in gpm.

Equation: **Pumping rate = Flow, gal/Time, min**

Pumping rate = 121.968 gal/375 min = **0.325 gpm**

50. **Water is flowing at a velocity of 1.75 ft/s in a 10.0-inch diameter pipe. If the pipe changes from the 10.0-inch to an 8.0-inch pipe, what will the velocity be in the 8.0-inch pipe?**

Flow in 10.0-inch pipe equals the flow in the 8.0-inch pipe as the flow must remain constant: $\mathbf{Q_1} = \mathbf{Q_2}$

Since Q, flow = (Area)(Velocity), it follows that: $\mathbf{(Area_1)(Velocity_1)} = \mathbf{(Area_2)(Velocity_2)}$

First, find the diameters in feet for the 10.0-inch and 8.0-inch pipes.

Diameter for 10.0-inch = 10.0-in.(1 ft/12-in.) = 0.833 ft

Diameter for 8.0-inch = 8.0-in.(1 ft/12-in.) = 0.667 ft

Then, determine the areas of each size pipe. Area = $(0.785)(Diameter)^2$

$Area_1$ (10.0-in.) = $(0.785)(0.833 \text{ ft})(0.833 \text{ ft}) = 0.545 \text{ ft}^2$

$Area_2$ (8.0-in.) = $(0.785)(0.667 \text{ ft})(0.667 \text{ ft}) = 0.349 \text{ ft}^2$

Lastly, substitute areas calculated and known velocity in 10.0-inch pipe.

$(0.545 \text{ ft}^2)(1.75 \text{ ft/s}) = (0.349 \text{ ft}^2)(x, \text{ ft/s})$ Solve for x, ft/s.

$x, \text{ ft/s} = \dfrac{(0.545 \text{ ft}^2)(1.75 \text{ ft/s})}{(0.349 \text{ ft}^2)} = \mathbf{2.7 \text{ ft/s in 8.0-in. pipe}}$

51. If a 29.8-ft diameter digester drops 1.75 inches in 0.25 hours, what is the pumping rate out of the tank in gpm?

First, convert 0.25 hours to minutes.

Number of minutes = (0.25 hr)(60 min/hr) = 15 min

Next, determine the amount in feet the tank level dropped.

Drop, ft = (1.75 in.)(1 ft/12 in.) = 0.1458 ft

Then, determine the volume in gallons for the drop in level of the tank.

Equation: **Volume, gal = (0.785)(Diameter)^2(Drop, ft)(7.48 gal/ft^3)**

Substitute values and solve.

Volume, gal = (0.785)(29.8 ft)(29.8 ft)(0.1458 ft)(7.48 gal/ft^3) = 760 gal

Now, calculate the pumping rate in gpm.

Equation: **Pumping rate, gpm = Flow, gal/Time, min**

Pumping rate = 760 gal/15 min = 50.67 gpm, round to **51 gpm**

52. **What is the flow velocity in feet per second (ft/s) for a trapezoidal channel, given the following data?**

Bottom width, w_1 = 8.15 ft
Water surface width, w_2 = 12.75 ft
Depth = 5.35 ft
Flow = 48 ft³/s

Equation: **Flow (Q), ft³/s** $= \dfrac{(w_1 + w_2)}{2}$ **(Depth, ft)(Velocity, ft/s)**

Rearrange the formula to solve for velocity in ft/s.

$$\text{Velocity, ft/s} = \frac{2\,(Q,\,\text{ft}^3/\text{s})}{(w_1 + w_2)(\text{Depth, ft})}$$

Substitute values and solve.

$$\text{Velocity, ft/s} = \frac{2\,(48\ \text{ft}^3/\text{s})}{(8.15\ \text{ft} + 12.75\ \text{ft})(5.35\ \text{ft})} = \frac{96\ \text{ft}^3/\text{s}}{(20.9\ \text{ft})(5.35\ \text{ft})} = \textbf{0.86 ft/s}$$

****

The following two problems involve flow through a pipeline that is **not** flowing full. The calculations are almost the same as determining flow in a full pipeline, except the multiplication factor of 0.785 is not used. A new factor is used and is based on the liquid level in the pipe divided by the pipe's diameter. These factors are presented in the depth/Diameter table in Appendix D. **Please note answers are only approximate.**

53. **What is the flow in gpm for a 36-inch sewage pipeline that is flowing at a velocity of 1.74 ft/s and where the depth of the sewage averages 18.2 inches?**

First, divide the depth of sewage flow by the diameter of the pipe. Converting inches to feet is not necessary.

Ratio = depth/Diameter = 18.2 in./36 in. = 0.5056

Next, determine the factor that needs to be used.

In Appendix D, look up 0.5056 under the column d/D. The number immediately to the right will be the factor that needs to be used. This number lies somewhere between 0.50 and 0.51. The number can be rounded, but in this problem, extrapolation will be used.

The factor for 0.50 is 0.3927 and for 0.51 is 0.4027.

To extrapolate, use the following equation:

$$\textbf{Percent Division Factor} = \frac{(\text{High d/D} - \text{Low d/D})(100\%)}{\text{Ratio d/D}}$$

$$\text{Division Factor} = \frac{(0.51 - 0.50)(100\%)}{0.5056} = 1.9778 \text{ Division factor}$$

Next, use the following equation to determine the Factor.

$$\text{Factor} = \frac{(0.4027 + 0.3927)}{1.9778} = 0.4022$$

Next, convert the pipe's diameter from inches to feet.

$$\text{Number of feet} = \frac{36 \text{ in.}}{12 \text{ in./ft}} = 3.0 \text{ ft}$$

Equation: **Flow, ft³/sec = (Area, ft²)(Velocity, ft/s)**

Where the area = (Factor)(Diameter)^2

Substitute values and solve.

Flow, ft³/s = (0.4022)(3.0 ft)(3.0 ft)(1.74 ft/s) = 6.2984 ft³/s

Now, convert ft³/s to gpm.

Flow, gpm = (6.2984 ft³/s)(60 s/min)(7.48 gal/ft³) = 2,826.72 gpm, round to **2,800 gpm**

54. **What is the gpm flow in a 48-inch sewage pipeline that is flowing at a velocity of 1.62 ft/s and where the depth of the sewage averages 30.5 inches?**

First, divide the depth of sewage flow by the diameter of the pipe. Converting inches to feet is not necessary.

Ratio = depth/Diameter = 30.5 in./48 in. = 0.6354, round to 0.64

Note: Again, extrapolation can also be used if more accuracy is required.

Next, determine the factor that needs to be used.

In Appendix D, look up 0.64 under the column d/D. The number immediately to the right will be the factor that needs to be used. In this case it is 0.5308. This will be the number used rather than 0.785.

Next, convert the pipe's diameter from inches to feet.

$$\text{Number of feet} = \frac{48 \text{ in.}}{12 \text{ in./ft}} = 4.0 \text{ ft}$$

Equation: **Flow, ft³/sec = (Area, ft²)(Velocity, ft/s)**

Where the area = (Factor)(Diameter)²

Substitute values and solve.

Flow, ft³/s = (0.5308)(4.0 ft)(4.0 ft)(1.62 ft/s) = 13.758 ft³/s

Now, convert ft³/s to gpm.

Flow, gpm = (13.758 ft³/s)(60 s/min)(7.48 gal/ft³) = 6,174.59 gpm, round to **6,200 gpm**

PUMPING CALCULATIONS

Operators need to understand pumping calculations, which helps in planning treatment processes and time it will take to complete the process; in determining how long a pump will take to discharge a certain amount of wastewater or chemical to treat the wastewater; and maybe in changing the size of a pump to fit the need better.

55. **Given the following data, how many gallons are being discharged from a lift station wet well?**

Wet well dimensions = 10.2 ft × 14.9 ft
Influent flow = 32 gpm
Well drops 4.2 inches in exactly 5 minutes

First, calculate the wet well's level drop in gpm.

$$\text{Level drop, gpm} = \frac{(10.2 \text{ ft})(14.9 \text{ ft})(4.2 \text{ in.})(1 \text{ ft}/12 \text{ in.})(7.48 \text{ gal/ft}^3)}{5 \text{ min}} = 79.58 \text{ gpm}$$

Next, determine the discharge rate in gpm. In this case, add the level drop because this is liquid that is being lost from the wet well via the discharge pump.

Equation: **Discharge rate, gpm = Influent flow, gpm + Level drop, gpm**

Discharge rate, gpm = 32 gpm + 79.58 gpm = 111.58 gpm, round to **110 gpm**

56. **Given the following data, how many gallons are being discharged from a lift station wet well?**

Wet well dimensions = 12.4 ft × 16.1 ft
Influent flow = 88.5 gpm
Well increases in level by 2.3 inches in exactly 10 minutes

First, calculate the wet well's level increase in gpm.

$$\text{Level increase, gpm} = \frac{(12.4 \text{ ft})(16.1 \text{ ft})(2.3 \text{ in.})(1 \text{ ft}/12 \text{ in.})(7.48 \text{ gal/ft}^3)}{10 \text{ min}} = 28.62 \text{ gpm}$$

Next, determine the discharge rate in gpm. In this case, subtract the level increase because this is liquid that is coming into the wet well and not being discharged via the discharge pump.

Equation: **Discharge rate, gpm = Influent flow, gpm − Level increase, gpm**

Discharge rate, gpm = 88.5 gpm − 28.62 gpm = 59.88 gpm, round to **60 gpm**

57. **A reclaim water tank has a radius of 42.1 ft. How long will it take in hours and minutes for a pump to fill a tank to the 24.1-ft level, if it already has a water level of 10.46 ft and the pumping rate is 685 gpm?**

First, determine the number of gallons in the tank if it were filled to the 24.1-ft level.

Equation: Volume, gal $= \pi(\text{radius})^2(\text{Depth, ft})(7.48 \text{ gal/ft}^3)$

Volume, gal $= 3.14(42.1 \text{ ft})(42.1 \text{ ft})(24.1 \text{ ft})(7.48 \text{ gal/ft}^3) = 1{,}003{,}258$ gal

Next, calculate the number of gallons already in the tank.

Volume, gal $= 3.14(42.1 \text{ ft})(42.1 \text{ ft})(10.46 \text{ ft})(7.48 \text{ gal/ft}^3) = 435{,}439$ gal

Next, determine the number of gallons that are required to fill the tank by subtracting the volume of water at the 4.25-ft level from the volume of water at the 19.5-ft level.

Number of gallons $= 1{,}003{,}258$ gal $- 435{,}439$ gal $= 567{,}819$ gal

Next, calculate the pump's discharge rate in gpm.

Equation: **Pumping time, min** $= \dfrac{\textbf{Discharge, gal}}{\textbf{Pump rate, gpm}}$

Substitute values and solve.

Pump's discharge rate, gal $= \dfrac{567{,}819 \text{ gal}}{685 \text{ gpm}} = 828.93$ min

Now, divide by 60 (60 min/hr) to determine the number of hours.

Number of hours $= (828.93)/60 \text{ min/hr}) = 13.8155$ hr

Now, determine the number of minutes in 0.8155 hours by multiplying by 60 (60 min/hr).

Number of minutes $= (0.8155 \text{ hr})(60 \text{ min/hr}) = 48.93$ min, round to 49 min

Thus, the tank will be filled to the 24.1-ft level in **13 hours and 49 minutes.**

58. **Given the following data, determine the rate a pump discharges from the clarifier in gpm:**

Duration pump operates = 14 hr and 29 minutes
Tank diameter = 49.7 ft
Wastewater level at beginning of pumping = 16.48 ft
Wastewater level at end of pumping = 0.86 ft

First, find the number of minutes the pump worked.

Number of min = (14 hr)(60 min/hr) + 29 min = 869 min

Next, calculate the change in level during pumping.

Level change, ft = 16.48 ft − 0.86 = 15.62 ft

Next, calculate the volume in gallons added to the tank by the pump.

Equation: Volume, gal = $(0.785)(\text{Diameter})^2(\text{Level change, ft})(7.48 \text{ gal/ft}^3)$

Volume, gal = $(0.785)(49.7 \text{ ft})(49.7 \text{ ft})(15.62 \text{ ft})(7.48 \text{ gal/ft}^3) = 226{,}550.52$ gal

Now, calculate the pump's discharge rate in gpm.

Equation: **Pump's discharge rate, gpm** $= \dfrac{\text{Discharge, gal}}{\text{Time, min}}$

Substitute values and solve.

Pump's discharge rate, gpm $= \dfrac{226{,}550.52 \text{ gal}}{869 \text{ min}} = 260.70$ gpm, round to **260 gpm**

59. **How long will it take in hours and minutes to empty a tanker truck containing a polymer, if the truck's pump unloads the polymer at 62.4 gpm and a total of 17,500 liters needs to be unloaded?**

First, determine the number of gallons in 17,500 liters.

$$\text{Number of gal} = \frac{17{,}500 \text{ liters}}{3.785 \text{ liters/gal}} = 4{,}623.51 \text{ gal}$$

Then, divide the number of gallons by the pumping rate.

Time to pump = 4,623.51 gal/62.4 gpm = 74.09 min

Divide by 60 min/hr.

74.09 min/60 min/hr = 1.2348 hr

Next, find how many minutes are in 0.2348 hr by multiplying by 60 min/hr.

(0.2348 hr)(60 min/hr) = 14.088 min, round to 14 min.

The unloading time = 1 hr and 14 min.

60. **A chemical truck's pump increases the level in a 14.5-ft diameter chemical storage tank by 41.25 inches in 1 hr and 12 minutes. What is the pumping rate of the pump in gpm?**

First, determine the number of minutes the pump worked.

Number of min = (1 hr)(60 min/hr) + 12 min = 72 min

Next, convert the level increase in inches to feet.

Number of ft = (41.25 in.)(1 ft/12 in.) = 3.4375 ft

Then, calculate the volume using the following equation.

Volume, gal = (0.785)(Diameter)²(Drop, ft)(7.48 gal/ft³)

Substitute values and solve.

Volume, gal = (0.785)(14.5 ft)(14.5 ft)(3.4375 ft)(7.48 gal/ft³) = 4,243.75 gal

Lastly, determine the pumping rate in gpm.

Equation: Pumping rate or Flow, gpm $= \dfrac{\text{Number of gal}}{\text{Time, min}}$

Substitute values and solve.

Flow, gpm $= \dfrac{4,243.75 \text{ gal}}{72 \text{ min}} = 58.94$ gpm, round to **59 gpm**

61. **A chemical pump delivers 846 mL in exactly 10 minutes. How much will the chemical tank level drop in in./day, if it is 13.85 ft in diameter and had a beginning level of 9.72 ft?**

First, convert the mL/min to gpd.

Chemical pump feed, gpd $= \dfrac{(846 \text{ mL})(1,440 \text{ min/day})}{(10 \text{ min})(3,785 \text{ mL/gal})} = 32.186$ gpd

Next, calculate the drop in feet for the chemical tank.

Equation: **Flow, gpd = (0.785)(Diameter)²(Drop, ft)(7.48 gal/ft³)**

32.186 gpd = (0.785)(13.85 ft)(13.85 ft)(Drop, ft)(7.48 gal/ft³)

Rearrange the problem to solve for "Drop in feet" for the tank.

Drop, in. $= \dfrac{(32.186 \text{ gpd})(12 \text{ in./ft})}{(0.785)(13.85 \text{ ft})(13.85 \text{ ft})(7.48 \text{ gal/ft}^3)} = $ **0.343 in.**

62. **What should the pumping rate be in gpm for waste activated sludge, given the following data?**

Solids to be wasted = 5,450 lb
Return activated sludge (RAS) suspended solids (SS) = 7,275 mg/L

First, calculate the flow in mgd.

Equation: **Solids, lb/day = (RAS SS, mg/L)(mgd)(8.34 lb/gal)**

Rearrange the equation to solve for mgd.

$$\text{mgd} = \frac{\text{Solids, lb/day}}{(\text{RAS SS, mg/L})(8.34\,\text{lb/gal})}$$

Substitute values and solve.

$$\text{mgd} = \frac{5,450\,\text{lb}}{(7,275\,\text{mg/L})(8.34\,\text{lb/gal})} = 0.089825\,\text{mgd}$$

Now, determine the pumping rate in gpm.

$$\text{Pumping rate, gpm} = \frac{(0.089825\,\text{mgd})(1,000,000/\text{mil})}{1,440\,\text{min/day}} = 62.38\,\text{gpm, round to } \mathbf{62.4\ gpm}$$

SOLIDS AND HYDRAULIC LOADING RATE CALCULATIONS

Solids and hydraulic loading rate calculations are used to determine the solids or hydraulic loading on clarifiers, trickling filters, and other processes. These calculations are important to know so that operators, for example, can determine when to discharge sludge from a clarifier or to know the contact time between organisms in a trickling filter and the food entering that trickling filter.

63. **What is the solids loading rate in lb/d/ft², if a 72.9-ft diameter gravity thickener receives 47,300 gpd of biosolids and the biosolids contain 1.77% solids?**

First, determine the surface area of the gravity thickener.

Surface area of gravity thickener, ft² = (0.785)(72.9 ft)(72.9 ft) = 4,171.81 ft²

Now, calculate the solids loading rate.

Equation: **Solids loading rate, lb/d/ft²** $= \dfrac{\text{(Percent solids)}\,\text{(Biosolids added, gpd)}\,(8.34\ \text{lb/gal})}{\text{(Surface area, ft}^2)}$

Substitute values and solve.

Solids loading rate, lb/d/ft² $= \dfrac{(1.77\%/100\%)\,(47{,}300\ \text{gpd})\,(8.34\ \text{lb/gal})}{(4{,}171.81\ \text{ft}^2)}$

Solids loading rate, lb/d/ft² $= \mathbf{1.67\ lb/d/ft^2}$

64. **What is the solids loading rate on a secondary clarifier with a diameter of 75.2 ft, if the primary effluent flow rate is 2,720,000 gpd, the return activated sludge (RAS) flow is 650,000 gpd, and there is a mixed liquor suspended solids (MLSS) of 3,375 mg/L?**

First, determine the area of the clarifier.

Equation: Area $= (0.785)(\text{Diameter})^2$

Area $= (0.785)(75.2\ \text{ft})(75.2\ \text{ft}) = 4{,}439.2\ \text{ft}^2$

Next, find the total flow by adding the primary effluent and the RAS flows.

Total flow, gpd $= 2{,}720{,}000 + 650{,}000 = 3{,}370{,}000\ \text{gpd}$

Next, convert gpd to mgd.

Number of mgd $= \dfrac{3{,}370{,}000\ \text{gpd}}{1{,}000{,}000\ \text{gal/mil}} = 3.37\ \text{mgd}$

Finally, calculate the solids loading rate.

Equation: **Solids loading rate** $= \dfrac{(\text{MLSS, mg/L})\,(\text{mgd})\,(8.34\ \text{lb/gal})}{\text{Area, ft}^2}$

Solids loading rate $= \dfrac{(3{,}375\ \text{mg/L})\,(3.37\ \text{mgd})\,(8.34\ \text{lb/gal})}{4{,}439.2\ \text{ft}^2}$

Solids loading rate $= 21.368$ lb of solids/d/ft², round to **21.4 lb of solids/d/ft²**

65. **Determine the solids loading rate in lb of solids/d/1,000 ft², given the following data:**

Secondary clarifier radius = 41.3 ft
Primary effluent flow = 1,745,000 gpd
Return of activated sludge is 0.863 mgd
Mixed liquor suspended solids (MLSS) = 2,614 mg/L
Specific gravity of the solids is 1.04

First, convert the primary effluent flow in gallons per day to mgd.

$$\text{mgd} = \frac{1,745,000 \text{ gpd}}{1,000,000/\text{mil}} = 1.745 \text{ mgd}$$

Next, determine the total flow.

Total flow = Primary flow + Return of activated sludge

Total flow = 1.745 mgd + 0.863 mgd = 2.608 mgd

Next, calculate the area of the clarifier.

Area = πr^2

Area = (3.14)(41.3 ft)(41.3 ft) = 5,355.87 ft²

Factor out 1,000 ft².

5,355.87 ft² = (5.35587)(1,000 ft²)

Next, determine the lb/gal of solids.

Solids, lb/gal = (8.34 lb/gal)(1.04 sp gr) = 8.6736 lb/gal

Next, calculate the solids loading rate.

Equation: **Solids loading rate** $= \dfrac{\text{(MLSS, mg/L)(mgd)(Solids, lb/gal)}}{\text{Area, 1,000 ft}^2}$

Solids loading rate $= \dfrac{(2,614 \text{ mg/L})(2.608 \text{ mgd})(8.6736 \text{ lb/gal})}{(5.35587)(1,000 \text{ ft}^2)} =$

Solids loading rate $= 11,040$ lb of solids/d/1,000 ft², round to **11,000 lb of solids/d/1,000 ft²**

66. **A trickling filter has a diameter of 150.2 ft. If the flow through the filter is 2.11 mgd and the recirculation rate is 20.5% of the flow rate, what is the hydraulic loading rate on a trickling filter in gallons per day per square foot (gpd/ft²)?**

First, determine the total flow in gallons per day (gpd) through the trickling filter.

Total flow, gal $= [2.11 \text{ mgd} + 2.11 \text{ mgd}(20.5\%/100\%)](1,000,000/\text{mil})$

Total flow, gal $= [2.11 \text{ mgd} + 0.43255 \text{ mgd}](1,000,000/\text{mil}) = 2,542,550$ gpd

Next, determine the surface area in ft² for the clarifier.

Area $= \pi r^2$ where r $=$ Diameter/2 $= 150.2$ ft/2 $= 75.1$ ft

Trickling filter surface area, ft² $= (3.14)(75.1 \text{ ft})(75.1 \text{ ft}) = 17,709.63$ ft²

Lastly, calculate the hydraulic loading rate.

Hydraulic loading rate $= \dfrac{\text{Total flow, gpd}}{\text{Surface area, ft}^2}$

Hydraulic loading rate, gpd/ft² $= \dfrac{2,542,550, \text{ gpd}}{17,709.63 \text{ ft}^2} = 143.57$ gpd/ft², round to **144 gpd/ft²**

67. What is the hydraulic loading rate for a pond that is 7.45 acre-ft in gpd/ft², if the flow into the pond is 1.02 mgd?

Since the problem asks for gpd, first convert the volume of the pond in acre-ft to gallons.

Know: 1 acre-ft = 43,560 ft² (see Appendix A)

Area of pond, ft² = (7.45 acre-ft)(43,560 ft²/acre-ft) = 324,522 ft²

Next, convert mgd to gallons.

Flow into pond, gal = (1.02 mgd)(1,000,000/mil) = 1,020,000 gpd

Lastly, divide the flow.

Equation: **Hydraulic loading rate** $= \dfrac{\text{Total flow, gpd}}{\text{Surface area, ft}^2}$

Hydraulic loading rate $= \dfrac{1,020,000, \text{gpd}}{324,522 \text{ ft}^2} = 3.143$ gpd/ft², round to **3.14 gpd/ft²**

68. Calculate the solids loading rate in lb of solids/d/ft², given the following data:

Secondary clarifier radius = 35.1 ft
Primary effluent flow = 1,950,000 gpd
Return of activated sludge is 0.875 mgd
MLSS = 2,574 mg/L
Specific gravity of the solids is 1.03

First, convert the primary effluent flow in gallons per day to mgd.

$\text{mgd} = \dfrac{1,950,000 \text{ gpd}}{1,000,000/\text{mil}} = 1.95$ mgd

Next, determine the total flow.

Total flow = Primary flow + Return of activated sludge

Total flow = 1.95 mgd + 0.875 mgd = 2.825 mgd

Next, calculate the area of the clarifier.

Area = πr^2

Area = (3.14)(35.1 ft)(35.1 ft) = 3,868.51 ft^2

Next, determine the lb/gal of solids.

Solids, lb/gal = (8.34 lb/gal)(1.03 sp gr) = 8.59 lb/gal

Next, calculate the solids loading rate.

Equation: **Solids loading rate** $= \dfrac{(\text{MLSS, mg/L})(\text{mgd})(8.34 \text{ lb/gal})}{\text{Area, ft}^2}$

Solids loading rate $= \dfrac{(2,574 \text{ mg/L})(2.825 \text{ mgd})(8.59 \text{ lb/gal})}{3,868.51 \text{ ft}^2} =$

Solids loading rate = 16.146 lb of solids/d/ft^2, round to **16.1 lb of solids/d/ft^2**

SLUDGE PUMPING PROBLEMS

Sludge pumping calculations are important for operators to determine so that they know how much sludge and solids are being loaded into a digester, so that underloading or overloading does not occur. Also, operators need to know how much sludge is being pumped to other sludge processing applications, such as sludge thickening, filter presses, or for land application.

69. **Given the following data, calculate the lb/day of solids pumped to a sludge thickener:**

Sludge sample = 2,005.629 grams (g)
Solids content after drying = 97.446 g
Pump operates exactly 8 minutes every 1.0 hour
Pump rate = 12.5 gpm
Specific gravity (sp gr) = 1.02
Clarifier effluent flow = 2.09 mgd

First, determine the percent solids in the sludge.

Equation: **Solids, % = (Dry solids, g)(100%) / Sludge sample, g**

$$\text{Solids, \%} = \frac{(97.446 \text{ g})(100\%)}{2,005.629 \text{ g}} = 4.859\% \text{ solids}$$

Now, calculate the solids pumped in lb/day.

Equation: **Solids, lb/day =**

(Pumping, min/day)(24 hr/day)(Pump rate, gpm)(8.34 lb/gal)(sp gr of sludge)(% solids)

Solids, lb/day = (8 min/hr)(24 hr/day)(12.5 gpm)(8.34 lb/gal)(1.02 sp gr)(4.859%/100%)

Solids, lb/day = 992.03 lb/day, round to **990 lb/day of solids**

70. **Given the following data, calculate the pumping rate for a sludge pump in gpd:**

Bore of pump diameter = 8.25 inches
Stroke setting = 2.33 inches
Strokes per minute = 55
Pump operates 5 min/hour
Pump is 88% efficient

First, convert the bore size and stroke setting from inches to feet.

Bore, ft = (8.25-in.)(1 ft/12 in.) = 0.6875 ft

Stroke setting, ft = (2.33-in.)(1 ft/12 in.) = 0.194 ft

Next, calculate the number of gallons per stroke.

Equation: **Number of gal per stroke = (0.785)(Bore diameter, ft)^2(Stroke, ft)(7.48 gal/ft^3)**

Number of gal per stroke = (0.785)(0.6875 ft)(0.6875 ft)(0.194 ft)(7.48 gal/ft^3) = 0.538 gal/stroke

Next, determine the total minutes the pump operates per day.

Total pump time, min/day = (6 min/hr)(24 hr/day) = 144 min/day pump operates

Now, determine the pumping rate in gpd

Pumping rate, gpd = (0.538 gal/stroke)(55 strokes/min)(144 min/day)(88%/100% efficiency)

Pumping rate, gpd = 3,749.64 gpd, round to **3,700 gpd**

71. **How many lb/day of solids were pumped to a digester, if a sludge pump operates exactly 8 minutes every hour at a rate of 15.5 gpm, the percent solids in the sludge was 5.92%, and the specific gravity of the sludge was 1.03?**

Solids, lb/day =

(Pumping, min/day)(24 hr/day)(Pump rate, gpm)(8.34 lb/gal)(sp gr, sludge)(% solids)

Solids, lb/day = (8 min/hr)(24 hr/day)(15.5 gpm)(8.34 lb/gal)(1.03 sp gr)(5.92%/100%)

Solids, lb/day = 1,513.41 lb/day, round to **1,510 lb/day of solids**

72. Given the following data, calculate the gallons of thickened sludge flow in gallons from a thickener.

Primary sludge pumped to thickener = 3,285 gallons
Primary sludge solids (PSS) = 4.85%
Thickened sludge solids (TSS) = 6.32%
Specific gravity of primary sludge = 1.02
Specific gravity of thickened sludge = 1.06

First, convert both the primary and thickened sludge specific gravity to lb/gal.

Primary sludge, lb/gal = (8.34 lb/gal)(1.02 sp gr) = 8.5068 lb/gal

Thickened sludge, lb/gal = (8.34 lb/gal)(1.06 sp gr) = 8.8404 lb/gal

Equation: **(Primary sludge, gal)(Primary sludge, lb/gal)(Percent PSS) =**

(x Thickened sludge, gal)(Thickened sludge, lb/gal)(Percent TSS)

Rearrange the equation to solve for the number of gallons of thickened sludge.

$$x \text{ Thickened sludge, gal} = \frac{(\text{Primary sludge, gal})\,(\text{Primary sludge, lb/gal})\,(\text{Percent PSS})}{(\text{Thickened sludge, lb/gal})\,(\text{Percent TSS})}$$

Substitute values and solve.

$$x \text{ Thickened sludge, gal} = \frac{(3{,}285 \text{ gal})\,(8.5068 \text{ lb/gal})\,(4.85\%)}{(8.8404 \text{ lb/gal})\,(6.32\%)}$$

x Thickened sludge, gal = 2,425.8 gal, round to **2,430 gal of Thickened sludge**

73. **Given the following parameters, calculate how long a primary sludge pump should operate in minutes per hour:**

Plant flow = 455 gpm
Sludge pump = 35.0 gpm
Influent suspended solids (SS) = 325 mg/L
Effluent SS = 125 mg/L
Sludge = 4.85% solids

First, convert gpm to mgd.

$$\text{Number of mgd} = \frac{(455 \text{ gpm})(1,440 \text{ min/day})}{1,000,000/\text{mil}} = 0.6552 \text{ mgd}$$

Equation: **Operating time, min/hr =**

$$\frac{(\text{Flow, mgd})(\text{Influent SS, mg/L} - \text{Effluent SS, mg/L})(100\%)}{(\text{Sludge pump, gpm})(\text{Percent Solids})(24 \text{ hr/day})}$$

Substitute values and solve.

$$\text{Operating time, min/hr} = \frac{(0.6552 \text{ mgd})(325 \text{ mg/L, SS} - 125 \text{ mg/L, SS})(100\%)}{(35.0 \text{ gpm})(4.85\%)(24 \text{ hr/day})}$$

$$\text{Operating time, min/hr} = \frac{(0.6552 \text{ mgd})(200 \text{ mg/L, SS})(100\%)}{(35.0 \text{ gpm})(4.85\%)(24 \text{ hr/day})}$$

Operating time, min/hr = **3.22 min/hr**

BIOSOLIDS PUMPING, PRODUCTION, AND RETENTION TIME PROBLEMS

Biosolids pumping calculations provide operators accurate process control data for the sedimentation process. Biosolids are mostly composed of water, with the biosolids ranging from only 3 to 7% by volume.

74. **A wastewater treatment plant produces the following biosolids. If the plant treated 3.38 mil gal in the same time period, what is the biosolids production in lb/mil gal for this time period?**

Biosolids Produced
Tuesday = 1,630 gallons
Wednesday = 1,560 gallons
Thursday = 1,580 gallons
Friday = 1,710 gallons

First, find the total amount of biosolids in gallons produced for these four days.

1,630 gallons
1,560 gallons
1,580 gallons
<u>1,710 gallons</u>
6,480 gallons

Now, calculate the biosolids produced in lb/mil gal.

Equation: **Biosolids, lb/mil gal** $= \dfrac{(\text{Biosolids, gal})\,(8.34\ \text{lb/gal})}{\text{Flow, mil gal}}$

Biosolids, lb/mil gal $= \dfrac{(6,480\ \text{gal})\,(8.34\ \text{lb/gal})}{3.38\ \text{mil gal}} = 15{,}989$ lb/mil gal, round to **16,000 lb/mil gal**

75. **What is the biosolids production in wet tons per year, if the plant flow averages 2.67 mgd and production of biosolids averages 18,250 gal/day?**

Equation: **Biosolids, wet tons/yr** $= \dfrac{(\text{Biosolids, lb/mil gal})\,(\text{mgd})\,(365\ \text{days/yr})}{2{,}000\ \text{lb/ton}}$

Substitute values and solve.

Biosolids, wet tons/yr $= \dfrac{(18{,}250\ \text{lb/mil gal})\,(2.67\ \text{mgd})\,(365\ \text{days/yr})}{2{,}000\ \text{lb/ton}}$

Biosolids, wet tons/yr $= 8{,}892.77$ wet tons/yr, round to **8,890 wet tons/yr Biosolids**

76. Given the following data, calculate the amount of solids and volatile solids removed in lb/day:

Pumping rate = 195 gpm
Pump frequency = 24 times/day
Pumping cycle = exactly 10 minutes per cycle
Solids = 4.20%
Volatile solids (VS) = 61.5%

First, determine the solids removal in lb/day.

Equation: **Solids, lb/day =**

(Time, min/cycle)(cycles/day)(Pump rate, gpm)(8.34 lb/gal)(Percent solids)

Substitute values and solve.

Solids, lb/day = (10 min/cycle)(24 cycles/day)(195 gpm)(8.34 lb/gal)(4.20%/100%)

Solids, lb/day = 16,393 lb/day, round to **16,400 lb/day Solids**

Next, calculate the amount of volatile solids removed in lb/day.

Equation: **VS, lb/day =**

(Time, min/cycle)(cycles/day)(Pump rate, gpm)(8.34 lb/gal)(Percent, solids)(Percent VS)

Substitute values and solve.

VS, lb/day = (10 min/cycle)(24 cycles/day)(195 gpm)(8.34 lb/gal)(4.20%/100%)(61.5%/100%)

VS, lb/day = 10,082 lb/day, round to **10,100 lb/day VS**

77. **What is the biosolids retention time (BRT) in days for a digester that is 49.9 ft in diameter, has a working level of 16.5 ft, and receives an average flow of 6.5 gpm?**

First, convert gpm to gpd.

Digester influent, gpm = (6.5 gpm)(1,440 min/day) = 9,360 gpd

Next, determine the working volume in gallons for the digester.

Equation: **Digester volume, gal = (0.785)(Diameter, ft)^2(Height, ft)(7.48 gal/ft^3)**

Digester volume, gal = (0.785)(49.9 ft)(49.9 ft)(16.5 ft)(7.48 gal/ft^3) = 241,244 gal

Lastly, calculate the BRT.

Equation: $$\textbf{BRT, days} = \frac{\textbf{Digester working volume, gal}}{\textbf{Influent flow, gpd}}$$

Substitute values and solve.

$$\text{BRT, days} = \frac{241,244 \text{ gal}}{9,360 \text{ gpd}} = 25.77 \text{ days, round to } \textbf{26 days}$$

78. **What is the estimated biosolids pumping rate for the following system?**

Plant flow = 2.59 mgd
Removed biosolids = 1.18%
Influent total suspended solids (TSS) = 317 mg/L
Effluent TSS = 128 mg/L
Weight of biosolids = 8.42 lb/gal

Equation: **Estimated pumping rate =**

$$\frac{(\text{Influent TSS, mg/L} - \text{Effluent TSS, mg/L})(\text{Flow, mgd})(8.34\ \text{lb/gal})}{(\text{Percent solids in sludge})(\text{Sludge, lb/gal})(1,440\ \text{min/day})}$$

Substitute values and solve.

$$\text{Estimated pumping rate} = \frac{(317\ \text{TSS mg/L} - 128\ \text{TSS, mg/L})(2.59\ \text{mgd})(8.34\ \text{lb/gal})}{(1.18\%/100\%)(8.42\ \text{lb/gal})(1,440\ \text{min/day})}$$

$$\text{Estimated pumping rate} = \frac{(189\ \text{TSS mg/L})(2.59\ \text{mgd})(8.34\ \text{lb/gal})}{(1.18\%/100\%)(8.42\ \text{lb/gal})(1,440\ \text{min/day})}$$

Estimated pumping rate = 28.53 gpm, round to **28.5 gpm**

79. **Determine the amount of solids and volatile solids removed in lb/day, given the following data:**

Pumping rate = 211,680 gal/day
Pump frequency = 24 times/day
Pumping cycle = 12 minutes exactly per cycle
Solids = 3.72%
Volatile solids (VS) = 59.6%

First, convert gal/day to gpm.

$$\text{Pumping rate, gpm} = \frac{211,680 \text{ gal/day}}{1,440 \text{ min/day}} = 147 \text{ gpm}$$

Next, determine the solids removal in lb/day.

Equation: **Solids, lb/day =**

(Time, min/cycle)(cycles/day)(Pump rate, gpm)(8.34 lb/gal)(Percent solids)

Substitute values and solve.

Solids, lb/day = (12 min/cycle)(24 cycles/day)(147 gpm)(8.34 lb/gal)(3.72%/100%)

Solids, lb/day = 13,135 lb/day, round to **13,100 lb/day Solids**

Next, calculate the amount of volatile solids removed in lb/day.

Equation: **VS, lb/day =**

(Time, min/cycle)(cycles/day)(Pump rate, gpm)(8.34 lb/gal)(Percent, solids)(Percent VS)

Substitute values and solve.

VS, lb/day = (12 min/cycle)(24 cycles/day)(147 gpm)(8.34 lb/gal)(3.72%/100%)(59.6%/100%)

VS, lb/day = 7,828 lb/day, round to **7,830 lb/day VS**

WASTE ACTIVATED SLUDGE PUMPING RATE CALCULATIONS

These calculations are used as a planning tool by the operator. The waste activated sludge (WAS) suspended solids (SS) are pumped out of the secondary clarifier and wasted or returned to the aeration tank. It is better to pump continuously rather than intermittently and not to change the amount by more than 15 percent from one day to the next.

80. **Determine the waste activated sludge (WAS) pumping rate in gal/hr, given the following data:**

Amount of WAS to be wasted = 4,650 lb/day
Suspended solids concentrations = 3,840 mg/L

First, use the "pounds" equation to solve for the number of mgd.

Equation: **Number of lb/day WAS = (WAS, mg/L)(Number of mgd)(8.34 lb/gal)**

Rearrange the equation to solve for mgd.

$$\text{Number of mgd} = \frac{\text{Number of lb/day WAS}}{(\text{Number of mg/L WAS})(8.34 \text{ lb/gal})}$$

Substitute values and solve.

$$\text{Number of mgd} = \frac{4,650 \text{ lb/day}}{(3,840 \text{ mg/L WAS})(8.34 \text{ lb/gal})} = 0.1452 \text{ mgd}$$

Now, convert mgd to gal/hr.

$$\text{WAS pumping rate, gal/hr} = \frac{(0.1452 \text{ mgd})(1,000,000/\text{mil})}{(24 \text{ hr/day})} = \textbf{6,050 gal/hr}$$

81. **What was the waste activated sludge (WAS) pumping rate in mgd, if the return-activated sludge (RAS) suspended solids (SS) concentration averaged that day were 7,360 mg/L and the solids removed from the system were 6,250 lb/day?**

Equation: **Solids, lb/day = (RAS mg/L)(mgd)(8.34 lb/gal)**

Rearrange the equation to solve for mgd.

$$\text{Number of mgd} = \frac{\text{Solids, lb/day}}{(\text{RAS mg/L})(8.34 \text{ lb/gal})}$$

$$\text{Number of mgd} = \frac{6{,}250 \text{ lb/day}}{(7{,}360 \text{ mg/L})(8.34 \text{ lb/gal})} = \mathbf{0.102 \text{ mgd}}$$

82. **Determine the waste activated sludge (WAS) flow in gpm, given the following data.**

Influent flow = 3.95 mgd
Clarifier radius = 45.1 ft
Clarifier depth = 16.2 ft
Aerator = 1.08 mil gal
MLSS = 2,565 mg/L
Return activated sludge (RAS) SS = 7,120 mg/L
Secondary effluent SS = 27.0 mg/L
Target solids retention time (SRT) = 10 days exactly

First, determine the volume in mil gal for the clarifier and add to the aerator volume.

$$\text{Equation: Clarifier, gal} = \frac{\pi(\text{radius})^2(\text{Depth, ft})(7.48 \text{ gal/ft}^3)}{1{,}000{,}000/\text{mil}}$$

$$\text{Clarifier, gal} = \frac{3.14(45.1 \text{ ft})(45.1 \text{ ft})(16.2 \text{ ft})(7.48 \text{ gal/ft}^3)}{1{,}000{,}000/\text{mil}} = 0.773926 \text{ gal}$$

Total volume = 0.773926 + 1.08 = 1.853926 mil gal

Equation: **Target SRT =**

$$\frac{(\text{MLSS mg/L})(\text{Clarifier, Aerator Volume, mil gal})(8.34\text{ lb/gal})}{(\text{RAS SS mg/L})(x\text{mgd})(8.34\text{ lb/gal}) + (\text{Effluent SS, mg/L})(\text{Flow, mgd})(8.34\text{ lb/gal})}$$

Substitute values and solve.

$$10\text{ days SRT} = \frac{(2,565\text{ mg/L})(1.853926\text{ mil gal})(8.34\text{ lb/gal})}{(7,120\text{ mg/L})(x\text{mgd})(8.34\text{ lb/gal}) + (27.0\text{ mg/L})(3.95\text{ mgd})(8.34\text{ lb/gal})}$$

$$10\text{ days SRT} = \frac{39,659.37\text{ lb MLSS}}{(7,120\text{ mg/L})(x\text{mgd})(8.34\text{ lb/gal}) + 889.461\text{ lb/day}}$$

Rearrange the equation so that x mgd is in the numerator.

$$(7,120\text{ mg/L})(x\text{mgd})(8.34\text{ lb/gal}) + 889.461\text{ lb/day} = \frac{39,659.37\text{ lb MLSS}}{10\text{ days SRT}}$$

$$(7,120\text{ mg/L})(x\text{mgd})(8.34\text{ lb/gal}) + 889.461\text{ lb/day} = 3,965.937\text{ lb/day}$$

Subtract 889.461 lb/day from both sides of the equation.

$$(7,120\text{ mg/L})(x\text{mgd})(8.34\text{ lb/gal}) = 3,076.476\text{ lb/day}$$

$$x\text{ mgd} = \frac{3,076.476\text{ lb/day}}{(7,120\text{ mg/L})(8.34\text{ lb/gal})} = 0.051809\text{ mgd WAS}$$

Lastly, convert mgd to gpm.

$$\text{WAS flow, gpm} = \frac{(0.051809\text{ mgd})(1,000,000/\text{mil})}{1,440\text{ min/day}} = 35.98\text{ gpm, round to }\textbf{36.0 gpm WAS}$$

83. Determine the waste activated sludge (WAS) flow in gpm, given the following data.

Influent flow = 3.12 mgd
Clarifier radius = 43.2 ft
Clarifier depth = 17.3 ft
Aerator = 1.25 mil gal
MLSS = 2,480 mg/L
RAS SS = 7,140 mg/L
Secondary effluent SS = 21.0 mg/L
Target solids retention time (SRT) = 8 days exactly

First, determine the volume in mil gal for the clarifier and add to the aerator volume.

Equation: $\text{Clarifier, gal} = \dfrac{\pi(\text{radius})^2(\text{Depth, ft})(7.48 \text{ gal/ft}^3)}{1{,}000{,}000/\text{mil}}$

$\text{Clarifier, gal} = \dfrac{3.14(43.2 \text{ ft})(43.2 \text{ ft})(17.3 \text{ ft})(7.48 \text{ gal/ft}^3)}{1{,}000{,}000/\text{mil}} = 0.7583 \text{ gal}$

Total volume = 0.7583 + 1.25 = 2.0083 mil gal

Equation: **Target SRT =**

$$\dfrac{(\text{MLSS mg/L})(\text{Clarifier, Aerator Volume, mil gal})(8.34 \text{ lb/gal})}{(\text{RAS SS mg/L})(x\text{mgd})(8.34 \text{ lb/gal}) + (\text{Effluent SS, mg/L})(\text{Flow, mgd})(8.34 \text{ lb/gal})}$$

Substitute values and solve.

$8 \text{ days SRT} = \dfrac{(2{,}480 \text{ mg/L})(2.0083 \text{ mil gal})(8.34 \text{ lb/gal})}{(7{,}140 \text{ mg/L})(x\text{mgd})(8.34 \text{ lb/gal}) + (21.0 \text{ mg/L})(3.12 \text{ mgd})(8.34 \text{ lb/gal})}$

$8 \text{ days SRT} = \dfrac{41{,}538.07 \text{ lb MLSS}}{(7{,}140 \text{ mg/L})(x\text{mgd})(8.34 \text{ lb/gal}) + 546.4368 \text{ lb/day}}$

Rearrange the equation so that x mgd is in the numerator.

$(7{,}140 \text{ mg/L})(x\text{mgd})(8.34 \text{ lb/gal}) + 546.4368 \text{ lb/day} = \dfrac{41{,}538.07 \text{ lb MLSS}}{8 \text{ days SRT}}$

$(7{,}140 \text{ mg/L})(x\text{mgd})(8.34 \text{ lb/gal}) + 546.4368 \text{ lb/day} = 5{,}192.26 \text{ lb/day}$

Subtract 546.4368 lb/day from both sides of the equation.

$$(7{,}140 \text{ mg/L})(x \text{ mgd})(8.34 \text{ lb/gal}) = 4{,}645.82 \text{ lb/day}$$

$$x \text{ mgd} = \frac{4{,}645.82 \text{ lb/day}}{(7{,}140 \text{ mg/L})(8.34 \text{ lb/gal})} = 0.078019 \text{ mgd WAS}$$

Lastly, convert mgd to gpm.

$$\text{WAS flow, gpm} = \frac{(0.078019 \text{ mgd})(1{,}000{,}000/\text{mil})}{1{,}440 \text{ min/day}} = 54.18 \text{ gpm, round to } \textbf{54.2 gpm WAS}$$

PUMPING HORSEPOWER, EFFICIENCY, AND COSTING CALCULATIONS

These types of calculations can be used for determining pump size, efficiency, and costing.

84. **If 395 horsepower (hp) is required to run a pump with a motor efficiency (ME) of 91.4% and a pump efficiency (PE) of 79.2%, what is the motor horsepower (mhp)?** *Note:* **The 395 hp in this problem is called the water horsepower (whp). The whp is the actual energy (horsepower) available to pump water.**

Equation: **Motor horsepower** $= \dfrac{\text{whp}}{(\text{ME})(\text{PE})}$

$$\text{mhp} = \frac{395 \text{ whp}}{(91.4\%/100\% \text{ ME})(79.2\%/100\% \text{ PE})}$$

$$\text{mhp} = \frac{395 \text{ whp}}{(0.914 \text{ ME})(0.792 \text{ PE})}$$

$\text{mhp} = 545.66 \text{ mhp, round to } \textbf{546 mhp}$

85. **Find the motor horsepower (mhp) for a pump with the following parameters:**

Motor efficiency (ME): 91.3%
Total head (TH): 49.2 ft
Pump efficiency (PE): 78.1%
Flow: 1.28 mgd

First, convert mgd to gpm.

Number of gpm = (1.28 mgd)(1,000,000/mil)(1 day/1,440 min) = 888.889 gpm

The equation for determining the mhp with the given data is different than the problem above.

Equation: $\mathbf{mhp} = \dfrac{(\text{Flow, gpm})(\text{TH, ft})}{(3,960)(\text{ME})(\text{PE})}$

Substitute values and solve.

$\text{mhp} = \dfrac{(888.889 \text{ gpm})(49.2 \text{ ft})}{(3,960)(91.3\%/100\% \text{ ME})(78.1\%/100\% \text{ PE})}$

mhp = 15.488 mhp, round to **15.5 mhp**

86. **Find the whp for the following system:**

Motor efficiency: 90.6
Pump efficiency: 77.8%
Mhp: 152

Equation: **Water horsepower = (mhp)(ME)(PE)**

Water horsepower = (152 mhp)(90.6%/100% ME)(77.8%/100 PE) =

Water horsepower = 107.14 whp, round to **107 whp**

87. **If the water horsepower (whp) is 91.2 and the brake horsepower is 112, what is the pump efficiency (PE)?**

Equation: **Brake horsepower = whp/PE**

Where PE is understood as pump efficiency/100%

Rearrange the equation to solve for pump efficiency.

$$PE = \frac{(whp)(100\%)}{Brake\ horsepower}$$

Substitute values and solve.

$$PE = \frac{(91.2\ whp)(100\%)}{112\ Brake\ hp} = 81.43\%\ PE,\ \text{round to } \mathbf{81.4\%\ PE}$$

88. **Given the following data, calculate the brake horsepower for a pump:**

Pumping rate = 530 gpm
Differential pressure = 82 psi
Pump efficiency = 81.3%

Equation: $\mathbf{bhp} = \dfrac{(\text{Flow, gpm})(\text{Differential pressure, psi})}{(1{,}714)(\text{Pump efficiency})}$

$$bhp = \frac{(530\ gpm)(82\ psi)}{(1{,}714)(81.3\%/100\%)} = 31.19\ hp,\ \text{round to } \mathbf{31\ hp}$$

89. **What is the brake horsepower (bhp) for a pump that is pumping sludge, given the following data?**

Specific gravity of the sludge = 1.05
Differential pressure = 12 psi
Pump efficiency (PE) = 80.7%
Sludge flow = 880 gpm

First, convert the differential pressure to total differential head (TDH).

Equation: $\textbf{TDH} = \dfrac{\text{(Differential pressure)}\,(2.31\ \text{ft/psi})}{\text{Specific gravity}}$

$\text{TDH} = \dfrac{(12\ \text{psi})\,(2.31\ \text{ft/psi})}{1.05\ \text{sp gr}} = 26.4\ \text{ft}$

Next, solve for bhp.

Equation: $\textbf{bhp} = \dfrac{\text{(Flow, gpm)}\,(\text{Head, ft})\,(\text{sp gr})}{(3{,}960)\,(\text{PE})}$

$\text{bhp} = \dfrac{(880\ \text{gal})\,(26.4\ \text{ft})\,(1.05\ \text{sp gr})}{(3{,}960)\,(80.7\%/100\%)} = 7.63\ \text{hp}$

90. **Given the following data, calculate the cost of running a pump in dollars per day:**

Flow = 972,000 gpd
TDH = 72.5 ft
Motor efficiency = 90.4%
Pump efficiency = 68.5%
Cost in kilowatt hours = $0.072

First, convert gpd to gpm.

$\text{Number of gpm} = \dfrac{(972{,}000\ \text{gpd})}{(1{,}440\ \text{min/day})} = 675\ \text{gpm}$

Next, determine the horsepower (hp).

Equation: $\textbf{Motor hp} = \dfrac{\text{(Flow, gpm)}\,(\text{TDH, ft})}{(3{,}960)\,(\text{Motor efficiency})\,(\text{Pump efficiency})}$

$$\text{Motor hp} = \frac{(675 \text{ gpm})(72.5 \text{ ft})}{(3,960)(90.4\%/100\%)(68.5\%/100\%)} = 19.96 \text{ hp}$$

Now, calculate the cost of running the pump in dollars.

Equation: **Cost, \$/day = (Motor hp)(24 hr/day)(0.746 kW/hp)(Cost/kW-hr)**

Cost, \$/day = (19.96 hp)(24 hr/day)(0.746 kW/hp)(\$0.072/kW-hr) = **\$25.73/day**

91. Given the following data, calculate the cost of running a pump in dollars per day:

Flow = 1.83 mgd
TDH = 129 ft
Motor efficiency = 88.3%
Pump efficiency = 63.7%
Cost in kilowatt hours = \$0.075

First, convert mgd to gpm.

$$\text{Number of mgd} = \frac{(1.83 \text{ mgd})(1,000,000/\text{mil})}{(1,440 \text{ min/day})} = 1,271 \text{ gpm}$$

Next, determine the horsepower (hp).

Equation: **Motor hp** $= \dfrac{\textbf{(Flow, gpm)(TDH, ft)}}{\textbf{(3,960)(Motor efficiency)(Pump efficiency)}}$

$$\text{Motor hp} = \frac{(1,271 \text{ gpm})(129 \text{ ft})}{(3,960)(88.3\%/100\%)(63.7\%/100\%)} = 73.61 \text{ hp}$$

Now, calculate the cost of running the pump in dollars.

Equation: **Cost, \$/day = (Motor hp)(24 hr/day)(0.746 kW/hp)(Cost/kW-hr)**

Cost, \$/day = (73.61 hp)(24 hr/day)(0.746 kW/hp)(\$0.075/kW-hr) = **\$98.84/day**

DOSAGE PROBLEMS

These calculations are used mainly for process control, which requires accurate determination before the chemical is actually applied to a particular process. By keeping accurate records of dosages and thus usage, operators can also plan ordering or costing.

92. **What are the Cl$_2$ and sulfur dioxide (SO$_2$) dosages in mg/L for a wastewater plant's effluent, given the following data?**

Pounds of chlorine used = 175 lb/day
Flow = 1,285 gpm
Chlorine demand = 9.86 mg/L
Assume SO$_2$ is 2.50 mg/L higher than the chlorine residual

Note: This additional amount of sulfur dioxide above the chlorine residual is applied as a safety factor and is typically started at 3 mg/L.

First, convert gpm to mgd.

$$\text{Number of mgd} = \frac{(1,285 \text{ gpm})(1,440 \text{ min/day})}{1,000,000/\text{mil}} = 1.8504 \text{ mgd}$$

Next, determine the chlorine dosage in mg/L

Equation: **Number of lb/day = (Dosage, mg/L)(Number of mgd)(8.34 lb/gal)**

Rearrange the equation to solve for the chlorine dosage.

$$\text{Chlorine dosage, mg/L} = \frac{\text{Number of lb/day}}{(\text{mgd})(8.34 \text{ lb/gal})}$$

Substitute values and solve.

$$\text{Chlorine dosage, mg/L} = \frac{175 \text{ lb/day}}{(1.8504 \text{ mgd})(8.34 \text{ lb/gal})} = 11.34 \text{ mg/L Cl}_2, \text{ round to } \textbf{11.4 mg/L Cl}_2$$

In order to determine the sulfur dioxide dosage, the chlorine residual needs to be known.

Chlorine residual = Chlorine dosage − Chlorine demand

Chlorine residual = 11.34 mg/L − 9.86 mg/L = 1.48 mg/L Cl_2 residual

Now, calculate the SO_2 dosage.

SO_2 dosage, mg/L = Chlorine residual, mg/L + 2.50 mg/L

SO_2 dosage, mg/L = 1.48 mg/L + 2.50 mg/L = **3.98 mg/L SO_2**

93. **What should the chemical feeder setting be in mL/min for a polymer solution, if the desired dosage is 3.90 mg/L and the treatment plant is treating 1,250 gpm? The specific gravity (sp gr) of the polymer is 1.34. Assume polymer is 100% pure.**

First, determine the lb/gal for the polymer.

Polymer, lb/gal = (8.34 lb/gal)(1.34 sp gr) = 11.1756 lb/gal

Next, convert gpm to mgd.

$$\text{Number of mgd} = \frac{(1,250 \text{ gpm})(1,440 \text{ min/day})}{1,000,000/\text{mil}} = 1.80 \text{ mgd}$$

Next, calculate the mL/min polymer using the following equation.

$$\text{Equation: } \textbf{mL/min} = \frac{(\text{mg/L})(3,785 \text{ mL/gal})(\text{mgd})(8.34 \text{ lb/gal})}{(1,440 \text{ min/day})(\text{Polymer, lb/gal})}$$

$$\text{Polymer, mL/min} = \frac{(3.90 \text{ mg/L})(3,785 \text{ mL/gal})(1.80 \text{ mgd})(8.34 \text{ lb/gal})}{(1,440 \text{ min/day})(11.1756 \text{ lb/gal})}$$

Polymer, mL/min = 13.77 mL/min, round to **13.8 mL/min Polymer**

94. **How many lb/day of chlorine gas are required to treat 1,785 gpm, given the following data?**

Chlorine demand = 9.3 mg/L
Chlorine residual = 1.25 mg/L?

First, determine the total chlorine dose in mg/L.

Equation: **Chlorine dose = Chlorine demand + Chlorine residual**

Chlorine dose = 9.3 mg/L + 1.25 mg/L = 10.55 mg/L

Next, convert gpd to mgd.

$$\text{Number of mgd} = \frac{(1,785 \text{ gpm})(1,440 \text{ min/day})}{1,000,000/\text{mil}} = 2.5704 \text{ mgd}$$

Lastly, calculate the lb/day of chlorine required.

Equation: **Number of lb/day = (Cl_2, mg/L)(Number of mgd)(8.34 lb/gal)**

Substitute values and solve.

Number of lb/day, Cl_2 = (10.55 mg/L)(2.5704 mgd)(8.34 lb/gal)

Number of lb/day, Cl_2 = 226.16 lb/day, round to **230 lb/day Cl_2**

95. **A wastewater treatment plant feeds 235 lb/day of chlorine. What must have been the dose if the plant treats 1,350 gpm?**

First, convert gpm to mgd.

$$\text{Number of mgd} = \frac{(1,350 \text{ gpm})(1,440 \text{ min/day})}{1,000,000/\text{mil}} = 1.944 \text{ mgd}$$

Equation: **Number of lb/day = (Dosage, mg/L)(Number of mgd)(8.34 lb/gal)**

Rearrange the equation to solve for dosage in mg/L.

$$\textbf{Chlorine dosage, mg/L} = \frac{\text{Number of lb/day}}{(\text{Number of mgd})(8.34\ \text{lb/gal})}$$

Substitute values and solve.

$$\text{Chlorine dosage, mg/L} = \frac{235\ \text{lb/day}}{(1.944\ \text{mgd})(8.34\ \text{lb/gal})} = 14.49\ \text{mg/L, round to } \textbf{14.5 mg/L } \textbf{Cl}_2$$

96. **A wastewater plant is treating a flow of 2,635,000 gpd with an alum dose of 2.95 mg/L. If the alum is 48.5% pure and has a specific gravity of 1.32, what is the alum feed in mL/min?**

First, convert gpd to mgd.

$$\text{Number of mgd} = \frac{2,635,000\ \text{gpd}}{1,000,000/\text{mil}} = 2.635\ \text{mgd}$$

Next, determine the lb/gal for the alum.

Number of lb/gal = (8.34 lb/gal)(1.32 sp gr) = 11.0088 lb/gal

$$\text{Equation: } \textbf{mL/min} = \frac{(\text{mg/L})(3,785\ \text{mL/gal})(\text{mgd})(8.34\ \text{lb/gal})}{(1,440\ \text{min/day})(\text{Percent purity}/100\%)(\text{Alum, lb/gal})}$$

$$\text{Alum, mL/min} = \frac{(2.95\ \text{mg/L})(3,785\ \text{mL/gal})(2.635\ \text{mgd})(8.34\ \text{lb/gal})}{(1,440\ \text{min/day})(48.5\%/100\%)(11.0088\ \text{lb/gal})}$$

Alum, mL/min = 31.91 mL/min, round to **31.9 mL/min of Alum**

97. Given the following data, calculate the polymer dosage in mg/L:

Wastewater flow = 2,160 gpm
Polymer flow = 48.5 mL/min
Polymer purity = 63.2%
Polymer specific gravity = 1.11

First, convert gpm to mgd.

$$\text{Number of mgd} = \frac{(2,160 \text{ gpm})(1,440 \text{ min/day})}{1,000,000/\text{mil}} = 3.1104 \text{ mgd}$$

Next, determine the lb/gal for the polymer.

Polymer, lb/gal = (8.34 lb/gal)(1.11 sp gr) = 9.2574 lb/gal

Next, calculate the polymer dosage in mg/L using the following equation.

$$\text{Equation: } \textbf{Polymer, mg/L} = \frac{(\text{mL/min})(1,440 \text{ min/day})(\text{Percent purity}/100\%)(\text{Polymer, lb/gal})}{(3,785 \text{ mL/gal})(\text{mgd})(8.34 \text{ lb/gal})}$$

$$\text{Polymer, mg/L} = \frac{(48.5 \text{ mL/min})(1,440 \text{ min/day})(63.2\%/100\%)(9.2574 \text{ lb/gal})}{(3,785 \text{ mL/gal})(3.1104 \text{ mgd})(8.34 \text{ lb/gal})}$$

Polymer, mg/L = **4.16 mg/L Polymer**

CHEMICAL FEED SOLUTION SETTINGS

As above, these calculations are used mainly for process control, which requires accurate determination before the chemical is actually applied to a particular process. Also as above, by keeping accurate records of dosages and thus usage, operators can plan ordering and costing.

98. **What should the chemical feeder be set on in mL/min, given the following data?**

Polymer dosage = 7.04 mg/L
Plant flow = 1,650 gpm
Sp gr of polymer = 12.78 lb/gal

First, convert gpm flow to mgd.

$$\text{Number of mgd} = \frac{(1{,}650 \text{ gpm})(1{,}440 \text{ min/day})}{1{,}000{,}000/\text{mil}} = 2.376 \text{ mgd}$$

Next, determine the lb/day of polymer using the pounds formula.

Polymer, lb/day = (Dosage, mg/L)(mgd)(8.34 lb/gal)

Polymer, lb/day = (7.04 mg/L)(2.376 mgd)(8.34 lb/gal) = 139.5 lb/day

Next, calculate the number of gallons polymer used.

$$\text{Polymer, gal} = \frac{(139.5 \text{ lb/day})}{(12.78 \text{ lb/gal})} = 10.92 \text{ gal}$$

Now, using the following equation, calculate the mL/min of polymer being used.

Equation: **Number of mL/min** $= \dfrac{(\text{Number of gallons used})(3{,}785 \text{ mL/gal})}{1{,}440 \text{ min/day}}$

Substitute values and solve.

$$\text{Polymer, mL/min} = \frac{(10.92 \text{ gal})(3{,}785 \text{ mL/gal})}{1{,}440 \text{ min/day}} = \textbf{28.7 mL/min Polymer}$$

99. What should the chemical feed pump be set on in gpd, given the following data?

Plant's treatment flow = 1,490,000 gpd
Alum dose = 8.75 mg/L
Alum = 5.37 lb/gal [since it is only 48.5% pure; thus (11.072 lb/gal)(48.5%) = 5.37 lb/gal]

First, convert gpd to mgd.

$$\text{Number of mgd} = \frac{1,490,000 \text{ gpd}}{1,000,000/\text{mil}} = 1.49 \text{ mgd}$$

Next, calculate the number of lb/day of alum required.

Equation: **lb/day = (Dose, mg/L)(mgd)(8.34 lb/gal)**

Alum, lb/day = (8.75 mg/L)(1.49 mgd)(8.34 lb/gal) = 108.733 lb/day of dry alum

Now, calculate the amount of liquid alum by dividing the amount of dry alum by 5.37 lb/gal.

$$\text{Alum, gpd} = \frac{108.733 \text{ lb/day}}{5.37 \text{ lb/gal}} = \textbf{20.2 gpd of Alum solution}$$

100. Given the following data, calculate the percent stroke setting on a chemical feed pump:

Maximum chemical feed by chemical pump (100% setting) = 104 gallons
Plant flow = 1,740 gpm
Alum solution strength = 48.3%
Alum dose = 9.45 mg/L
Alum = 11.065 lb/gal

First, convert the flow in gpm to mgd.

$$\text{Number of mgd} = \frac{(1,740 \text{ gpm})(1,440 \text{ min/day})}{1,000,000/\text{mil}} = 2.5056 \text{ mgd}$$

Next, convert the alum percent strength to lb/gal (since it is only 48.3% pure)

$$\text{Alum, lb/gal} = (48.3\%/100\%)(11.065 \text{ lb/gal}) = 5.344 \text{ lb/gal}$$

Next, calculate the number of lb/day of alum that is fed for given dosage.

$$\text{Alum, lb/day} = (9.45 \text{ mg/L})(2.5056 \text{ mgd})(8.34 \text{ lb/gal}) = 197.4739 \text{ lb/day}$$

Next, calculate how many gpd the pump needs to feed for this dosage.

Equation: **Alum feed, gpd** $= \dfrac{\textbf{Chemical feed, lb/day}}{\textbf{Chemical solution, lb/gal}}$

$$\text{Alum feed, gpd} = \frac{197.4739 \text{ lb/day}}{5.344 \text{ lb/gal}} = 36.95 \text{ gpd}$$

Lastly, calculate the percent stroke setting.

Equation: **Percent stroke setting** $= \dfrac{\textbf{(Chemical feed, gal/day)(100\%)}}{\textbf{Maximum feed, gpd}}$

$$\text{Percent stroke setting} = \frac{(36.95 \text{ gpd})(100\%)}{104 \text{ gpd}} = \textbf{35.5\% Stroke setting}$$

101. **Given the following data, calculate the percent stroke setting on a chemical feed pump:**

Maximum chemical feed by chemical pump (100% setting) = 115 gallons
Flow = 1,150 gpm
Polymer solution strength = 32.3%
Polymer dose = 12.4 mg/L
Polymer = 9.75 lb/gal

First, convert the flow in gpm to mgd.

$$\text{Number of mgd} = \frac{(1{,}150 \text{ gpm})(1{,}440 \text{ min/day})}{1{,}000{,}000/\text{mil}} = 1.656 \text{ mgd}$$

Next, convert the polymer percent strength to lb/gal.

Polymer, lb/gal = (32.3%/100%)(9.75 lb/gal) = 3.149 lb/gal

Next, calculate the number of lb/day of polymer that is fed for given dosage.

Polymer, lb/day = (12.4 mg/L)(1.656 mgd)(8.34 lb/gal) = 171.257 lb/day

Next, calculate the how many gpd the pump needs to feed for this dosage.

$$\text{Equation: } \textbf{Polymer feed, gpd} = \frac{\text{Chemical feed, lb/day}}{\text{Chemical solution, lb/gal}}$$

$$\text{Polymer feed, gpd} = \frac{171.257 \text{ lb/day}}{3.149 \text{ lb/gal}} = 54.38 \text{ gpd}$$

Lastly, calculate the percent stroke setting.

Equation: **Percent stroke setting** $= \dfrac{(\text{Chemical feed, gal/day})(100\%)}{\text{Maximum feed, gpd}}$

Percent stroke setting $= \dfrac{(54.38 \text{ gal/day})(100\%)}{115 \text{ gpd}} =$ **47.3% Stroke setting**

DRY CHEMICAL FEED SETTINGS

As with liquid dosing, accuracy in dosing dry chemicals is important, too. The more accurate the dosage calculation is, the more probability there will be for an operator to control a treatment process and the better the records for future referral.

102. **If a sample collection bowl collected 2,468.4 grams in exactly 10 minutes, what was the feed rate of the dry chemical in lb/day?**

Know: 454 grams = 1 pound

First, determine the number of g/min.

Number of grams = 2,468.4 g/10.0 min = 246.84 g/min

Equation: **Chemical, lb/day** $= \dfrac{(\text{Number of g/min})(1{,}440 \text{ min/day})}{454 \text{ g/lb}}$

Substitute values and solve.

Chemical, lb/day $= \dfrac{(246.84 \text{ g/min})(1{,}440 \text{ min/day})}{454 \text{ g/lb}} = 782.93$ lb/day, round to **783 lb/day**

103. What must have been the setting of a dry chemical feeder in grams/min, if the number of lb/day was 235.8?

Equation: **Chemical, lb/day** $= \dfrac{(\text{Number of g/min})(1{,}440 \text{ min/day})}{454 \text{ g/lb}}$

Rearrange to solve for the feeder setting in g/min.

Number of g/min $= \dfrac{(\text{Chemical, lb/day})(454 \text{ g/lb})}{1{,}440 \text{ min/day}}$

Substitute values and solve.

Number of g/min $= \dfrac{(235.8 \text{ lb/day})(454 \text{ g/lb})}{1{,}440 \text{ min/day}} = 74.34$ g/min, round to **74.3 g/min**

104. Determine the feed rate for a dry polymer in lb/day, if the drawdown in exactly 10 minutes was 86.3 grams (g) and the flow was 1,057,000 gpd.

First, determine the number of grams used per minute.

Polymer, g $= 86.3$ g/10 min $= 8.63$ g/min

Equation: **Polymer, lb/day** $= \dfrac{(\text{Number of g/min})(1{,}440 \text{ min/day})}{454 \text{ g/lb}}$

Substitute values and solve.

Polymer, lb/day $= \dfrac{(8.63 \text{ g/min})(1{,}440 \text{ min/day})}{454 \text{ g/lb}} = 27.37$ lb/day, round to **27.4 lb/day Polymer**

105. What is the dosage in mg/L for magnesium hydroxide and the feed rate in grams/min (g/min), if the wastewater plant is treating 1,370 gpm and the magnesium feed rate is 63.4 lb/day?

First, calculate the feed rate in g/min by converting lb/day to g/min.

$$\text{Feed rate, g/min} = \frac{(63.4 \text{ lb/day})(454 \text{ g/lb})}{1,440 \text{ min/day}} = 19.9886 \text{ g/min, round to } \mathbf{20.0 \text{ g/min}}$$

Next, convert gpm to mgd.

$$\text{Number of mgd} = \frac{(1,370 \text{ gpm})(1,440 \text{ min/day})}{1,000,000/\text{mil}} = 1.9728 \text{ mgd}$$

Next, find the dosage in mg/L by using the "pounds" equation.

Equation: **Chemical, lb/day = (Dosage, mg/L)(mgd)(8.34 lb/gal)**

Rearrange the "pounds" equation and solve for dosage in mg/L.

$$\text{Dosage, mg/L} = \frac{\text{Chemical, lb/day}}{(\text{mgd})(8.34 \text{ lb/gal})}$$

Substitute values and solve.

$$\text{Dosage, mg/L} = \frac{63.4 \text{ lb/day}}{(1.9728 \text{ mgd})(8.34 \text{ lb/gal})} = \mathbf{3.85 \text{ mg/L}}$$

106. How many pounds of a dry polymer (39.4% active) are required to make exactly 200 gallons of a solution that is exactly 5.00% and what feed rate will be required in mL/min for a dosage of 6.85 mg/L, if the plant is treating 2.47 mgd?

First, convert 250 gallons to be mixed to mil gal.

$$\text{Number of mil gal} = \frac{200 \text{ gallons}}{1,000,000/\text{mil}} = 0.000200 \text{ mil gal}$$

Know: 1% = 10,000 mg/L

Therefore, 5.00% = 50,000 mg/L

Next, calculate the number of lb of dry polymer required.

Equation: **Dry polymer, lb** $= \dfrac{\textbf{(Dose, mg/L)(mil gal)(8.34 lb/gal)}}{\textbf{Percent purity}}$

Substitute values and solve.

$$\text{Dry polymer, lb} = \frac{(50,000 \text{ mg/L})(0.000200 \text{ mil gal})(8.34 \text{ lb/gal})}{39.4\%/100\%}$$

Dry polymer, lb = **211.68 lb**

Next, determine the number of pounds of polymer per gallon in this solution.

$$\text{Dry polymer, lb/gal} = \frac{211.68 \text{ lb}}{200 \text{ gal}} = 1.0584 \text{ lb/gal}$$

Now, calculate the mL/min required for a dosage of 6.85 mg/L.

Equation: **Dosage desired, mg/L** $= \dfrac{\textbf{(mL/min)(1,440 min/day)(lb/gal)}}{\textbf{(3,785 mL/gal)(8.34 lb/gal)(mgd)}}$

Rearrange the formula to solve for mL/min.

$$\text{Dry polymer feed, mL/min} = \frac{\textbf{(Dosage, mg/L)(3,785 mL/gal)(8.34 lb/gal)(mgd)}}{\textbf{(1,440 min/day)(lb/gal)}}$$

Substitute values and solve.

$$\text{Dry polymer feed, mL/min} = \frac{(6.85 \text{ mg/L})(3,785 \text{ mL/gal})(8.34 \text{ lb/gal})(2.47 \text{ mgd})}{(1,440 \text{ min/day})(1.0584 \text{ lb/gal})}$$

Dry Polymer feed = 350.43 mL/min, round to **350 mL/min Dry polymer**

SLUDGE PRODUCTION CALCULATIONS

Sludge production calculations are important for costing and disposal purposes. Plants that use processes like digestion or heat treatment produce less sludge compared to plants that use chemical addition to treat wastes because more of the sludge is destroyed.

107. **What is the amount of dry solids produced in lb/day, if a wastewater treatment plant has an influent flow of 2,140 gpm, has primary influent suspended solids of 272 mg/L, and the secondary suspended solids are 98 mg/L?** *Note:* **The specific gravity of the suspended solids is 1.03.**

First, determine the number of mg/L of suspended solids (SS) removed.

SS removed, mg/L = 272 mg/L, influent − 98 mg/L effluent = 174 mg/L, SS removed

Next, convert gpm to mgd.

$$\text{Number of mgd} = \frac{(2,140 \text{ gpm})(1,440 \text{ min/day})}{1,000,000/\text{mil}} = 3.0816 \text{ mgd}$$

Then, determine the lb/gal of sludge.

Sludge, lb/gal = (8.34 lb/gal)(1.03 sp gr) = 8.59 lb/gal

Next, calculate the dry solids produced per day.

Equation: **SS removed, lb/day = (SS removed, mg/L)(Number of mgd)(Sludge lb/gal)**

SS removed, lb/day = (174 mg/L, SS)(3.0816 mgd)(8.59 lb/gal)

SS removed, lb/day = 4,605.94 lb/day, round to **4,600 lb/day**

108. Given the following data, determine the amount of dry solids produced in lb/day:

Flow = 1,335,000 gpd
BOD_5 = 228 mg/L
Influent suspended solids (SS) = 341 mg/L
Primary effluent suspended solids = 106 mg/L
Specific gravity (sp gr) = 1.03

First, determine the number of mg/L of suspended solids (SS) removed.

SS removed, mg/L = 341 mg/L, influent − 106 mg/L effluent = 235 mg/L, SS removed

Next, convert gpd to mgd.

$$\text{Number of mgd} = \frac{1,335,000 \text{ gpd}}{1,000,000/\text{mil}} = 1.335 \text{ mgd}$$

SS removed, lb/day = (SS removed, mg/L)(Number of mgd)(8.34 lb/gal)(SS sp gr)

SS removed, lb/day = (235 mg/L, SS)(1.335 mgd)(8.34 lb/gal)(1.03 sp gr)

SS removed, lb/day = 2,694.96 lb/day, round to **2,690 lb/day**

109. Given the following data, determine the primary effluent suspended solids in mg/L.

Plant flow = 880 gpm
Primary effluent suspended solids removed = 2,480 lb/day
Specific gravity (sp gr) = 1.04

First, convert gpm to mgd.

$$\text{Number of mgd} = \frac{(880 \text{ gpm})(1,440 \text{ min/day})}{1,000,000/\text{mil}} = 1.2672 \text{ mgd}$$

Next, determine the primary suspended solids (SS).

Equation: **SS removed, lb/day = (SS removed, mg/L)(mgd)(8.34 lb/gal)(SS sp gr)**

Rearrange to solve for mg/L, SS removed.

$$\text{SS, mg/L} = \frac{\text{SS removed, lb/day}}{(\text{mgd})(8.34 \text{ lb/gal})(\text{SS sp gr})}$$

$$\text{SS, mg/L} = \frac{2{,}480 \text{ lb/day removed}}{(1.2672 \text{ mgd})(8.34 \text{ lb/gal})(1.04)} = 225.64 \text{ mg/L, round to } \textbf{230 mg/L, SS}$$

110. Given the following data, estimate the dry sludge solids produced by a secondary clarifier in lb/day:

Secondary influent flow = 1,660 gpm
Influent BOD$_5$ = 216 mg/L
Effluent BOD$_5$ = 20.4 mg/L
Bacterial growth rate for this plant = 0.395 lb SS/lb BOD$_5$
Suspended solids = 337 mg/L

First, determine the BOD$_5$ removal.

$$\text{BOD}_5 \text{ removal, mg/L} = 216 \text{ mg/L} - 20.4 \text{ mg/L} = 195.6 \text{ mg/L}$$

Next, convert gpm to mgd.

$$\text{Number of mgd} = \frac{(1{,}660 \text{ gpm})(1{,}440 \text{ min/day})}{1{,}000{,}000/\text{mil}} = 2.39 \text{ mgd}$$

Next, determine the BOD$_5$ removal in lb/day.

Equation: **BOD$_5$ removal, lb/day = (BOD$_5$, mg/L)(Flow, mgd)(8.34 lb/gal)**

$$\text{BOD}_5 \text{ removal, lb/day} = (195.6 \text{ mg/L})(2.39 \text{ mgd})(8.34 \text{ lb/gal}) = 3{,}898.82 \text{ lb/day}$$

Lastly, use the bacterial growth rate to calculate the estimated lb/day of solids produced.

$$\frac{x \text{ lb/day Solids produced}}{3{,}898.82 \text{ lb/day BOD}_5 \text{ removed}} = \frac{0.395 \text{ lb SS/lb BOD}_5}{1 \text{ lb BOD}_5 \text{ removed}}$$

$$x \text{ lb/day Solids produced} = \frac{(3{,}898.82 \text{ lb/day BOD}_5 \text{ removed})(0.395 \text{ lb SS/lb BOD}_5)}{1 \text{ lb BOD}_5 \text{ removed}}$$

$$x \text{ lb/day Solids produced} = 1{,}540.03 \text{ lb/day, round to } \textbf{1,540 lb/day Solids produced}$$

SLUDGE AGE (GOULD) CALCULATIONS

Operators need to understand sludge age calculations because this will help them maintain an appropriate amount of activated sludge in an aeration tank. The age of the sludge refers to the average solids retention time (usually in days) that the solids remain in the aeration tank. The sludge age is controlled by the sludge wasting rate, which affects the sludge yield in the system. This calculation is similar to detention time. See Figure E-8 in Appendix E for one type of wastewater plant using an oxidation ditch.

111. Given the following data, determine the sludge age in days at an oxidation ditch wastewater treatment plant:

Mixed liquor suspended solids (MLSS) = 3,673 mg/L
Solids added = 535 lb/day
Average top width of ditch at water surface = 14.8 ft
Average depth = 5.23 ft
Average bottom width = 8.57 ft
Length of ditch = 195 ft
Diameter of half circles = 112 ft

First, determine the number of mil gal in the ditch.

Equation: $\dfrac{[(b_1 + b_2)\,\text{Depth}]}{2}$ **(Length of 2 sides + Length of 2 half circles)(7.48 gal/ft³)**

Where b_1 = bottom width of ditch and b_2 = top width of ditch at water surface, and the length of the two half circles is π(Diameter)

Substitute values and solve.

$$\text{Volume, gal} = \frac{[(8.57\ \text{ft} + 14.8\ \text{ft})(5.23\ \text{ft})]}{2}[(2)(195\ \text{ft}) + (3.14)(112\ \text{ft})](7.48\ \text{gal/ft}^3)$$

Volume, gal = (61.11 ft²)(390 ft + 351.68 ft)(7.48 gal/ft³)

Volume, gal = (61.11 ft²)(741.68 ft)(7.48 gal/ft³)

Volume, gal = 339,024 gal

Next, convert gallons to mil gal.

Number of mil gal $= \dfrac{339{,}024}{1{,}000{,}000/\text{mil}} = 0.339024$ mil gal

Next, calculate the amount of solids under aeration.

Equation: **Solids under aeration, lb = (Ditch volume, mil gal)(MLSS, mg/L)(8.34 lb/gal)**

Solids under aeration, lb $= (0.339024$ mil gal$)(3{,}673$ mg/L$)(8.34$ lb/gal$) = 10{,}385$ lb

Next, determine the sludge age in days.

Equation: **Sludge age, days** $= \dfrac{\text{Solids under aeration, lb}}{\text{Solids added, lb/day}}$

Sludge age, days $= \dfrac{10{,}385 \text{ lb}}{535 \text{ lb/day}} = 19.41$ days, round to **19.4 days**

112. **The flow through an oxidation ditch is 750,000 gpd. If the mixed liquor suspended solids (MLSS) is 3,170 mg/L, the SS is 142 mg/L, and the capacity of the oxidation ditch is 1.888 acre-ft, what is the sludge age in days?**

First, determine the volume of the oxidation ditch by converting acre-ft to mil gal.

Know: 1 acre foot $= 325{,}829$ gal

Number of mil gal $= \dfrac{(1.888 \text{ acre-ft})(325.829 \text{ gal/acre-ft})}{1{,}000{,}000/\text{mil}} = 0.61517$ mil gal

Convert gpd to mgd

Number of mil gal $= \dfrac{750{,}000 \text{ gpd}}{1{,}000{,}000/\text{mil}} = 0.75$ mgd

Next, calculate the amount of sludge age in days.

Know: **Sludge age, days** $= \dfrac{(\text{MLSS, mg/L})(\text{Volume, mil gal})(8.34 \text{ lb/gal})}{(\text{SS, mg/L})(\text{Flow, mgd})(8.34 \text{ lb/gal})}$

Substitute values and solve.

Sludge age, days $= \dfrac{(3{,}170 \text{ mg/L})(0.61517 \text{ mil gal})(8.34 \text{ lb/gal})}{(142 \text{ mg/L})(0.75 \text{ mgd})(8.34 \text{ lb/gal})} = 18.31$ days, round to **18.3 days**

ORGANIC LOADING RATE CALCULATIONS

Organic loading rate calculations tell the operator the amount of food entering the plant. These calculations are used for wastewater treatment ponds, rotating biological contactors, or trickling filters. See figures in Appendix E for the types of wastewater plants using these processes.

113. **What is the organic loading rate for a trickling filter that is 125 ft in diameter and 5.25 ft deep in lb BOD$_5$/d/1,000 ft^3, if the primary effluent flow is 1,710 gpm and the BOD$_5$ is 186 mg/L?**

First, determine the volume of the tricking filter in ft^3.

Equation: **Volume, ft^3 = (0.785)(Diameter)^2(Depth, ft)**

Volume, ft^3 = (0.785)(125 ft)(125 ft)(5.25 ft) = 64,394.53 ft^3

Next, factor out 1,000 ft^3 from the volume = (64.39453)(1,000 ft^3)

Next, convert gpm to mgd.

$$\text{Number of mgd} = \frac{(1,710 \text{ gpm})(1,440 \text{ min/day})}{1,000,000/\text{mil}} = 2.4624 \text{ mgd}$$

Next, determine the pounds of BOD$_5$/d/1,000 ft^3 using a modified version of the "pounds" equation.

$$\textbf{Organic loading rate, lb BOD}_5\textbf{/d/1,000 ft}^3 = \frac{\textbf{(BOD}_5\textbf{, mg/L)(Flow, mgd)(8.34 lb/gal)}}{\textbf{Volume of trickling filter, ft}^3\textbf{/1,000 ft}^3}$$

$$\text{Organic loading rate, lb BOD}_5\text{/d/1,000 ft}^3 = \frac{(186 \text{ mg/L BOD}_5)(2.4624 \text{ mgd})(8.34 \text{ lb/gal})}{(64.39453)(1,000 \text{ ft}^3)}$$

Organic loading rate, lb BOD$_5$/d/1,000 ft^3 = 59.32, round to **59 lb BOD$_5$/d/1,000 ft^3**

114. **A wastewater pond receives an influent flow of 135,000 gpd. Given the following data, calculate the organic loading rate in lb BOD$_5$/d/acre for this pond:**

Pond size = 310 by 125 ft and averages 4.88 ft in depth
BOD$_5$ content = 183 mg/L

First, calculate the volume of the pond in acres.

Know: **Volume, acres** $= \dfrac{(\text{Length})(\text{Width})(\text{Depth})}{43{,}560 \text{ ft}^3/\text{acre}}$

Volume, acres $= \dfrac{(310 \text{ ft})(125 \text{ ft})(4.88 \text{ ft})}{43{,}560 \text{ ft}^3/\text{acre}} = 4.34 \text{ acres}$

Next, convert gpd to mgd.

Number of mgd $= (135{,}000 \text{ gpd})/(1{,}000{,}000/\text{mil}) = 0.135 \text{ mgd}$

Organic loading rate, lb BOD$_5$/d/acres $= \dfrac{(\text{BOD}_5, \text{mg/L})(\text{Flow, gpd})(8.34 \text{ lb/gal})}{\text{Volume of pond, acres}}$

Organic loading rate, lb BOD$_5$/d/acres $= \dfrac{(183 \text{ mg/L BOD}_5)(0.135 \text{ mgd})(8.34 \text{ lb/gal})}{4.34 \text{ acres}}$

Organic loading rate, lb BOD$_5$/d/acres $= 47.47$, round to **47 lb BOD$_5$/d/acres**

115. **Given the following data on a wastewater treatment pond, calculate the organic loading rate in lb BOD$_5$/d/acre:**

Influent flow = 193,000 gpd
Surface area of pond = 4.88 acre-ft
Influent BOD$_5$ concentration = 208 mg/L

First, convert gallons per day to mgd.

Number of mgd $= (193{,}000 \text{ gpd}) / (1{,}000{,}000/\text{mil}) = 0.193 \text{ mgd}$

Next, determine the pounds of BOD$_5$/d/acre using a modified version of the "pounds" equation.

Organic loading rate, lb BOD$_5$/d/acre $= \dfrac{(\text{BOD}_5, \text{mg/L})(\text{Flow, mgd})(8.34 \text{ lb/gal})}{\text{Surface area of pond, acre-ft}}$

Organic loading rate, lb BOD$_5$/d/acre $= \dfrac{(208 \text{ mg/L BOD}_5)(0.193 \text{ mgd})(8.34 \text{ lb/gal})}{4.88 \text{ acre-ft}}$

Organic loading rate, lb BOD$_5$/d/acre $= 68.61$ lb BOD$_5$/d/acre, round to **68.6 lb BOD$_5$/d/acre**

116. **What is the organic loading rate for a rotating biological contactor (RBC) in lb BOD$_5$/d/1,000 ft^2, given the following data?**

Surface area of RBC = 575,800 ft^2
BOD$_5$ = 182 mg/L
Flow = 2,815,000 gpd

First, convert gallons per day to mgd.

Number of mgd = (2,815,000 gpd)/(1,000,000/mil) = 2.815 mgd

Next, factor out 1,000 ft^2 from the RBC surface area = (575.8)(1,000 ft^2)

Organic loading rate, lb BOD$_5$/d/1,000 ft^2 $= \dfrac{(\text{BOD}_5, \text{mg/L})(\text{Flow, mgd})(8.34 \text{ lb/gal})}{(\text{Surface area of RBC})(1,000 \text{ ft}^2)}$

Organic loading rate, lb BOD$_5$/d/1,000 ft^2 $= \dfrac{(182 \text{ mg/L BOD}_5)(2.815 \text{ mgd})(8.34 \text{ lb/gal})}{(575.8)(1,000 \text{ ft}^2)}$

Organic loading rate, lb BOD$_5$/d/1,000 ft^2 = **7.42 lb BOD$_5$/d/1,000 ft^2**

SUSPENDED SOLIDS LOADING CALCULATIONS

Operators use suspended solids loading for evaluating process control.

117. **What is the amount of suspended solids (SS) entering a trickling filter in lb/day, if the influent flow is 1,080 gpm and the amount of SS is 438 mg/L?**

First, convert gallons per day to mgd.

Number of mgd $= \dfrac{(1,080 \text{ gpm})(1,440 \text{ min/day})}{1,000,000/\text{mil}} = 1.5552 \text{ mgd}$

Then, calculate the lb/day of suspended solids.

Equation: **Number of lb/day = (SS, mg/L)(Number of mgd)(8.34 lb/gal)**

Substitute values and solve.

Number of lb/day SS = (438 mg/L, SS)(1.5552 mgd)(8.34 lb/gal)

Number of lb/day SS = 5,681.02 lb/day, round to **5,680 lb/day SS**

118. What are the suspended solids (SS) entering a trickling filter in mg/L, if the influent flow to a trickling filter is 965 gpm and the SS loading is 3,790 lb/day?

First, determine the number of mgd.

$$\text{Number of mgd} = \frac{(965 \text{ gpm})(1,440 \text{ min/day})}{1,000,000/\text{mil}} = 1.3896 \text{ mgd}$$

Equation: **Number of lb/day SS = (SS, mg/L)(Number of mgd)(8.34 lb/gal)**

Rearrange the equation to solve for mg/L, SS.

$$\text{Number of mg/L, SS} = \frac{\text{Number of lb/day}}{(\text{Number of mgd})(8.34 \text{ lb/gal})}$$

Substitute values and solve.

$$\text{Number of mg/L, SS} = \frac{3,790 \text{ lb/day}}{(1.3896 \text{ mgd})(8.34 \text{ lb/gal})} = \textbf{327 mg/L, SS}$$

119. **What must have been the flow in gpm to a trickling filter, if the suspended solids (SS) loading was 3,950 lb/day and the SS was 372 mg/L?**

Equation: **Number of lb/day = (SS, mg/L)(Number of mgd)(8.34 lb/gal)**

Rearrange the equation to solve for mg/L, SS.

$$\text{Number of mgd} = \frac{\text{Number of lb/day SS}}{(\text{SS, mg/L})(8.34\ \text{lb/gal})}$$

Substitute values and solve.

$$\text{Number of mg/L, SS} = \frac{3{,}950\ \text{lb/day SS}}{(372\ \text{mg/L})(8.34\ \text{lb/gal})} = 1.273\ \text{mgd}$$

Lastly, convert mgd to gpm.

$$\text{Number of gpm} = \frac{(1.273\ \text{mgd})(1{,}000{,}000/\text{mil})}{1{,}440\ \text{min/day}} = \textbf{884 gpm}$$

120. **The flow through a trickling filter is 1,475 gpm. If the influent suspended solids (SS) are 276 mg/L and the effluent suspended solids are 73 mg/L, how many lb/day of suspended solids are removed?**

First, determine the amount in mg/L of suspended solids removed by the trickling filter.

SS removed, mg/L = 276 mg/L − 73 mg/L = 203 mg/L, SS removed

Next, convert gpm to mgd.

$$\text{Number of mgd} = \frac{(1{,}475\ \text{gpm})(1{,}440\ \text{min/day})}{1{,}000{,}000/\text{mil}} = 2.124\ \text{mgd}$$

Now, calculate the lb/day of suspended solids removed.

Equation: **SS removed, lb/day = (SS, mg/L, SS)(mgd)(8.34 lb/gal)**

SS removed, lb/day = (203 mg/L, SS)(2.124 mgd)(8.34 lb/gal)

SS removed, lb/day = 3,595.97 lb/day, round to **3,600 lb/day SS removed**

BIOCHEMICAL OXYGEN DEMAND LOADING CALCULATIONS

Biochemical oxygen demand is the demand for oxygen made by bacteria, as they decompose organic matter in wastewater or in the natural environment. This calculation sometimes is helpful in evaluating treatment pond processes. The BOD_5 is a five-day test. See Figures E-1 and E-6 in Appendix E for two types of wastewater plants using a trickling filter.

121. **What is the BOD_5 loading on a wastewater plant in lb/day, if the influent flow is 5.18 mgd, the raw influent BOD_5 is 264 mg/L, and the MLSS is 2,735 mg/L?**

Equation: **Number of lb/day BOD_5 = (BOD_5, mg/L)(mgd)(8.34 lb/gal)**

Substitute values and solve.

Number of lb/day BOD_5 = (264 mg/L)(5.18 mgd)(8.34 lb/gal)

Number of lb/day BOD_5 = 11,405 lb/day, round to **11,400 lb/day BOD_5**

122. Given the following data, calculate the biochemical oxygen demand (BOD_5) loading on an aerator in lb/day/1,000 ft³:

Influent flow = 1,629,000 gallons per day (gpd)
Influent BOD_5 = 195 mg/L
Effluent BOD_5 = 68 mg/L
Aerator dimensions = 225 ft by 48 ft with a depth of 17.5 ft

First, convert gallons to mgd.

$$\text{Number of mgd} = \frac{1,629,000 \text{ gpd}}{1,000,000/\text{mil}} = 1.629 \text{ mgd}$$

Then, calculate the BOD_5 loading in lb/day/1,000 ft³.

Equation: **Number of lb/day/1,000 ft³** $= \dfrac{(BOD_5, \text{mg/L})(\text{Number of mgd})(8.34 \text{ lb/gal})}{(\text{Length})(\text{Width})(\text{Depth})}$

Substitute values and solve.

$$BOD_5, \text{lb/day/1,000 ft}^3 = \frac{(195 \text{ mg/L } BOD_5)(1.629 \text{ mgd})(8.34 \text{ lb/gal})}{(225 \text{ ft})(48 \text{ ft})(17.5 \text{ ft})}$$

$$BOD_5, \text{lb/day/1,000 ft}^3 = \frac{2,649.24 \text{ lb/day}}{189,000 \text{ ft}^3}$$

Next, factor out 1,000 ft³ from the denominator.

$$BOD_5, \text{lb/day/1,000 ft}^3 = \frac{2,649.24 \text{ lb/day}}{(189)(1,000 \text{ ft}^3)} = \textbf{14 lb/day/1,000 ft}^3 \textbf{ of BOD}_5$$

123. What is the influent flow to a trickling filter in gpm, if the BOD_5 loading is 3,072 lb/day and the BOD_5 is 194 mg/L?

Equation: **Number of lb/day = (BOD_5, mg/L)(Number of mgd)(8.34 lb/gal)**

Rearrange the equation to solve for mgd.

$$\textbf{(Number of mgd)} = \frac{\text{Number of lb/day}}{(BOD_5, \text{mg/L})(8.34 \text{ lb/gal})}$$

Substitute values and solve.

$$\text{Number of mgd flow} = \frac{3{,}072 \text{ lb/day}}{(194 \text{ mg/L})(8.34 \text{ lb/gal})} = 1.8987 \text{ mgd}$$

Lastly, convert mgd to gpm.

$$\text{Number of gpm} = \frac{(1.8987 \text{ mgd})(1{,}000{,}000/\text{mil})}{1{,}440 \text{ min/day}} = 1{,}318.5 \text{ gpm, round to } \mathbf{1{,}320 \text{ gpm}}$$

124. Given the following data, calculate the biochemical oxygen demand (BOD$_5$) loading on a trickling filter in lb/day/1,000 ft³:

Influent flow = 1,095 gpm
Influent BOD$_5$ = 173 mg/L
Effluent BOD$_5$ = 18 mg/L
Trickling filter diameter 125 ft
Trickling filter depth 7.50 ft

First, determine the number of mgd.

$$\text{Number of mgd} = \frac{(1{,}095 \text{ gpm})(1{,}440 \text{ min/day})}{1{,}000{,}000/\text{mil}} = 1.5768 \text{ mgd}$$

Equation: **Number of lb/day/1,000 ft³** $= \dfrac{(\text{BOD}_5, \text{mg/L})(\text{Number of mgd})(8.34 \text{ lb/gal})}{(0.785)(\text{Diameter, ft})^2(\text{Depth, ft})}$

Substitute values and solve.

$$\text{BOD}_5, \text{ lb/day/1,000 ft}^3 = \frac{(173 \text{ mg/L BOD}_5)(1.5768 \text{ mgd})(8.34 \text{ lb/gal})}{(0.785)(125 \text{ ft})(125 \text{ ft})(7.50 \text{ ft})}$$

$$\text{BOD}_5, \text{ lb/day/1,000 ft}^3 = \frac{2{,}275.04 \text{ lb/day}}{91{,}992.19 \text{ ft}^3}$$

Next, factor out 1,000 ft³ from the denominator.

$$\text{BOD}_5, \text{ lb/day/1,000 ft}^3 = \frac{2{,}275.04 \text{ lb/day}}{(91.99219)(1{,}000 \text{ ft}^3)} = \mathbf{25 \text{ lb/day/1,000 ft}^3}$$

125. **Calculate the influent flow to a trickling filter in gpm, if the BOD$_5$ loading is 2,715 lb/day and the BOD$_5$ is 196 mg/L.**

Equation: **Number of gpm** $= \dfrac{(\text{Number of lb/day})(1,000,000/\text{mil})}{(\text{BOD}_5, \text{mg/L})(\text{Number of mgd})(8.34 \text{ lb/gal})(1,440 \text{ min/day})}$

Substitute values and solve.

Number of gpm $= \dfrac{(2,715 \text{ lb/day})(1,000,000/\text{mil})}{(196 \text{ mg/L})(8.34 \text{ lb/gal})(1,440 \text{ min/day})} = 1,153.41$ gpm, round to **1,150 gpm**

SOLUBLE AND PARTICULATE BIOCHEMICAL OXYGEN DEMAND CALCULATIONS

Biochemical oxygen demand (BOD$_5$) measures the amount of organic matter that is present in water. Bacteria break down this organic matter by natural decomposition and in the process utilize oxygen. Thus, the more organic matter present in the wastewater, the more demand for oxygen by the bacteria. Operators need to know how to do BOD$_5$ calculations because there are strict regulations for the amount of BOD$_5$ that can be discharged to a natural water body from the treated plant. The K value in these problems is the portion of suspended solids in the wastewater that are organic suspended solids. Domestic water usually has about 50 to 70% of the suspended solids as organic suspended solids, which is usually written in decimal form (0.5 to 0.7).

126. **What is the soluble BOD$_5$, if the total BOD$_5$ is 216 mg/L, the suspended solids (particulates) are 183 mg/L, and the K factor is 0.67?**

Equation: **Total BOD$_5$ = (Particulate BOD$_5$)(K factor) + Soluble BOD$_5$**

Rearrange the equation to solve for soluble BOD$_5$.

Soluble BOD$_5$ = Total BOD$_5$ − (Particulate BOD$_5$)(K factor)

Substitute values and solve.

Soluble BOD$_5$ = 216 mg/L BOD$_5$ − (183 mg/L BOD$_5$)(0.67 K factor)

Soluble BOD$_5$ = 216 mg/L BOD$_5$ − 122.61 mg/L BOD$_5$

Soluble BOD$_5$ = 93.39 mg/L, round to **93 mg/L Soluble BOD$_5$**

127. How many lb/day of soluble BOD$_5$ enters a rotating biological contactor (RBC) each day, if the flow is 1,935 gpm, total BOD$_5$ is 266 mg/L, the particulate BOD$_5$ is 147 mg/L, and the K factor is 0.56?

Equation: **Total BOD$_5$ = (Particulate BOD$_5$)(K factor) + Soluble BOD$_5$**

Rearrange the equation to solve for soluble BOD$_5$ by subtracting (Particulate BOD)(K factor) from each side.

Soluble BOD$_5$ = Total BOD$_5$ − (Particulate BOD$_5$)(K factor)

Substitute values and solve.

Soluble BOD$_5$ = 266 mg/L BOD$_5$ − (147 mg/L BOD$_5$)(0.56 K factor)

Soluble BOD$_5$ = 266 mg/L BOD$_5$ − 82.32 mg/L BOD$_5$

Soluble BOD$_5$ = 183.68 mg/L Soluble BOD$_5$

Next, convert gpm to mgd.

$$\text{Number of mgd} = \frac{(1,935 \text{ gpm})(1,440 \text{ min/day})}{1,000,000/\text{mil}} = 2.7864 \text{ mgd}$$

Next, determine the Soluble BOD$_5$.

Equation: **Soluble BOD$_5$, lb/day = (Soluble BOD$_5$, mg/L)(mgd)(8.34 lb/gal)**

Soluble BOD$_5$, lb/day = (183.68 mg/L)(2.7864 mgd)(8.34 lb/gal)

Soluble BOD$_5$, lb/day = 4,268.46 lb/day, round to **4,300 lb/day Soluble BOD$_5$**

128. Given the following data, calculate the number of lb/day of soluble BOD_5 that enters a rotating biological contactor (RBC):

Flow to the RBC = 1,165 gpm
Total BOD_5 = 253 mg/L
Particulate BOD_5 = 129 mg/L
The K factor = 0.675?

Equation: **Total BOD_5 = (Particulate BOD_5)(K factor) + Soluble BOD_5**

Rearrange the equation to solve for soluble BOD_5 by subtracting (Particulate BOD_5)(K factor) from each side.

Soluble BOD_5 = Total BOD_5 − (Particulate BOD_5)(K factor)

Substitute values and solve.

Soluble BOD_5 = 253 mg/L BOD_5 − (129 mg/L BOD_5)(0.675 K factor)

Soluble BOD_5 = 253 mg/L BOD_5 − 87.075 mg/L BOD_5

Soluble BOD_5 = 165.925 mg/L Soluble BOD_5

Next, convert gpm to mgd.

$$\text{Number of mgd} = \frac{(1,165 \text{ gpm})(1,440 \text{ min/day})}{1,000,000/\text{mil}} = 1.6776 \text{ mgd}$$

Next, determine the Soluble BOD_5.

Equation: **Soluble BOD_5, lb/day = (Soluble BOD_5, mg/L)(mgd)(8.34 lb/gal)**

Soluble BOD_5, lb/day $= (165.925 \text{ mg/L})(1.6776 \text{ mgd})(8.34 \text{ lb/gal})$

Soluble BOD_5, lb/day $= 2,321.49$ lb/day, round to **2,320 lb/day Soluble BOD_5**

CHEMICAL OXYGEN DEMAND LOADING CALCULATIONS

Chemical oxygen demand is a measure of the capacity of water to consume oxygen, when organic matter and the oxidation of inorganic matter are decomposed. Because it also measures the decomposition of inorganic matter such as nitrate and ammonia, it is only an indirect measure of the organic matter in water.

129. Determine the chemical oxygen demand (COD) in lb/day that is applied to an aeration tank, if the flow is 1,680,000 gallons and the COD concentration is 127 mg/L.

First, convert the gpd to mgd.

Number of gpd $= (1,680,000 \text{ gpd})/(1,000,000/\text{mil}) = 1.68$ mgd

Equation: **Number of lb/day $=$ (COD, mg/L)(Number of mgd)(8.34 lb/gal)**

Substitute values and solve.

Number of lb/day $= (127 \text{ mg/L})(1.68 \text{ mgd})(8.34 \text{ lb/gal})$

Number of lb/day $= 1,779.42$ lb/day, round to **1,780 lb/day**

130. **What is the influent flow to an aeration tank in gpm, if the COD loading is 2,443 lb/day and the COD is 154 mg/L?**

First, determine the number of mgd.

Equation: **Number of lb/day COD = (COD, mg/L)(Number of mgd)(8.34 lb/gal)**

Rearrange the equation to solve for mg/L.

$$(\text{Number of mgd}) = \frac{\text{Number of lb/day COD}}{(\text{COD, mg/L})(8.34\ \text{lb/gal})}$$

Substitute values and solve.

$$\text{Number of mgd flow} = \frac{2,443\ \text{lb/day}}{(154\ \text{mg/L})(8.34\ \text{lb/gal})} = 1.902\ \text{mgd}$$

Then, convert mgd to gpm.

$$\text{Number of gpm} = \frac{(1.902\ \text{mgd})(1,000,000/\text{mil})}{1,440\ \text{min/day}} = 1,320.83\ \text{gpm, round to } \textbf{1,320 gpm}$$

HYDRAULIC DIGESTION TIME CALCULATIONS

Hydraulic digestion time tells the operator how long the process will take to complete, and is thus used for planning purposes. See Figures E-2, E-4, E-5, and E-6 in Appendix E for four types of wastewater plants using a digester.

131. **What is the hydraulic digestion time for a digester that is 28.0 ft in radius, has a level of 16.3 ft, and has a sludge flow of 8,075 gallons per day (gpd)?**

First, determine the volume of the digester in gallons.

Volume, gal = πr^2(Depth, ft)(7.48 gal/ft³)

Volume, gal = (3.14)(28.0 ft)(28.0 ft)(16.3 ft)(7.48 gal/ft³) = 300,147.63 gal

Next, calculate the digestion time in days.

Equation: **Digestion time, days** $= \dfrac{\text{Number of gallons}}{\text{Influent sludge flow, gal/day}}$

Substitute values and solve.

Digestion time, days $= \dfrac{300{,}147.63 \text{ gal}}{8{,}075 \text{ gal/day}} = 37.17$ days, round to **37.2 days**

132. What is the hydraulic digestion time for a 50.1-ft diameter digester with a level of 19.2 ft and sludge flow of 10.5 gpm?

First, convert gpm to gpd.

Flow, gpd $= (10.5 \text{ gpm})(60 \text{ min/hr})(24 \text{ hr/day}) = 15{,}120$ gpd

Next, calculate the digestion time.

Equation: **Digestion time, days** $= \dfrac{(0.785)(\text{Diameter})^2(\text{Depth, ft})(7.48 \text{ gal/ft}^3)}{\text{Influent sludge flow, gal/day}}$

Substitute values and solve.

Digestion time, days $= \dfrac{(0.785)(50.1 \text{ ft})(50.1 \text{ ft})(19.2 \text{ ft})(7.48 \text{ gal/ft}^3)}{15{,}120 \text{ gpd}}$

Digestion time, days $= 18.715$ days, round to **18.7 days**

DIGESTER LOADING RATE CALCULATIONS

Digester loading rate calculations tell the operator how much volatile solids are stabilized per cubic foot of digester space. It is used for evaluating process control.

133. If the sludge flow into a digester is 82,250 gpd, the digester is 25.8 ft in radius, the sludge level is 17.9 ft, the sludge is 5.34% solids with a specific gravity of 1.04, and the sludge has 62.5% volatile solids, what is the loading on a digester in lb volatile solids (VS)/day/ft³ ?

Equation: **Digester loading, lb VS/day/ft³ =**

$$\frac{(\text{Flow, gpd})(8.34\ \text{lb/gal})(\text{sp gr})(\text{Percent sludge})(\text{Percent volatile solids})}{\pi(\text{radius})^2(\text{Sludge level})}$$

Substitute values and solve.

$$\text{Digester loading, lb VS/day/ft}^3 = \frac{(82,250\ \text{gpd})(8.34\ \text{lb/gal})(1.04)(5.34\%/100\%)(62.5\%/100\%)}{(3.14)(25.8\ \text{ft})(25.8\ \text{ft})(17.9\ \text{ft})}$$

$$\text{Digester loading, lb VS/day/ft}^3 = \frac{23,809.85\ \text{lb VS/day}}{37,412.96\ \text{ft}^3}$$

$$\text{Digester loading, lb VS/day/ft}^3 = \textbf{0.636 lb VS/day/ft}^3$$

134. Given the following data, calculate the digester loading in lb volatile solids (VS)/day/1,000 ft³:

Digester radius = 29.8 ft
Sludge level = 18.5 ft
Influent sludge flow = 35.5 gpm
Percent sludge solids = 4.95%
Percent volatile solids = 67.2%
Specific gravity of sludge = 1.03

First, convert gpm to gpd.

Number of gpd = (35.5 gpm)(60 min/hr)(24 hr/day) = 51,120 gpd

Equation: **Digester loading, lb VS/day/1,000 ft³ =**

$$\frac{(\text{Flow, gpd})(8.34 \text{ lb/gal})(\text{sp gr})(\text{Percent sludge})(\text{Percent volatile solids})}{\pi(\text{radius})^2(\text{Sludge level})}$$

Substitute values and solve.

Digester loading, lb VS/day/1,000 ft³ =

$$\frac{(51,120 \text{ gpd})(8.34 \text{ lb/gal})(1.03 \text{ sp gr})(4.95\%/100\%)(67.2\%/100\%)}{(3.14)(29.8 \text{ ft})(29.8 \text{ ft})(18.5 \text{ ft})}$$

Digester loading, lb VS/day/1,000 ft³ $= \dfrac{14,607 \text{ lb VS/day}}{51,586.24 \text{ ft}^3}$

Factor out 1,000 ft³ from the denominator and do not divide by 1,000 ft³, as it will become part of the units and not part of the calculation.

Digester loading, lb VS/day/1,000 ft³ $= \dfrac{14,607 \text{ lb VS/day}}{51.5862 \times 1,000 \text{ ft}^3}$

Digester loading, lb VS/day/1,000 ft³ $= \dfrac{283.16 \text{ lb VS/day}}{1,000 \text{ ft}^3}$, round to **283 lb VS/day/1,000 ft³**

135. Given the following data, calculate the digester loading in lb VS/day/1,000 ft³.

Digester diameter = 41.5 ft
Sludge level = 18.75 ft
Influent sludge flow = 11.5 gpm
Percent sludge solids = 5.27%
Percent volatile solids = 66.2%
Specific gravity of sludge = 1.03

First, covert gpm to gpd.

Number of gpd = (11.5 gpm)(60 min/hr)(24 hr/day) = 16,560 gpd

Equation: **Digester loading, lb VS/day/1,000 ft³** =

$$\frac{(\text{Flow, gpd})(8.34\,\text{lb/gal})(\text{sp gr})(\text{Percent sludge})(\text{Percent volatile solids})}{(0.785)(\text{Diameter})^2(\text{Sludge level})}$$

Substitute values and solve.

Digester loading, lb VS/day/1,000 ft³ =

$$\frac{(16,560\,\text{gpd})(8.34\,\text{lb/gal})(1.03\,\text{sp gr})(5.27\%/100\%)(66.2\%/100\%)}{(0.785)(41.5\,\text{ft})(41.5\,\text{ft})(18.75\,\text{ft})}$$

Digester loading, lb VS/day = $\dfrac{4{,}962.86\,\text{lb VS/day}}{25{,}349.37\,\text{ft}^3}$

Next, factor out 1,000 ft³ from the denominator.

Digester loading, lb VS/day = $\dfrac{4{,}962.86\,\text{lb VS/day}}{(25.34937)(1{,}000\,\text{ft}^3)}$

Digester loading, lb VS/day/1,000 ft³ = $\dfrac{195.78\,\text{lb VS/day}}{1{,}000\,\text{ft}^3}$, round to **196 lb VS/day/1,000 ft³**

MEAN CELL RESIDENCE TIME (SOLIDS RETENTION TIME) CALCULATIONS

The Mean Cell Residence Time (MCRT) is the average time the activated-sludge solids are in an activated biosolids system. The MCRT is an important design and operating parameter for operators to use in the activated-sludge process and is normally expressed in days. This calculation is used for operational process control. See Figure E-2 in Appendix E for one type of wastewater plant using the activated-sludge process.

136. Calculate the mean cell residence time (MCRT), given the following data:

Flow = 1,860 gpm
Aeration tanks volume = 571,000 gallons
Clarifier tank volume = 288,000 gallons
Mixed liquor suspended solids (MLSS) = 2,815 mg/L
Waste rate = 31,300 gpd
Waste activated sludge (WAS) = 6,880 mg/L
Effluent TSS = 15.5 mg/L

First, convert the flow rate in gpm to mgd.

$$\text{Number of mgd} = \frac{(1,860)\,(1,440\,\text{min/day})}{1,000,000/\text{mil}} = 2.6784\,\text{mgd}$$

Next, convert the volumes for the tanks to mil gal.

$$\text{Aeration tank, mgd} = \frac{571,000\,\text{gal}}{1,000,000/\text{mil}} = 0.571\,\text{mil gal}$$

$$\text{Clarifier tank, mgd} = \frac{288,000\,\text{gal}}{1,000,000/\text{mil}} = 0.288\,\text{mil gal}$$

Next, convert the waste rate from gpd to mgd.

$$\text{Waste rate, mgd} = \frac{31,300\,\text{gal}}{1,000,000/\text{mil}} = 0.0313\,\text{mgd}$$

Next, calculate the MCRT.

Equation: **MCRT, days =**

$$\frac{(\text{MLSS, mg/L})\,(\text{Aeration tank mil gal} + \text{Clarifier tank mil gal})\,(8.34\,\text{lb/gal})}{(\text{WAS, mg/L})\,(\text{Waste rate, mgd})\,(8.34\,\text{lb/gal}) + (\text{TSS, mg/L})\,(\text{Flow, mgd})\,(8.34\,\text{lb/gal})}$$

MCRT, days =

$$\frac{(2,815\,\text{mg/L MLSS})\,(0.571\,\text{mil gal} + 0.288\,\text{mil gal})\,(8.34\,\text{lb/gal})}{(6,880\,\text{mg/L})\,(0.0313\,\text{mgd})\,(8.34\,\text{lb/gal}) + (15.5\,\text{mg/L TSS})\,(2.6784\,\text{mgd})\,(8.34\,\text{lb/gal})}$$

$$\text{MCRT, days} = \frac{(2,815\,\text{mg/L MLSS})\,(0.859\,\text{mil gal})\,(8.34\,\text{lb/gal})}{1,795.97\,\text{lb/day} + 346.24\,\text{lb/day}}$$

$$\text{MCRT, days} = \frac{20,166.83\,\text{lb}}{2,142.21\,\text{lb/day}} = \textbf{9.41 days}$$

137. What is the mean cell residence time (MCRT), given the following data:

Flow = 3.96 mgd
Aeration tank volume = 584,000 gallons
Clarifier tank volume = 301,000 gallons
MLSS = 2,382 mg/L
Waste rate = 36,200 gpd
Waste activated sludge (WAS) = 6,770 mg/L
Effluent TSS = 14.5 mg/L

First, convert the volumes for the tanks to mil gal.

$$\text{Aeration tank, mgd} = \frac{584,000 \text{ gal}}{1,000,000/\text{mil}} = 0.584 \text{ mil gal}$$

$$\text{Clarifier tank, mgd} = \frac{301,000 \text{ gal}}{1,000,000/\text{mil}} = 0.301 \text{ mil gal}$$

Next, convert the waste rate from gpd to mgd.

$$\text{Waste rate, mgd} = \frac{36,200 \text{ gal}}{1,000,000/\text{mil}} = 0.0362 \text{ mgd}$$

Next, calculate the MCRT.

Equation: **MCRT, days =**

$$\frac{(\text{MLSS, mg/L})(\text{Aeration tank mil gal} + \text{Clarifier tank mil gal})(8.34 \text{ lb/gal})}{(\text{WAS, mg/L})(\text{Waste rate, mgd})(8.34 \text{ lb/gal}) + (\text{TSS, mg/L})(\text{Flow, mgd})(8.34 \text{ lb/gal})}$$

$$\text{MCRT, days} = \frac{(2,382 \text{ mg/L MLSS})(0.584 \text{ mil gal} + 0.301 \text{ mil gal})(8.34 \text{ lb/gal})}{(6,770 \text{ mg/L})(0.0362 \text{ mgd})(8.34 \text{ lb/gal}) + (14.5 \text{ mg/L TSS})(3.96 \text{ mgd})(8.34 \text{ lb/gal})}$$

$$\text{MCRT, days} = \frac{(2,382 \text{ mg/L MLSS})(0.885 \text{ mil gal})(8.34 \text{ lb/gal})}{2,043.917 \text{ lb/day} + 478.883 \text{ lb/day}}$$

$$\text{MCRT, days} = \frac{17,581.3 \text{ lb}}{2,522.8 \text{ lb/day}} = \textbf{6.97 days}$$

138. If the mean cell residence time (MCRT) desired were 6.75 days, what would the waste rate be for the following system in lb/day?

Flow = 1,645 gpm
Aeration tank (AT) volume = 0.502 mil gal
Clarifier tank (CT) volume = 0.285 mil gal
Mixed liquor suspended solids (MLSS) = 2,725 mg/L
Effluent TSS = 17.1 mg/L

First, convert the flow in gpm to mgd.

$$\text{Flow, mgd} = \frac{(1,645)(1,440 \text{ min/day})}{1,000,000/\text{mil}} = 2.369 \text{ mgd}$$

Equation: **Waste rate, lb/day =**

$$\frac{\text{MLSS, mg/L} \, [\text{AT, mil gal} + \text{CT, mil gal}] \, (8.34 \text{ lb/gal})}{\text{Desired MCRT}} - \textbf{(TSS, mg/L)(Flow, mgd)(8.34 lb/gal)}$$

Waste rate, lb/day =

$$\frac{2,725 \text{ mg/L} \, [0.502 \text{ mil gal} + 0.285 \text{ mil gal}] \, (8.34 \text{ lb/gal})}{6.75 \text{ days, Desired MCRT}} - (17.1 \text{ mg/L TSS})(2.369 \text{ mgd})(8.34 \text{ lb/gal})$$

$$\text{Waste rate, lb/day} = \frac{(2,725 \text{ mg/L})(0.787 \text{ mil gal})(8.34 \text{ lb/gal})}{6.75 \text{ days, Desired MCRT}} - 337.853 \text{ lb/day}$$

Waste rate, lb/day = 2649.74 lb/day − 337.853 lb/day

Waste rate, lb/day = 2,311.887 lb/day, round to **2,310 lb/day**

139. If the mean cell residence time (MCRT) desired was 7.40 days, what would the waste rate be for the following system in lb/day and gpm?

Flow = 4.28 mgd
Aeration tank (AT) volume = 982,000 gal
Clarifier tank (CT) volume = 425,000 gal
Mixed liquor suspended solids (MLSS) = 2,820 mg/L
Waste activated sludge (WAS) = 6,310 mg/L
Effluent TSS = 18.5 mg/L

First, convert the volumes for the aeration and clarifier tanks from gallons to mil gal.

$$\text{Aeration tank, mil gal} = \frac{982,000 \text{ gal}}{1,000,000/\text{mil}} = 0.982 \text{ mil gal}$$

$$\text{Clarifier tank, mil gal} = \frac{425,000 \text{ gal}}{1,000,000/\text{mil}} = 0.425 \text{ mil gal}$$

Equation: **Waste rate, lb/day =**

$$\frac{\text{MLSS, mg/L} \, [\text{AT, mil gal} + \text{CT, mil gal}] \, (8.34 \text{ lb/gal})}{\text{Desired MCRT}} - (\text{TSS, mg/L})(\text{Flow, mgd})(8.34 \text{ lb/gal})$$

Waste rate, lb/day =

$$\frac{(2,820 \text{ mg/L}) \, [0.982 \text{ mil gal} + 0.425 \text{ mil gal}] \, (8.34 \text{ lb/gal})}{7.40 \text{ days, Desired MCRT}} - (18.5 \text{ mg/L TSS})(4.28 \text{ mgd})(8.34 \text{ lb/gal})$$

$$\text{Waste rate, lb/day} = \frac{(2,820 \text{ mg/L}) \, (1.407 \text{ mil gal}) \, (8.34 \text{ lb/gal})}{7.40 \text{ days, Desired MCRT}} - 660.36 \text{ lb/day}$$

Waste, lb/day = 4471.75 lb − 660.36 lb

Waste, lb/day = 3,811.39 lb/day, round to **3,810 lb/day**

Now, the waste rate in gpm can be calculated because the concentration of WAS is known.

$$\text{Equation: } \textbf{Waste, mgd} = \frac{\text{Waste, lb/day}}{(\text{WAS, mg/L})(8.34 \text{ lb/gal})}$$

Substitute values, in this case, before rounding and solve.

$$\text{Waste, mgd} = \frac{3{,}811.39\ \text{lb/day}}{(6{,}310\ \text{mg/L})(8.34\ \text{lb/gal})} = 0.07242\ \text{mgd}$$

Lastly, convert mgd to gpm.

$$\text{Waste, gpm} = \frac{(0.07242\ \text{mgd})(1{,}000{,}000/\text{mil})}{1{,}440\ \text{min/day}} = \mathbf{50.3\ gpm}$$

DIGESTER VOLATILE SOLIDS LOADING RATIO CALCULATIONS

This calculation compares the volatile solids added to the volatile solids in the digester. It is used for evaluating process control.

140. **What is the ratio of volatile solids loading on a digester, if the digester has 41,150 kg of volatile solids (VS), 1,085 lb/day are pumped into it, percentage total solids (TS) are 4.59%, and percentage volatile solids (VS) are 65.8%?**

First, convert the number of kg to lb.

Know: 1 kg = 2.205 lb

Number of lb = (41,150 kg)(2.205 lb/kg) = 90,736 lb

Use expanded equation with percentages.

$$\text{Equation: } \textbf{Digester VS ratio} = \frac{\text{VS added lb/day}}{(\text{lb VS, digester})(\text{TS \%}/100\%)(\text{VS \%}/100\%)}$$

Substitute values and solve.

$$\text{Digester VS ratio} = \frac{1{,}085\ \text{lb/day}}{(90{,}736\ \text{lb})(4.59\%/100\%\ \text{TS})(65.8\%/100\%\ \text{VS})}$$

Digester VS ratio = **0.396 VS ratio**

141. Given the following data, calculate the volatile solids (VS) loading ratio on a digester:

Digester capacity = 99,950 gallons
Total solids (TS) in digester = 5.88%
VS in digester = 63.4%
Sludge pumped to digester = 43,350 lb/day
Total solids (TS) pumped = 5.35%
VS pumped = 71.6%
Specific gravity of sludge in digester = 1.03
Specific gravity of sludge pumped to digester = 1.01

First, convert the specific gravity (sp gr) to lb/gal.

sp gr in digester = (8.34 lb/gal)(1.03 sp gr) = 8.59 lb/gal

sp gr of pumped sludge = (8.34 lb/gal)(1.01 sp gr) = 8.42 lb/gal

Use the following expanded equation with percentages to solve for digester volatile solid ratio.

Digester VS ratio =

$$\frac{\text{(VS added lb/day)(Pumped sludge, lb/gal)(Pumped TS \%/100\%)(Pumped VS \%/100\%)}}{\text{(Digester, gal)(Digester sludge, lb/gal)(Digester TS \%/100\%)(Digester VS \%/100\%)}}$$

Substitute values and solve.

$$\text{Digester VS ratio} = \frac{(43,350 \text{ lb/day})(8.42 \text{ lb/gal})(5.35\%/100\%)(71.6\%/100\%)}{(99,950 \text{ gal})(8.59 \text{ lb/gal})(5.88\%/100\%)(63.4\%/100\%)}$$

Digester VS ratio = 0.4368, round to **0.437 VS ratio**

DIGESTER GAS PRODUCTION PROBLEMS

Operators calculate the amount of gases produced per pound of volatile solids destroyed to determine the effectiveness of the digestion process. Also, it is important to know the gas production because in some cases it is used as a fuel for other plant processes.

142. What is the daily gas production in ft³/lb VS destroyed, given the following data?

Gas production = 14,350 ft³/day
Pumped volatile solids (VS) to digester = 1,660 lb/day
Percent VS reduction = 51.5%

Equation: **Gas produced, ft³/lb VS destroyed** $= \dfrac{\text{Gas production, ft}^3/\text{day}}{(\text{VS added, lb/day})(\text{Percent VS reduction})}$

Substitute values and solve.

Gas produced, ft³/lb VS destroyed $= \dfrac{14{,}350 \text{ ft}^3/\text{day}}{(1{,}660 \text{ lb/day})(51.5\%/100\%)}$

Gas produced, ft³/lb VS destroyed = 16.79 ft³/lb, round to **16.8 ft³/lb**

143. Given the following data, determine the gas produced by a digester in cubic meters (m³) per pound of volatile solids (VS) destroyed:

Digester gas production = 12,370 ft³/day
Pumped volatile solids to digester = 1,490 lb/day
Percent VS reduction = 53.2%
1 cubic meter = 35.3 cubic feet

Equation: **Gas produced, m³/lb VS destroyed** =

$$\dfrac{\text{Gas production, ft}^3/\text{day}}{(\text{VS destroyed, lb/day})(\text{Percent VS reduction})(35.3 \text{ ft}^3/\text{m}^3)}$$

Gas produced, m³/lb VS destroyed $= \dfrac{12{,}370 \text{ ft}^3/\text{day}}{(1{,}490 \text{ lb/day})(53.2\%/100\%)(35.3 \text{ ft}^3/\text{m}^3)}$

Gas produced, m³/lb VS destroyed = **0.442 m³/lb of VS destroyed**

144. **What must have been the gas production by a digester in ft³/day, given the following data?**

Volatile solids added = 395 kg/day
Gas produced in ft³/lb VS destroyed = 15.7 ft³/lb
Percent VS reduction = 55.8%

First, convert kg/day to lb/day.

Number of lb/day = (395 kg/day)(2.205 lb/kg) = 870.975 lb/day

Equation: **Gas produced, ft³/lb VS destroyed** $= \dfrac{\text{Gas production, ft}^3/\text{day}}{(\text{VS added, lb/day})(\text{Percent VS reduction})}$

Rearrange to solve for gas production in ft³/day.

Gas production, ft³/day = (Gas produced, ft³/lb VS destroyed)(VS added, lb/day)(VS % reduction)

Substitute values and solve.

Gas production, ft³/day = (15.7 ft³/lb)(870.975 lb/day)(55.8% VS/100%)

Gas production, ft³/day = **7,630 ft³/day**

DIGESTER SOLIDS BALANCE

Digester solids balance calculations are used to confirm the many calculations used to evaluate process control and maximize the digester process.

145. **Calculate the solids balance for a digester, given the following data:**

<u>Sludge Before Digestion</u>
Influent raw sludge = 32,500 lb/day
Percent solids in raw sludge = 6.75%
Percent volatile solids in raw sludge = 65.0%

<u>**Digested Sludge**</u>
Percent solids = 4.90%
Percent volatile solids (VS) = 50.5%

First, determine the solids entering the digester unit in lb/day.

Equation: **Total solids, lb/day = (Raw sludge, lb/day)(Percent solids)**

Total solids, lb/day = (32,500 lb/day)(6.75%/100%)

Total solids, lb/day = 2,193.75 lb/day, round to **2,190 lb/day Total solids**

Next, calculate the VS entering the digester unit in lb/day

VS, lb/day = (2,193.75 lb/day)(65.0%/100%) = 1,425.94 lb/day, round to **1,430 lb/day VS**

Next, calculate the fixed solids (residue left over after drying at 550°C) entering the digester unit in lb/day.

Equation: **Fixed solids, lb/day = Total solids, lb/day − VS, lb/day**

Fixed solids, lb/day = 2,193.75 lb/day − 1,425.94 lb/day

Fixed solids, lb/day = 767.81 lb/day, round to **768 lb/day Fixed solids**

Next, determine the amount of water entering the digester unit in lb/day.

Equation: **Water in sludge, lb/day = Sludge, lb/day − Total solids, lb/day**

Water, lb/day = 32,500 lb/day − 2,193.75 lb/day

Water, lb/day = 30,306.25 lb/day, round to **30,300 lb/day Water**

Next, calculate the volatile solids reduction (VSR).

Equation: **Percent VSR** $= \dfrac{(\text{In} - \text{Out})(100\%)}{\text{In} - (\text{In})(\text{Out})}$

First, convert percentage to decimal form.

Decimal form for 65.0% = 65.0%/100% = 0.650

Decimal form for 50.5% = 50.5%/100% = 0.505

Percent VSR $= \dfrac{(0.650 - 0.505)(100\%)}{0.650 - (0.650)(0.505)}$

Reduce and simplify.

Percent VSR $= \dfrac{(0.145)(100\%)}{0.650 - 0.32825} = \dfrac{14.5\%}{0.32175} = 45.07\%$, round to **45.1% VSR**

Next, determine the gas produced by the digester in lb/day.

Know: Pounds VSR = Pounds gas produced

Equation: **Gas produced, lb/day = (Effluent VS, lb/day)(Percent VSR)**

Gas produced, lb/day = (1,425.94 lb/day VS)(45.07%/100% VSR)

Gas produced, lb/day = 642.67 lb/day, round to **643 lb/day Gas production**

Next, calculate the VS in the digested sludge in lb/day.

Equation: **VS in digested sludge, lb/day = Influent VS, lb/day − Destroyed VS, lb/day**

VS in digested sludge, lb/day = 1,425.94 lb/day VS − 642.67 lb/day Destroyed VS

VS in digested sludge, lb/day = 783.27, lb/day, round to **783 lb/day VS in digested sludge**

Next, calculate the total solids in digested sludge (lb/day).

Equation: **Total digested solids, lb/day** $= \dfrac{\text{VS Digested, lb/day}}{\text{Percent digested VS}}$

Total digested solids, lb/day $= \dfrac{783.27\ \text{lb/day}}{50.5\%/100\%\ \text{digested VS}}$

Total digested solids, lb/day $= 1{,}551.03$ lb/day, round to **1,550 lb/day Total digested solids**

Next, calculate the fixed solids in the digested sludge in lb/day.

Equation: **Fixed solids, lb/day = Total digested solids, lb/day − VS digested, lb/day**

Fixed solids, lb/day $= 1{,}551.03$ lb/day $- 783.27$ lb/day

Fixed solids, lb/day $= 767.76$ lb/day, round to **768 lb/day Fixed solids**

Next, calculate the digested sludge in lb/day

Equation: **Digested sludge, lb/day** $= \dfrac{\text{Total digested solids, lb/day}}{\text{Digested sludge percent solids}}$

Digested sludge, lb/day $= \dfrac{1{,}551.03\ \text{lb/day}}{4.90\%/100\%\ \text{Digested solids}}$

Digested sludge, lb/day $= 31{,}653.67$ lb/day, round to **31,700 lb/day**

Next, calculate the lb/day of water in the digested sludge.

Equation: **Water in digested sludge, lb/day = Sludge, lb/day − Total digested solids, lb/day**

Water in digested sludge, lb/day $= 31{,}653.67$ lb/day $- 1{,}551.03$ lb/day

Water in digested sludge, lb/day $= 30{,}102.64$ lb/day, round to **30,100 lb/day Water**

Finally do a comparison analysis of the results.

COMPARISON ANALYSES OF DIGESTER SOLIDS MASS BALANCE			
Sludge Entering the Digester		**Digested Sludge Exiting Digester**	
Total Solids	2,193.75 lb/day	Total Solids	1,551.03 lb/day
Volatile Solids	(1,425.94 lb/day)	Volatile Solids	(783.27 lb/day)
Fixed Solids	(767.81 lb/day)	Fixed Solids	(767.76 lb/day)
Water in Sludge	30,306.25 lb/day	Water in Sludge	30,102.64 lb/day
		Gas Produced	642.67 lb/day
Final rounded total	**32,500 lb/day**	Final rounded total	**32,296 lb/day**

Note: Numbers in parenthesis are not added.

146. Calculate the solids balance for a digester, given the following data:

Sludge Before Digestion
Influent raw sludge = 28,300 lb/day
Percent solids in raw sludge = 6.10%
Percent volatile solids in raw sludge = 70.8%

Digested Sludge
Percent solids = 4.23%
Percent volatile solids (VS) = 56.5%

First, determine the solids entering the digester unit in lb/day.

Equation: **Total solids, lb/day = (Raw sludge, lb/day)(Percent solids)**

Total solids, lb/day = (28,300 lb/day)(6.10%/100%)

Total solids, lb/day = 1,726.3 lb/day, round to **1,730 lb/day Total solids**

Next, calculate the VS entering the digester unit in lb/day.

VS, lb/day = (1,726.3 lb/day)(70.8%/100%) = 1,222.22 lb/day, round to **1,220 lb/day VS**

Next, calculate the fixed solids (residue left over after drying at 550°C) entering the digester unit in lb/day.

Equation: **Fixed solids, lb/day = Total solids, lb/day − VS, lb/day**

Fixed solids, lb/day = 1,726.3 lb/day − 1,222.22 lb/day

Fixed solids, lb/day = 504.08 lb/day, round to **504 lb/day Fixed solids**

Next, determine the amount of water entering the digester unit in lb/day.

Equation: **Water in sludge, lb/day = Sludge, lb/day − Total solids, lb/day**

Water, lb/day = 28,300 lb/day − 1,726.3 lb/day

Water, lb/day = 26,573.7 lb/day, round to **26,600 lb/day Water**

Next, calculate the volatile solids reduction (VSR).

Equation: **Percent VSR** $= \dfrac{(\text{In} - \text{Out})(100\%)}{\text{In} - (\text{In})(\text{Out})}$

First, convert percentage to decimal form

Decimal form for 70.8% = 70.8%/100% = 0.708

Decimal form for 56.5% = 56.5%/100% = 0.565

Percent VSR $= \dfrac{(0.708 - 0.565)(100\%)}{0.708 - (0.708)(0.565)}$

Reduce and simplify.

Percent VSR $= \dfrac{(0.143)(100\%)}{0.708 - 0.40002} = \dfrac{14.3\%}{0.30798} = 46.43\%$, round to **46.4% VSR**

Next, determine the gas produced by the digester in lb/day.

Know: Pounds VSR = Pounds gas produced

Equation: **Gas produced, lb/day = (Effluent VS, lb/day)(Percent VSR)**

Gas produced, lb/day = (1,222.22 lb/day VS)(46.43%/100% VSR)

Gas produced, lb/day = 567.48 lb/day, round to **567 lb/day Gas production**

Next, calculate the VS in the digested sludge in lb/day.

Equation: **VS in digested sludge, lb/day = Influent VS, lb/day − Destroyed VS, lb/day**

VS in digested sludge, lb/day = 1,222.22 lb/day VS − 567.48 lb/day Destroyed VS

VS in digested sludge, lb/day = 654.74, lb/day, round to **655 lb/day VS in digested sludge**

Next, calculate the total solids in digested sludge (lb/day).

Equation: **Total digested solids, lb/day** $= \dfrac{\text{VS Digested, lb/day}}{\text{Percent digested VS}}$

Total digested solids, lb/day $= \dfrac{654.74\,\text{lb/day}}{56.5\%/100\%\text{ digested VS}}$

Total digested solids, lb/day = 1,158.83 lb/day, round to **1,160 lb/day Total digested solids**

Next, calculate the fixed solids in the digested sludge in lb/day.

Equation: **Fixed solids, lb/day = Total digested solids, lb/day − VS digested, lb/day**

Fixed solids, lb/day = 1,158.83 lb/day − 654.74 lb/day

Fixed solids, lb/day = 504.09 lb/day, round to **504 lb/day Fixed solids**

Next, calculate the digested sludge in lb/day.

Equation: **Digested sludge, lb/day** $= \dfrac{\text{Total digested solids, lb/day}}{\text{Digested sludge percent solids}}$

Digested sludge, lb/day $= \dfrac{1,158.83\ \text{lb/day}}{4.23/100\%\ \text{Digested solids}}$

Digested sludge, lb/day $= 27,395.51$ lb/day, round to **27,400 lb/day**

Next, calculate the lb/day of water in the digested sludge.

Equation: **Water in digested sludge, lb/day $=$ Sludge, lb/day $-$ Total solids, lb/day**

Water in digested sludge, lb/day $= 27,395.51$ lb/day $- 1,158.83$ lb/day

Water in digested sludge, lb/day $= 26,236.68$ lb/day, round to **26,200 lb/day Water**

Finally do a comparison analysis of the results.

COMPARISON ANALYSES OF DIGESTER SOLIDS MASS BALANCE			
Sludge Entering the Digester		**Digested Sludge Exiting Digester**	
Total Solids	1,726.3 lb/day	Total Solids	1,158.83 lb/day
Volatile Solids	(1,222.22 lb/day)	Volatile Solids	(654.74 lb/day)
Fixed Solids	(504.08 lb/day)	Fixed Solids	(504.09 lb/day)
Water in Sludge	26,573.7 lb/day	Water in Sludge	26,236.68 lb/day
		Gas Produced	567.48 lb/day
Final rounded total	**28,300 lb/day**	Final rounded total	**27,963 lb/day**

Note: Numbers in parenthesis are not added.

SOLIDS (MASS) BALANCE CALCULATIONS

Solids balance calculations are used to confirm the many calculations used to evaluate process control and maximize the wastewater treatment process.

Note: **Systems within 10% are considered in balance.**

MATH FOR WASTEWATER TREATMENT OPERATORS GRADES 3 AND 4

147. Given the following data, calculate the mass balance for the following activated biosolids system that has primary treatment using BOD_5 removal. Is there a problem with this system? If so, discuss.

Plant flow = 3.25 mgd
Plant influent BOD_5 = 244 mg/L
Influent BOD_5 = 131 mg/L
Plant effluent flow = 3.24 mgd
Effluent system activated biosolids BOD_5 = 22.5 mg/L
Effluent system activated biosolids TSS = 25 mg/L
Waste concentration = 6,845 mg/L
Waste flow = 0.0585 mgd

Given: This activated biosolids system has lb solids/lb BOD_5 = 0.68 lb solids/lb BOD_5

First, calculate the BOD_5 influent and effluent in lb/day.

Influent BOD_5, lb/day = (244 mg/L)(3.25 mgd)(8.34 lb/gal) = 6,613.62 lb/day

Effluent BOD_5, lb/day = (22.5 mg/L)(3.25 mgd)(8.34 lb/gal) = 609.8625 lb/day

Then, determine the difference in lb/day BOD_5 removal.

BOD_5 removed, lb/day = 6,613.62 lb/day − 609.8625 lb/day = 6,003.76 lb/day

Next, determine the solids produced in lb/day.

Equation: **Solids produced, lb/day = (BOD$_5$ removed, lb/day)(0.68 lb solids/lb BOD$_5$)**

Solids produced, lb/day = (6,003.76 lb/day)(0.68 lb/lb BOD$_5$) = 4,082.56 lb/day

Next, calculate the solids and sludge removed.

Effluent TSS, lb/day = (25 mg/L)(3.25 mgd)(8.34 lb/gal) = 677.62 lb/day

Solids removed, lb/day = (6,845 mg/L)(0.0585 mgd)(8.34 lb/gal) = 3,339.61 lb/day

Next, calculate the total solids removed.

Total solids removed, lb/day = 677.62 lb/day + 3,339.61 lb/day = 4,017.23 lb/day

Now, determine the difference in lb/day.

Balance difference, lb/day = 4,082.56 lb/day − 4,017.23 lb/day = 65.33 lb/day

Lastly, calculate the percent difference.

Percent difference $= \dfrac{(4,082.56 \text{ lb/day} - 4,017.23 \text{ lb/day})(100\%)}{4,082.56 \text{ lb/day}} =$ **1.60%**

This system is in balance.

148. Calculate the mass balance for the following conventional biological system. Is there a problem with this system? If so, discuss.

Influent waste flow = 1.73 mgd
Influent BOD_5 = 248 mg/L
Total suspended solids (TSS) = 296 mg/L
Effluent flow = 1.65 mgd
Effluent BOD_5 = 23 mg/L
Effluent TSS = 42 mg/L
Waste flow = 0.0294 mgd
Waste TSS = 8,310 mg/L

Given: This conventional activated biosolids system without primary has a lb solids/lb BOD_5 = 0.85 lb solids/lb BOD_5

First, calculate the BOD_5 influent and then the effluent in lb/day.

Influent BOD_5, lb/day = (248 mg/L)(1.73 mgd)(8.34 lb/gal) = 3,578.19 lb/day

Effluent BOD_5, lb/day = (23 mg/L)(1.73 mgd)(8.34 lb/gal) = 331.85 lb/day

BOD_5 removed, lb/day = 3,578.19 lb/day − 331.85 lb/day = 3,246.34 lb/day

Next, determine the solids produced in lb/day.

Equation: **Solids produced, lb/day = (BOD$_5$ removed, lb/day)(0.85 lb solids/lb BOD$_5$)**

Solids produced, lb/day = (3,246.34 lb/day)(0.85 lb/lb BOD$_5$) = 2,759.39 lb/day

Next, calculate the solids and sludge removed.

Effluent TSS, lb/day = (42 mg/L)(1.73 mgd)(8.34 lb/gal) = 605.98 lb/day

Effluent sludge, lb/day = (8,310 mg/L)(0.0294 mgd)(8.34 lb/gal) = 2,037.58 lb/day

Next, calculate the total solids removed.

Total solids removed, lb/day = 605.98 lb/day + 2,037.58 lb/day = 2,643.56 lb/day

Now, calculate the percent mass balance of the system.

Equation: **Percent mass balance** $= \dfrac{(\text{Solids produced, lb/day} + \text{Solids removed, lb/day})(100\%)}{\text{Solids produced, lb/day}}$

Percent mass balance $= \dfrac{(2{,}759.39\ \text{lb/day} - 2{,}643.58\ \text{lb/day})(100\%)}{2.759.39\ \text{lb/day}} = $ **4.2%**

This system is within balance.

149. **Calculate the mass balance for the following conventional biological system. Is there a problem with this system? If so. discuss.**

Influent waste flow = 1.47 mgd
Influent BOD_5 = 226 mg/L
Total suspended solids (TSS) = 281 mg/L
Effluent flow = 1.42 mgd
Effluent BOD_5 = 21.5 mg/L
Effluent TSS = 39.5 mg/L
Waste flow = 0.0208 mgd
Waste TSS = 8,445 mg/L

Given: This activated biosolids without primary, extended air system has a lb solids/lb BOD_5 = 0.65 lb solids/lb BOD_5

First, calculate the BOD_5 influent and then the effluent in lb/day.

Influent BOD_5, lb/day = (226 mg/L)(1.47 mgd)(8.34 lb/gal) = 2,770.71 lb/day

Effluent BOD_5, lb/day = (21.5 mg/L)(1.47 mgd)(8.34 lb/gal) = 263.59 lb/day

BOD_5 removed, lb/day = 2,770.71 lb/day − 263.59 lb/day = 2,507.12 lb/day

Next, determine the solids produced in lb/day.

Equation: **Solids produced, lb/day = (BOD_5 removed, lb/day)(0.65 lb solids/lb BOD_5)**

Solids produced, lb/day = (2,507.12 lb/day)(0.65 lb/lb BOD_5) = 1,629.63 lb/day

Next, calculate the solids and sludge removed.

Effluent solids, lb/day $= (39.5 \text{ mg/L})(1.47 \text{ mgd})(8.34 \text{ lb/gal}) = 484.26 \text{ lb/day}$

Effluent sludge, lb/day $= (8{,}445 \text{ mg/L})(0.0208 \text{ mgd})(8.34 \text{ lb/gal}) = 1{,}464.97 \text{ lb/day}$

Next, calculate the total solids removed.

Total solids removed, lb/day $= 484.26 \text{ lb/day} + 1{,}464.97 \text{ lb/day} = 1{,}949.23 \text{ lb/day}$

Now, calculate the percent mass balance of the system.

Equation: **Percent mass balance** $= \dfrac{(\text{Solids produced, lb/day} + \text{Solids removed, lb/day})(100\%)}{\text{Solids produced, lb/day}}$

Percent mass balance $= \dfrac{(1{,}629.63 \text{ lb/day} - 1{,}949.23 \text{ lb/day})(100\%)}{1{,}629.63 \text{ lb/day}} = \mathbf{19.6\%}$

This system is *not* in balance, since more solids are being removed than produced by an amount greater than 10%. *Note:* The negative sign can be dropped.

Now, calculate the waste rate in gpd based on the mass balance results.

Equation: **Waste rate, gpd** $= \dfrac{(\text{Solids produced, lb/day})(1{,}000{,}000/\text{mil})}{(\text{Waste TSS, mg/L})(8.34 \text{ lb/gal})}$

Waste rate, gpd $= \dfrac{(1{,}629.63 \text{ lb/day})(1{,}000{,}000/\text{mil})}{(8{,}445 \text{ mg/L TSS})(8.34 \text{ lb/gal})} = 23{,}138 \text{ gpd}$, round to **23,100 gpd**

Current waste rate is 0.0208 mil gal, which is 20,800 gpd. This is less than the calculated waste rate of 23,138 gpd.

Analyzing these results indicate that less wasting is required because more solids are being removed than added based on the mass balance results. The waste rate result indicates more solids need to be wasted. These conflicting results indicate the system is out of balance. In addition, problems with this system could be accentuated or be caused by improper sampling and by laboratory analytical error(s).

150. Given the following data, calculate whether a gravity thickener's sludge blanket will increase, remain the same, or decrease, and how much that change will be in lb/day solids, if it increases or decreases:

Pumped sludge to thickener = 124 gpm
Pumped sludge from thickener = 49 gpm
Thickener's effluent suspended solids (SS) = 743 mg/L
Primary sludge solids = 3.38%
Thickened sludge solids = 7.91%

First, calculate the thickener's influent and effluent solids in lb/day.

Equation: **Influent solids, lb/day =**

(Influent flow gpm)(1,440 min/day)(Percent sludge solids)(8.34 lb/gal)

Influent solids, lb/day = (124 gpm)(1,440 min/day)(3.38%/100%)(8.34 lb/gal)

Influent solids, lb/day = 50,334.64 lb/day

Effluent solids, lb/day = (49 gpm)(1,440 min/day)(7.91%/100%)(8.34 lb/gal)

Effluent solids, lb/day = 46,548.01 lb/day

Next, calculate the flow leaving the thickener.

Effluent flow = 124 gpm − 49 gpm = 75 gpm

Next, convert thickened sludge solids to percent.

Know: 1% = 10,000 mg/L

Percent sludge solids = (743 mg/L)(1%/10,000 mg/L) = 0.0743%

Now, calculate the sludge solids that are leaving the thickener.

Sludge solids, lb/day = (75 gpm)(1,440 min/day)(0.0743%/100%)(8.34 lb/gal)

Sludge solids, lb/day = 669.235 lb/day

Now, calculate whether the sludge blanket will increase, remain the same, or decrease.

Total solids, lb/day = 50,334.64 lb/day − 46,548.01 lb/day − 669.235 lb/day

Total solids, lb/day = 3,117.395 lb/day, round to **3,120 lb/day**

The sludge blanket will increase because the total solids are positive.

VOLATILE-ACIDS-TO-ALKALINITY RATIO PROBLEMS

The first phase of anaerobic digestion is acid fermentation, which is dependent upon new volatile solids entering the digester. The second stage is methane fermentation. These two processes need to be in delicate balance with each other for the anaerobic digestion process to proceed properly. Different treatment plants have different ratios, but typically the ratio is less than 0.1.

151. **What is the ratio of volatile acids to alkalinity, if the alkalinity in an anaerobic digester is 1,462 mg/L and the volatile acid concentration of the sludge is 181 mg/L?**

Equation: **Ratio = Volatile acids/Alkalinity**

Substitute values and solve.

$$\text{Ratio} = \frac{181 \text{ mg/L}}{1,462 \text{ mg/L}} = 0.1238, \text{ round to } \mathbf{0.124 \text{ Volatile-acids-to-alkalinity ratio}}$$

152. **What must have been the volatile acid concentration in an anaerobic digester, if the alkalinity was 1,603 mg/L and the ratio of volatile acids to alkalinity was 0.134?**

Equation: **Volatile acids = (Alkalinity)(Ratio)**

Substitute values and solve.

Volatile acids = (1,603 mg/L)(0.134) = 214.8 mg/L, round to **215 mg/L of Volatile acids**

153. If the ratio of volatile solids (VS) added to volatile solids already in an anaerobic digester is 0.153 and the amount of VS already in the digester is 19,447 lb, what is the amount of volatile solids in lb/day that must have been added to a digester?

Equation: **Digester VS ratio** $= \dfrac{\text{VS added lb/day}}{\text{lb VS in digester}}$

Rearrange the problem to solve for volatile solids added.

VS added lb/day = (Digester VS ratio)(lb VS in digester)

Substitute values and solve.

VS added lb/day = (0.153 VS ratio)(19,447 lb VS in digester)

VS added lb/day = 2,975.39 lb/day, round to **2,980 lb/day VS added**

LIME NEUTRALIZATION PROBLEMS

When the sludge in an anaerobic digester becomes acidic, it is called a sour digester. A sour digester occurs when the volatile acid-to-alkalinity ratio increases above 0.8. It is not always possible to wait for a digester to naturally correct itself because of the digester's capacity or time constraints. Under these circumstances, it is necessary to neutralize the acid conditions in the digester with lime. The following problems show how operators calculate the appropriate dosage of lime. The lime dosage is based on the amount of volatile acids in the sludge and is in a 1-to-1 ratio, that is, 1 mg/L of lime will neutralize 1 mg/L of volatile acid.

154. How many pounds of lime will be required to neutralize a sour digester that is 42.6 ft in diameter, has a sludge level of 16.9 ft, and has a volatile acid content of 2,805 mg/L?

Know: 1 mg/L of lime will neutralize 1 mg/L of volatile acids

First, calculate the volume of the digester in gallons.

Number of gallons $=$ (0.785)(42.6 ft)(42.6 ft)(16.9 ft)(7.48 gal/ft³) $=$ 180,085 gal

Next, convert gallons to mil gal.

Number of mil gal $= \dfrac{180,085 \text{ gal}}{1,000,000/\text{mil}} = 0.180085$ mil gal

Equation: **Number of lb Lime $=$ (Volatile acids, mg/L)(mil gal)(8.34 lb/gal)**

Lime, lb $=$ (2,805 mg/L)(0.180085 mil gal)(8.34 lb/gal) $=$ 4,212.85 lb, round to **4,210 lb of Lime**

155. Given the following data, calculate the amount of lime in lb that are needed to neutralize a sour digester:

Digester diameter $=$ 40.3 ft
Digester fluid level $=$ 15.1 ft
Volatile acids $=$ 2,287 mg/L

Know: 1 mg/L of lime will neutralize 1 mg/L of volatile acids

First, calculate the volume of the digester in gallons.

Number of gallons $=$ (0.785)(40.3 ft)(40.3 ft)(15.1 ft)(7.48 gal/ft³) $=$ 143,998.6 gal

Next, convert gallons to mil gal.

Number of mil gal $=$ (143,998.6 gal) / (1,000,000/mil) $=$ 0.144 mil gal

Equation: **Number of lb Lime $=$ (Volatile acids, mg/L)(mil gal)(8.34 lb/gal)**

Lime, lb $=$ (2,287 mg/L)(0.144 mil gal)(8.34 lb/gal) $=$ 2,746.596 lb, round to **2,750 lb of Lime**

156. **What must have been the concentration of volatile acids in mg/L for a sour digester with a radius of 27.5 ft and a sludge level of 17.1 ft, if the number of pounds of lime to neutralize the volatile acids was 2,930 lb?**

Know: 1 mg/L of lime will neutralize 1 mg/L of volatile acids

First, determine the sludge volume in gallons that is in the digester.

Number of gallons $= \pi(r^2)(\text{Height})(7.48 \text{ gal/ft}^3)$

Number of gallons $= 3.14(27.5 \text{ ft})(27.5 \text{ ft})(17.1 \text{ ft})(7.48 \text{ gal/ft}^3) = 303{,}734$ gal

First, convert the digester's volume from gallons to mil gal.

Number of mil gal $= (303{,}734 \text{ gal}) / (1{,}000{,}000/\text{mil}) = 0.303734$ mil gal

Equation: **Number of lb Lime $=$ (Volatile acids, mg/L)(mil gal)(8.34 lb/gal)**

Rearrange the equation to solve for the concentration of volatile acids.

$$\text{Volatile acids, mg/L} = \frac{\text{Number of lb, lime}}{(\text{mil gal})(8.34 \text{ lb/gal})}$$

Substitute values and solve.

$$\text{Volatile acids, mg/L} = \frac{2{,}930 \text{ lb, lime}}{(0.303734 \text{ mil gal})(8.34 \text{ lb/gal})}$$

Volatile acids, mg/L $= 1{,}156.67$ mg/L, round to **1,160 mg/L Volatile acids**

POPULATION LOADING CALCULATIONS

These calculations are used for wastewater treatment ponds. They are based on the number of people per acre of pond, and it is a helpful tool in evaluating process control of ponds.

157. Given the following data, calculate the population loading in people per acre on four ponds.

Pond 1 = 6.49 acres
Pond 2 = 5.28 acres
Pond 3 = 4.61 acres
Pond 4 = 3.55 acres
Services = 27,800
Average persons per service = 3.28

First, add the area in acres for each pond to get the total acres.

Total area of ponds = 6.49 acres + 5.28 acres + 4.61 acres + 3.55 acres = 19.93 acres

Next, determine the number of people served.

Number of people = (27,800 services)(3.28 people/service) = 91,184 people served

Next, using the following equation, determine the population loading.

$$\textbf{Population loading, people/acre} = \frac{\text{Number of people served}}{\text{Area of pond(s), acres}}$$

Substitute values and solve.

$$\text{Population loading, people/acre} = \frac{91,184 \text{ people served}}{19.93 \text{ acres}}$$

Population loading, people/acre = 4,575 people/acre, round to **4,580 people/acre**

158. What is the population loading in people/acre, if there are 22,675 services with 3.09 people per service, given the following data?

Wastewater pond 1 = 13.44 acres
Wastewater pond 2 = 8.67 acres
Wastewater pond 3 = 6.88 acres
Wastewater pond 4 = 5.92 acres

First, add the area in acres for each pond to get the total acres.

Total area of ponds = 13.44 acres + 8.67 acres + 6.88 acres + 5.92 acres = 34.91 acres

Next, determine the number of people served.

Number of people = (22,675 services)(3.09 people/service) = 70,066 people served

Next, using the following equation determine the population loading.

$$\text{Population loading, people/acre} = \frac{\text{Number of people served}}{\text{Area of pond(s), acres}}$$

Substitute values and solve.

$$\text{Population loading, people/acre} = \frac{70,066 \text{ people served}}{34.91 \text{ acres}}$$

Population loading, people/acre = 2007.05 people/acre, round to **2,010 people/acre**

POPULATION EQUIVALENT CALCULATIONS

Wastewater discharge from industries or commercial sources usually has a higher organic content than domestic wastewaters. Operators use population equivalent calculations to compare domestic wastewater to wastewater from these former sources. This is important in determining the loading that will be placed on a wastewater system when a new industry wants to connect to a system. What is needed is the flow from this industry in mgd and the BOD_5 concentration in mg/L. Domestic wastewater systems usually contain a range of 0.17 to 0.20 pounds of BOD_5 per day, which the wastewater plant should have already determined. Also, population equivalent calculations are require for designing proper size wastewater treatment plants, pump stations, and pipe sizes, because the volumetric flow that is expected to be treated and pumped needs to be estimated.

159. Given the following data, determine the population equivalent:

Average wastewater flow for the day $= 3.28$ ft³/s
BOD_5/person $= 0.213$ lb/day
BOD_5 concentration in the wastewater $= 2,443$ mg/L

First, convert ft³/s to mgd.

Number of mgd $= (3.28$ ft³/s$)(0.6463$ mgd/ ft³/s$) = 2.1199$ mgd

Use the following equation to solve this problem.

$$\textbf{Number of people} = \frac{(BOD_5, \text{mg/L})(\text{mgd})(8.34 \text{ lb/gal})}{\text{lb/day of } BOD_5/\text{person}}$$

$$\text{Number of people} = \frac{(2,443 \text{ mg/L } BOD_5)(2.1199 \text{ mgd})(8.34 \text{ lb/gal})}{0.213 \text{ lb/day}}$$

Number of people $= 202,780$ people, round to **203,000 people**

160. A wastewater treatment plant has an influent flow of 4.16 ft³/s. If the BOD_5 is 2,308 mg/L and the average BOD_5 per person is 0.237 lb/day, what is the population equivalent that this plant is currently treating?

First, convert ft³/s to mgd.

Number of mgd $= (4.16$ ft³/s$)(0.6463$ mgd/ ft³/s$) = 2.689$ mgd

$$\text{Equation: } \textbf{Number of people} = \frac{(BOD_5, \text{mg/L})(\text{mgd})(8.34 \text{ lb/gal})}{\text{lb/day of } BOD_5/\text{person}}$$

$$\text{Number of people} = \frac{(2,308 \text{ mg/L } BOD_5)(2.689 \text{ mgd})(8.34 \text{ lb/gal})}{0.237 \text{ lb/day}}$$

Number of people $= 218,396$ people, round to **218,000 people**

SOLIDS UNDER AERATION

Solids under aeration calculations are used by operators for evaluating process control.

161. **Calculate the number of pounds of suspended solids (SS) contained in an aeration tank, if the tank is 255 ft by 39.5 ft, with an average depth of 14.9 ft, the concentration of SS is 2,270 mg/L, and the specific gravity is 1.03.**

First, determine the volume in mil gal contained in the aeration tank.

$$\text{Aeration tank, mil gal} = \frac{(255\ \text{ft})\,(39.5\ \text{ft})\,(14.9\ \text{ft})\,(7.48\ \text{gal/ft}^3)}{1,000,000/\text{mil}} = 1.123\ \text{mil gal}$$

Next, determine the number of lb/gal for the SS.

Number of lb/gal, SS $= (8.34\ \text{lb/gal})(1.03\ \text{sp gr}) = 8.59$

Next, determine the pounds of SS under aeration using a modified version of the "pounds" equation because the SS weighs more than water.

Equation: **Number of lb SS $=$ (SS, mg/L)(Number of mil gal)(SS lb/gal)**

Substitute values and solve.

Number of lb SS $=$ (2,270 mg/L, SS)(1.123 mil gal)(8.59 lb/gal)

Number of lb SS $=$ 21,898 lb, round to **21,900 lb SS**

162. How many pounds of MLSS are being aerated, if the aeration tank is 47.5 ft in diameter, with a sludge height of 16.8 ft, the concentration of MLSS is 2,649 mg/L, and the specific gravity of the MLSS is 1.03?

First, determine how many gallons are in the aeration tank.

Number of gallons = (0.785)(Diameter)^2(Height, ft)(7.48 gal/ft^3)

Number of gallons = (0.785)(47.5 ft)(47.5 ft)(16.8 ft)(7.48 gal/ft^3) = 222,571 gal

Next, convert gallons to mil gal.

$$\text{Number of mil gal} = \frac{222,571 \text{ gal}}{1,000,000/\text{mil}} = 0.22257 \text{ mil gal}$$

Next, determine the lb/gal for the MLSS.

Number of lb/gal, MLSS = (8.34 lb/gal)(1.03 sp gr) = 8.59 lb/gal

Next, determine the pounds of MLSS under aeration using a modified version of the "pounds" equation because the MLSS weighs more than water.

Equation: **Number of lb MLSS = (MLSS, mg/L)(Number of mil gal)(MLSS lb/gal)**

Substitute values and solve.

Number of lb MLSS = (2,649 mg/L MLSS)(0.22257 mil gal)(8.59 lb/gal)

Number of lb MLSS = 5,064.56 lb, round to **5,060 lb MLSS**

SUSPENDED SOLIDS REMOVAL

The suspended solids removal calculations are used by operators as a sign for the efficiency of the treatment process in question. Typically, the suspended solids removed from wastewater systems ranges from 100 to 350 mg/L.

163. **Calculate the number of pounds of mixed liquor suspended solids (MLSS) being aerated by an aeration tank that is 48.75 ft in diameter, with a sludge height of 17.3 ft; the concentration of MLSS is 2,335 mg/L, and the specific gravity of the MLSS is 1.04.**

First, determine how many gallons are in the aeration tank.

Number of gallons = (0.785)(Diameter)^2(Height, ft)(7.48 gal/ft^3)

Number of gallons = (0.785)(48.75 ft)(48.75 ft)(17.3 ft)(7.48 gal/ft^3) = 241,416 gal

Next, convert gallons to mil gal.

Number of mil gal = $\dfrac{241,416 \text{ gal}}{1,000,000/\text{mil}}$ = 0.241416 mil gal

Next, determine the lb/gal for the MLSS.

Number of lb/gal, MLSS = (8.34 lb/gal)(1.04 sp gr) = 8.6736 lb/gal

Next, determine the pounds of MLSS under aeration using a modified version of the "pounds" equation because the MLSS weighs more than water.

Equation: **Number of lb MLSS = (MLSS, mg/L)(Number of mil gal)(MLSS lb/gal)**

Substitute values and solve.

Number of lb MLSS = (2,335 mg/L MLSS)(0.241416 mil gal)(8.6736 lb/gal)

Number of lb MLSS = 4,889.36 lb, round to **4,890 lb MLSS**

164. **What must have been the average influent concentration of suspended solids (SS) in mg/L, if a wastewater treatment plant's clarifier had a flow of 1,110 gpm and removed 2,020 lb/day of SS?**

First, convert gpm to mgd.

$$\text{Number of mgd} = \frac{(1{,}110 \text{ gpm})(1{,}440 \text{ min/day})}{1{,}000{,}000/\text{mil}} = 1.5984 \text{ mgd}$$

Equation: **Number of lb/day SS = (SS, mg/L)(Number of mgd)(8.34 lb/gal)**

Rearrange the equation, substitute values, and solve.

$$\textbf{SS removed, mg/L} = \frac{\text{Number of lb/day SS}}{(\text{Number of mgd})(8.34 \text{ lb/gal})}$$

$$\text{SS removed, mg/L} = \frac{2{,}020 \text{ lb/day}}{(1.5984 \text{ mgd})(8.34 \text{ lb/gal})} = 151.53 \text{ mg/L, round to } \textbf{152 mg/L, SS removed}$$

165. **What is the quantity of suspended solids (SS) in lb/day that was removed from a primary clarifier, if the average influent flow for the day was 3.44 ft³/s, the suspended solids was 186 mg/L, and the specific gravity of the SS 1.04?**

Know: 1 ft³/s = 0.6463 mgd/ ft³/s

First, convert ft³/s to mgd.

Number of mgd = (3.44 ft³/s)(0.6463 mgd/ ft³/s) = 2.223 mgd

Next, determine the lb/gal for the SS.

Number of lb/gal, MLSS = (8.34 lb/gal)(1.04 sp gr) = 8.6736 lb/gal

Next, determine the pounds of SS under aeration using a modified version of the "pounds" equation because the SS weighs more than water.

Equation: **Number of lb/day SS = (SS, mg/L)(Number of mgd)(SS lb/gal)**

Substitute values and solve.

Number of lb/day SS removed = (186 mg/L, SS)(2.223 mgd)(8.6736 lb/gal)

Number of lb/day SS removed = 3,586 lb/day, round to **3,590 lb/day SS removed**

BIOCHEMICAL OXYGEN DEMAND REMOVAL CALCULATIONS

The biochemical oxygen demand (BOD$_5$) removal calculations are used to inform operators about the efficiency of the treatment process for a pond or trickling filter. The BOD$_5$ is an empirical test that informs the operator on the relative oxygen requirements of a wastewater, and is an indicator of how much food is in the wastewater. The BOD$_5$ is a five-day test.

166. Given the following data, determine the BOD$_5$ removal in lb/day from a trickling filter.

Plant influent flow = 2.68 ft^3/s
Influent BOD$_5$ concentration = 346 mg/L
Effluent BOD$_5$ concentration = 124 mg/L

First, determine the amount of BOD$_5$ removed in mg/L by subtracting the influent BOD$_5$ from the effluent BOD$_5$.

BOD$_5$ removed, mg/L = (Influent BOD$_5$, mg/L − Effluent BOD$_5$, mg/L)

BOD$_5$ removed, mg/L = 346 mg/L − 124 mg/L = 222 mg/L BOD$_5$ removed

Next, convert ft^3/s to mgd.

Number of mgd = (2.68 ft^3/s)(0.6463 mgd/ ft^3/s) = 1.732 mgd

Next, solve for the amount of BOD$_5$ removed in lb/day by using the "pounds" formula.

Equation: **Number of lb/day BOD$_5$ = (BOD$_5$, mg/L)(Number of mgd)(8.34 lb/gal)**

Substitute values and solve.

Number of lb/day BOD$_5$ removed = (222 mg/L BOD$_5$)(1.732 mgd)(8.34 lb/gal)

Number of lb/day BOD$_5$ removed = 3,206.76 lb/day, round to **3,210 lb/day BOD$_5$ removed**

167. Given the following data, determine the BOD$_5$ removal in lb/day from a trickling filter.

Plant influent flow = 3.08 ft^3/s
Influent BOD$_5$ concentration = 312 mg/L
Effluent BOD$_5$ concentration = 117 mg/L
Trickling filter diameter = 41.5 ft
Trickling filter media depth = 7.75 ft

First, determine the amount of BOD$_5$ removed in mg/L by subtracting the influent BOD$_5$ from the effluent BOD$_5$.

BOD$_5$ removed, mg/L = (Influent BOD$_5$, mg/L − Effluent BOD$_5$, mg/L)

BOD$_5$ removed, mg/L = 312 mg/L − 117 mg/L = 195 mg/L BOD$_5$ removed

Next, convert ft^3/s to mgd.

Number of mgd = (3.08 ft^3/s)(0.6463 mgd/ ft^3/s) = 1.99 mgd

Next, solve for the amount of BOD$_5$ removed in lb/day by using the "pounds" formula.

Equation: **Number of lb/day BOD$_5$ = (BOD$_5$, mg/L)(Number of mgd)(8.34 lb/gal)**

Substitute values and solve.

Number of lb/day BOD$_5$ removed = (195 mg/L BOD$_5$)(1.99 mgd)(8.34 lb/gal)

Number of lb/day BOD$_5$ removed = 3,236.34 lb/day, round to **3,240 lb/day BOD$_5$ removed**

168. **What must have been the daily flow to a trickling filter in mgd, if the influent BOD$_5$ was 284 mg/L, the effluent BOD$_5$ was 91 mg/L, and the BOD$_5$ removed was 2,765 lb/ day?**

First, the amount of BOD$_5$ removed must still be determined by subtracting the influent BOD$_5$ from the effluent BOD$_5$.

BOD$_5$ removed, mg/L $= 284$ mg/L $- 91$ mg/L $= 193$ mg/L BOD$_5$ removed

Next, solve for the amount of BOD$_5$ removed in lb/day by using the "pounds" formula.

Equation: **Number of lb/day BOD$_5$ $=$ (BOD$_5$, mg/L)(Number of mgd)(8.34 lb/gal)**

Rearrange the equation, substitute values, and solve.

$$\textbf{Number of mgd} = \frac{\textbf{Number of lb/day BOD}_5}{\textbf{(BOD}_5 \textbf{ removed, mg/L)(8.34 lb/gal)}}$$

Substitute values and solve.

$$\text{Number of mgd} = \frac{2,765 \text{ lb/day}}{(193 \text{ mg/L BOD}_5 \text{ removed})(8.34 \text{ lb/gal})} = \textbf{1.7 mgd}$$

FOOD-TO-MICROORGANISM RATIO CALCULATIONS

A properly operated activated sludge process has a balance between the food entering the system and the microorganisms in the aeration tank. The best ratio varies because it depends on the activated-sludge process and the characteristics of the wastewater being treated. It measures the pounds of food coming in divided by the pounds of microorganisms present. The ratio is a process control number because it helps the operator determine the proper number of microorganisms for the system in question. *Note:* The day in mgd flow can de dropped in these types of problems.

169. What is the food-to-microorganism (F/M) ratio for an aeration tank that has a radius of 30.5 ft, liquid level of 16.3 ft, if the primary effluent flow averages 1,675 gpm, the mixed liquor volatile suspended solids (MLVSS) are 2,240 mg/L, and the BOD_5 is 345 mg/L?

First, calculate the number of gallons in the aeration tank.

Number of gallons $= \pi(\text{radius})^2(\text{Height, ft})(7.48 \text{ gal/ft}^3)$

Number of gallons $= (3.14)(30.5 \text{ ft})(30.5 \text{ ft})(16.3 \text{ ft})(7.48 \text{ gal/ft}^3) = 356{,}138 \text{ gal}$

Next, convert the number of gallons in the aeration tank to mil gal.

$\text{Number of mil gal} = \dfrac{356{,}138 \text{ gal}}{1{,}000{,}000/\text{mil}} = 0.356138 \text{ mil gal}$

Next, convert the effluent flow in gpm to mgd.

$\text{Number of mgd} = \dfrac{(1{,}675 \text{ gpm})(1{,}440 \text{ min/day})}{1{,}000{,}000/\text{mil}} = 2.412 \text{ mgd}$

Next, write the equation.

$$\text{F/M} = \frac{(BOD_5, \text{ mg/L})(\text{Flow, mgd})}{(\text{mg/L MLVSS})(\text{Volume of tank, mil gal})}$$

Substitute values and solve.

$$\text{F/M} = \frac{(345 \text{ mg/L } BOD_5)(2.412 \text{ mgd})}{(2{,}240 \text{ mg/L MLVSS})(0.356138 \text{ mil gal})} = \textbf{1.04 F/M ratio}$$

170. Given the following data on an aeration tank, calculate the current food-to-microorganism (F/M) ratio.

Primary effluent flow $= 2.43 \text{ ft}^3/\text{s}$
Aeration tank $= 172$ ft by 38.5 ft with a depth of 11.8 ft
MLVSS $= 2{,}014$ mg/L
$BOD_5 = 339$ mg/L

First, calculate the aeration tank volume in gallons.

Volume, gal = (Length, ft)(Width, ft)(Depth, ft)(7.48 gal/ft^3)

Substitute values and solve.

Volume, gal = (172 ft)(38.5 ft)(11.8 ft)(7.48 gal/ft^3) = 584,484 gal

Now, convert the volume of wastewater in the aeration tank to mil gal.

Number of mil gal = 584,484 gal/1,000,000/mil = 0.5845 mil gal

Next, convert the primary effluent flow from gpm to mgd.

Number of mgd = (2.43 ft^3/s)(0.6463 mgd/ft^3/s) = 1.57 mgd

Next, write the equation.

$$\text{F/M} = \frac{(\text{BOD}_5, \text{mg/L})(\text{Flow, mgd})(8.34 \text{ lb/gal})}{(\text{mg/L MLVSS})(\text{Volume in tank, mil gal})(8.34 \text{ lb/gal})}$$

The 8.34 lb/gal cancels, leaving the following equation.

$$\text{F/M} = \frac{(\text{BOD}_5, \text{mg/L})(\text{Flow, mgd})}{(\text{mg/L MLVSS})(\text{Volume of tank, mil gal})}$$

Substitute values and solve.

$$\text{F/M} = \frac{(339 \text{ mg/L BOD}_5)(1.57 \text{ mgd})}{(2,014 \text{ mg/L MLVSS})(0.5845 \text{ mil gal})} = \textbf{0.45 F/M ratio}$$

WASTE RATES USING FOOD/MICROORGANISM RATIO CALCULATIONS

These calculations are used in determining the wasting rate from aeration tanks. Because these calculations concentrate on organics, it is important to also use the mean cell residence time determinations, as it also includes inorganics. In addition, since the F/M ratio varies from day to day, the seven-day moving average should be used when estimating wasting rates with the F/M ratio calculation.

171. **Calculate the required waste rate from an aeration tank in mgd and gpm, given the following data:**

Aeration tank volume = 0.895 mil gal
Desired COD lb/MLVSS lb = 0.185
Primary effluent flow = 2.83 mgd
Primary effluent COD = 131 mg/L
Mixed liquor volatile suspended solids (MLVSS) = 3,391 mg/L
Waste volatile solids (WVS) concentration = 4,615 mg/L

First, find the existing MLVSS in pounds.

Equation: **Existing MLVSS, lb = (MLVSS, mg/L)(Aeration tank, mil gal)(8.34 lb/gal)**

Existing MLVSS, lb = (3,391 mg/L)(0.895 mil gal)(8.34 lb/gal) = 25,311 lb MLVSS

Next, determine the desired MLVSS in pounds.

Equation: $\textbf{Desired MLVSS, lb} = \dfrac{\textbf{(Primary effluent COD, mg/L)(mgd)(8.34 lb/gal)}}{\textbf{Desired COD lb/MLVSS lb}}$

$\text{Desired MLVSS, lb} = \dfrac{(131 \text{ mg/L})(2.83 \text{ mgd})(8.34 \text{ lb/gal})}{0.185 \text{ COD lb/MLVSS lb}} = 16{,}713 \text{ lb MLVSS}$

Next, subtract the existing MLVSS from the desired MLVSS to find the waste in pounds.

Waste, lb = 25,311 lb − 16,713 lb = 8,598 lb

Next, calculate the waste rate in mgd.

Equation: $\textbf{Waste rate, mgd} = \dfrac{\textbf{Waste, lb}}{\textbf{(WVS concentration, mg/L)(8.34 lb/gal)}}$

$\text{Waste rate, mgd} = \dfrac{8{,}598 \text{ lb}}{(4{,}615 \text{ mg/L})(8.34 \text{ lb/gal})} = 0.223388 \text{ mgd, round to } \textbf{0.223 mgd}$

Lastly, calculate the waste rate in gpm.

$\text{Waste rate, gpm} = \dfrac{(0.223388 \text{ mgd})(1{,}000{,}000 \text{ gpd/mgd})}{1{,}440 \text{ min/day}} = 155.13 \text{ gpm, round to } \textbf{155 gpm}$

172. **Calculate the required waste rate from an aeration tank in mgd and gpm, given the following data:**

Aeration tank volume = 0.798 mil gal
Desired COD lb/MLVSS lb = 0.180
Primary effluent flow = 2.60 mgd
Primary effluent COD = 129 mg/L
Mixed liquor volatile suspended solids (MLVSS) = 3,684 mg/L
Waste volatile solids (WVS) concentration = 4,378 mg/L

First, find the existing MLVSS in pounds.

Equation: **Existing MLVSS, lb = (MLVSS, mg/L)(Aeration tank, mil gal)(8.34 lb/gal)**

Existing MLVSS, lb = (3,684 mg/L)(0.798 mil gal)(8.34 lb/gal) = 24,518 lb MLVSS

Next, determine the desired MLVSS in pounds.

Equation: $$\textbf{Desired MLVSS, lb} = \frac{\textbf{(Primary effluent COD, mg/L)(mgd)(8.34 lb/gal)}}{\textbf{Desired COD lb/MLVSS lb}}$$

$$\text{Desired MLVSS, lb} = \frac{(129 \text{ mg/L})(2.60 \text{ mgd})(8.34 \text{ lb/gal})}{0.180 \text{ COD lb/MLVSS lb}} = 15,540 \text{ lb MLVSS}$$

Next, subtract the existing MLVSS from the desired MLVSS to find the waste in pounds.

Waste, lb = 24,518 lb − 15,540 lb = 8,978 lb

Next, calculate the waste rate in mgd.

Equation: $$\textbf{Waste rate, mgd} = \frac{\textbf{Waste, lb}}{\textbf{(WVS concentration, mg/L)(8.34 lb/gal)}}$$

$$\text{Waste rate, mgd} = \frac{8,978 \text{ lb}}{(4,378 \text{ mg/L})(8.34 \text{ lb/gal})} = 0.245888 \text{ mgd, round to } \textbf{0.246 mgd}$$

Lastly, calculate the waste rate in gpm.

$$\text{Waste rate, gpm} = \frac{(0.245888 \text{ mgd})(1,000,000 \text{ gpd/mgd})}{1,440 \text{ min/day}} = 170.756 \text{ gpm, round to } \textbf{171 gpm}$$

SEED SLUDGE PROBLEMS

This calculation is required for determining how much seed sludge in gallons to use for starting a new digester.

173. Given the following data, determine the seed sludge required in gallons:

Digester radius = 25.3 ft
Liquid level in digester = 16.5 ft
Requires 17.8% seed sludge

First, determine the number of gallons in the digester.

Volume, gal = π(radius)^2(Depth, ft)(7.48 gal/ft^3)

Volume, gal = (3.14)(25.3 ft)(25.3 ft)(16.5 ft)(7.48 gal/ft^3) = 248,060 gal

Next, use the following equation.

$$\textbf{Seed sludge, gal} = \frac{(\text{Capacity of digester})(\text{Percent seed sludge required})}{100\%}$$

$$\text{Seed sludge, gal} = \frac{(248,060 \text{ gallons})(17.8\%)}{100\%}$$

Seed sludge, gal = 44,154.68 gal, round to **44,200 gal of Seed sludge**

174. A digester with a diameter of 41.2 ft and a sludge level of 14.2 ft has a seed sludge requirement of 16.3% of the digester capacity. How many gallons of seed sludge will be needed?

First, determine the number of gallons in the digester.

Number of gallons = (0.785)(Diameter)^2(Height, ft)(7.48 gal/ft^3)

Number of gallons = (0.785)(41.2 ft)(41.2 ft)(14.2 ft)(7.48 gal/ft^3) = 141,532 gal

Next, use the following equation.

$$\text{Seed sludge, gal} = \frac{(\text{Capacity of digester})(\text{Percent seed sludge required})}{100\%}$$

$$\text{Seed sludge, gal} = \frac{(141,532 \text{ gallons})(16.3\%)}{100\%}$$

Seed sludge, gal = 23,069.72 gal, round to **23,100 gal of seed sludge**

GRAVITY THICKENER SOLIDS LOADING PROBLEMS

Gravity thickeners use large tanks that separate suspended solid and mineral matter from the liquid by gravity. The gravity thickener concentrates the sludge to reduce the load on processes that follow (conditioning, dewatering, and digestion) and produces a clear liquid, which is decanted. Flocculants are used to speed up the settling process. Operators can calculate the solids loading in lb/d/ft^2 or the hydraulic loading in gal/day/ft^2. The hydraulic loading calculation is used by operators to determine if the process is being overloaded or underloaded. See Figures E-4, E-5, and E-6 in Appendix E for three types of wastewater plants using the thickening process.

175. **Given the following data, determine the hydraulic loading on a gravity thickener in lb/d/ft^2:**

Gravity thickener radius = 30.2 ft
Influent flow = 36.5 gpm
Percent solids = 4.08%

First, determine the area of the gravity thickener.

Know: Area = $\pi(\text{radius})^2$ or πr^2

Area = (3.14)(30.2 ft)(30.2 ft) = 2,863.81 ft^2

Equation: **Hydraulic loading, gal/d/ft^2** $= \dfrac{(\text{Flow, gpm})(1,440 \text{ min/day})(\text{Percent solids})}{(\text{Gravity thickener area})(100\%)}$

Substitute values and solve.

$$\text{Hydraulic loading, gal/d/ft}^2 = \frac{(36.5 \text{ gpm})(1,440 \text{ min/day})(4.08\%)}{(2,863.81 \text{ ft}^2)(100\%)}$$

Hydraulic loading, gal/d/ft^2 = **0.749 gal/d/ft^2**

176. **What are the solids loading on the gravity thickener, if the percent solids is 4.46%, the influent flow is 31.3 gpm, and the gravity thickener has a radius of 29.8 ft?**

First, convert the gpm to gpd.

Number of gpd = (31.3 gpm)(1,440 min/day) = 45,072 gpd

Know: Area of gravity thickener = πr^2 where $\pi = 3.14$

Equation: **Solids loading, lb/d/ft^2** $= \dfrac{\text{(Flow, gpd)(8.34 lb/gal)(Percent solids)}}{\text{(Gravity thickener area)(100\%)}}$

Substitute values and solve.

Solids loading, lb/d/ft^2 $= \dfrac{(45,072 \text{ gpd})(8.34 \text{ lb/gal})(4.46\%)}{(3.14)(29.8 \text{ ft})(29.8 \text{ ft})(100\%)}$

Solids loading, lb/d/ft^2 = **6.01 lb/d/ft^2**

DISSOLVED AIR FLOTATION: THICKENER SOLIDS LOADING PROBLEMS

The dissolved air flotation technique is used to thicken sludge. These types of calculations are used for evaluating process control.

177. **Given the following data, determine the solids loading in lb/d/ft^2 on a dissolved air flotation (DAF) thickener unit:**

DAF unit length = **85.25 ft**
DAF unit width = **32.4 ft**
Waste-activated sludge (WAS) = **7,280 mg/L**
Sludge flow = **136 gpm**
Sludge specific gravity = **1.03 lb/gal**

First, determine the area of the DAF.

DAF area, ft^2 = (Length, ft)(Width, ft)

DAF area, ft^2 = (85.25 ft)(32.4 ft) = 2,762.1 ft^2

Next, convert gpm to mgd.

$$\text{Number of gpd} = \frac{(136 \text{ gpm})(1,440 \text{ min/day})}{1,000,000/\text{mil}} = 0.19584 \text{ mgd}$$

Next, determine the lb/gal for the sludge using the specific gravity.

Sludge, lb/gal = (8.34 lb/gal)(1.03 sp gr) = 8.59 lb/gal

Equation: **Solids loading, lb/d/ft^2** $= \dfrac{\textbf{(WAS, mg/L)(mgd)(Sludge, lb/gal)}}{\textbf{DAF area, ft}^2}$

$$\text{Solids loading, lb/d/ft}^2 = \frac{(7,280 \text{ mg/L, WAS})(0.19584 \text{ mgd})(8.59 \text{ lb/gal, Sludge})}{2,762.1 \text{ ft}^2 \text{ DAF}}$$

Solids loading, lb/d/ft^2 = **4.43 lb/d/ft^2**

178. **What are the solids loading for a dissolved air flotation (DAF) unit in lb/hr/ft² that is 48.9 ft by 25.2 ft, where sludge flow averages 95 gpm, with a waste-activated sludge (WAS) concentration of 6,085 mg/L, and the sludge has a specific gravity of 1.05?**

First, determine the area of the DAF unit in ft².

DAF area, ft² = (48.9 ft)(25.2 ft) = 1,232.28 ft²

Next, convert gpm to mgd.

$$\text{Number of mgd} = \frac{(95 \text{ gpm})(1,440 \text{ min/day})}{1,000,000/\text{mil}} = 0.1368 \text{ mgd}$$

Next, calculate the weight of the sludge in lb/gal.

Sludge, lb/gal = (8.34 lb/gal)(1.05 sp gr) = 8.757 lb/gal

Next, calculate the solids loading.

Equation: **Solids loading, lb/d/ft²** $= \dfrac{(\textbf{WAS, mg/L})(\textbf{mgd})(\textbf{Sludge, lb/gal})}{\textbf{DAF area, ft}^2}$

$$\text{Solids loading, lb/d/ft}^2 = \frac{(6,085 \text{ mg/L, WAS})(0.1368 \text{ mgd})(8.757 \text{ lb/gal, Sludge})}{(1,232.28 \text{ ft}^2 \text{ DAF})(24 \text{ hr/day})}$$

Solids loading, lb/d/ft² = 0.246 lb/hr/ft², round to 0.25 **lb/hr/ft²**

DISSOLVED AIR FLOTATION: AIR-TO-SOLIDS RATIO CALCULATIONS

Air-to-solids ratio calculations are used to determine the efficiency of the process, as the air flotation thickener and the solids in the system must be in balance. Typically the ratio ranges from 0.01 to 0.1.

179. **What is the air-to-solids ratio for a dissolved air flotation (DAF) unit that has an air flow rate equal to 8.75 ft³/min, a solids concentration of 0.78%, and a flow of 0.204 mgd?**

Know: Air $= 0.0807$ lb/ft^3 at standard temperature, pressure, and average composition

Equation: **Air-to-solids ratio** $= \dfrac{(\text{Air flow, ft}^3/\text{min})(\text{Air, lb/ft}^3)}{(\text{gpm})(\text{Percent solids}/100\%)(8.34 \text{ lb/gal})}$

Because equation requires gpm, first convert mgd to gpm.

Number of gpm $= \dfrac{(0.204 \text{ mgd})(1,000,000/\text{mil})}{1,440 \text{ min}/\text{day}} = 141.667$ gpm

Now, using the air-to-solids equation, substitute values and solve.

Air-to-solids ratio $= \dfrac{(8.75 \text{ ft}^3/\text{min})(0.0807 \text{ lb/ft}^3)}{(141.667 \text{ gpm})(0.78\%/100\%)(8.34 \text{ lb/gal})} = $ **0.077 Air-to-solids ratio**

180. Given the following data, determine the air-to-solids ratio for a DAF unit.

DAF influent flow $= 0.205$ mgd
Air flow $= 7.65$ ft^3/min
Solids concentration $= 0.786\%$
Solids specific gravity $= 1.05$

Know: Air $= 0.0807$ lb/ft^3 at standard temperature, pressure, and average composition

Equation: **Air-to-solids ratio** $= \dfrac{(\text{Air flow, ft}^3/\text{min})(\text{Air, lb/ft}^3)}{(\text{gpm})(\text{Percent solids}/100\%)(\text{Solids, lb/gal})}$

Since equation requires gpm, first convert mgd to gpm.

Number of gpm $= \dfrac{(0.205 \text{ mgd})(1,000,000/\text{mil})}{1,440 \text{ min}/\text{day}} = 142.36$ gpm

Then, the number of lb/gal is required because the solids weigh more than water.

Solids, lb/gal $= (8.34 \text{ lb/gal})(1.05 \text{ sp gr}) = 8.757$ lb/gal

Now, using the air-to-solids equation, substitute values and solve.

Air-to-solids ratio $= \dfrac{(7.65 \text{ ft}^3/\text{min})(0.0807 \text{ lb/ft}^3)}{(142.36 \text{ gpm})(0.786\%/100\%)(8.757 \text{ lb/gal})} = $ **0.063 Air-to-solids ratio**

DISSOLVED AIR FLOTATION: AIR RATE FLOW CALCULATIONS

Operators use air rate flow calculations for evaluating process control.

181. **How many lb/hr of air does a DAF unit receive, if the unit receives air at an average rate of 0.282 m³/min?**

First, convert m³/min to ft³/min.

Air flow, ft³/min $=$ (0.282 m³/min)(35.3 ft³/m³) $=$ 9.9546 ft³/min

Know: Air $=$ 0.0807 lb/ft³ at standard temperature, pressure, and average composition

Equation: **Air, lb/hr $=$ (Air flow, ft³/min)(60 min/hr)(0.0807 lb/ft³, Air)**

Substitute values and solve.

Air, lb/hr $=$ (9.9546 ft³/min)(60 min/hr)(0.0807 lb/ft³)

Air, lb/hr $=$ **48.2 lb/hr of Air**

182. **If a DAF unit receives air at an average rate of 0.198 m³/min, how many lb/day of air does it receive?**

First, convert m³/min to ft³/min.

Air flow, ft³/min $=$ (0.198 m³/min)(35.3 ft³/m³) $=$ 6.9894 ft³/min

Know: Air $=$ 0.0807 lb/ft³ at standard temperature, pressure, and average composition

Equation: **Air, lb/day $=$ (Air flow, ft³/min)(1,440 min/day)(0.0807 lb/ft³, Air)**

Substitute values and solve.

Air, lb/day $= (6.9894 \text{ ft}^3/\text{min})(1{,}440 \text{ min/day})(0.0807 \text{ lb/ft}^3)$

Air, lb/day $= 812.22$ lb/day, round to **812 lb/day of Air**

CENTRIFUGE THICKENING PROBLEMS

Centrifuges are used to dewater sludge usually after applying gravity thickening. They apply forces that are a thousand times greater than gravity. Polymers may be applied to the influent of the centrifuge to facilitate solids thickening.

183. Given the following data, determine the feed time for a centrifuge in minutes:

Capacity of basket centrifuge = 28.5 ft³
Influent sludge flow = 78 gpm
Influent sludge solids concentration = 6,880 mg/L
Average solids concentration in the basket = 8.15%

First, convert the influent sludge solids concentration to percent.

Know: 1% = 10,000 ppm or mg/L

Percent sludge solids concentration $= \dfrac{6{,}880 \text{ mg/L}}{10{,}000 \text{ mg/L}/1\%} = 0.688\%$

Equation: **Feed time, min =**

$$\frac{(\textbf{Centrifuge capacity})(\textbf{7.48 gal/ft}^3)(\textbf{8.34 lb/gal})(\textbf{Percent solids conc.})}{(\textbf{Sludge flow, gpm})(\textbf{8.34 lb/gal})(\textbf{Sludge solids conc.})}$$

Feed time, min $= \dfrac{(28.5 \text{ ft}^3)(7.48 \text{ gal/ft}^3)(8.34 \text{ lb/gal})(8.15\%/100\%)}{(78 \text{ gpm})(8.34 \text{ lb/gal})(0.688\%/100\%)}$

Feed time, min = 32.38 min, round to **32 min**

184. Given the following data, what is the feed time in minutes for a basket centrifuge thickener?

Basket centrifuge thickener capacity = 24.9 ft³
Sludge flow rate = 39,875 gpd
Solids concentration = 7,160 mg/L
Percent solids = 5.17%

Know: 1% = 10,000 ppm or mg/L

First, convert the solids concentration in mg/L to percent.

$$\text{Percent solids} = \frac{7{,}160 \text{ mg/L}}{10{,}000 \text{ mg/L/1\%}} = 0.716\%$$

Next, convert gpd to gpm.

$$\text{Number of gpm} = \frac{39{,}875 \text{ gpd}}{1{,}440 \text{ min/day}} = 27.69 \text{ gpm}$$

Now, calculate the feed time in minutes.

Equation: **Feed time, min** = $\dfrac{(\text{Capacity, ft}^3)(\text{Solids, \%/100\%})(7.48 \text{ gal/ft}^3)(8.34 \text{ lb/gal})}{(\text{Flow, gpm})(\text{Solids concentration, \%/100\%})(8.34 \text{ lb/gal})}$

Simplify equation by canceling out the 8.34 lb/gal and the 100%.

Feed time, min = $\dfrac{(\text{Capacity, ft}^3)(\text{Solids, \%})(7.48 \text{ gal/ft}^3)}{(\text{Flow, gpm})(\text{Solids concentration, \%})}$

$$\text{Feed time, min} = \frac{(24.9 \text{ ft}^3)(5.17\%)(7.48 \text{ gal/ft}^3)}{(27.69 \text{ gpm})(0.716\%)} = 48.57 \text{ min, round to } \textbf{48.6 min}$$

SAND DRYING BED PROBLEMS

By knowing how thick the sludge was when applied and later measuring the thickness of the dried sludge, an operator can use these calculations to determine the efficiency of the drying bed process.

185. Given the following data, what is the total amount of sludge in gallons applied to these drying beds?

Drying bed 1 = 227 ft by 60.5 ft
Drying bed 2 = 196 ft by 56.3 ft
Drying bed 1 had 2.15 inches of sludge applied to it
Drying bed 2 had 3.42 inches of sludge applied to it

First, convert the inches of sludge applied to each drying bed to feet.

Drying bed 1, ft = 2.15 in./12 in./ft = 0.1792 ft

Drying bed 2, ft = 3.42 in./12 in./ft = 0.285 ft

Next, determine the volume in ft^3 sent to drying bed 1 and 2.

Drying bed 1, ft^3 = (227 ft)(60.5 ft)(0.1792 ft) = 2,461 ft^3

Drying bed 2, ft^3 = (196 ft)(56.3 ft)(0.285 ft) = 3,145 ft^3

Lastly, calculate the volume in gallons sent to both sand drying beds.

Number of gal = (2,461 ft^3 + 3,145 ft^3)(7.48 gal/ft^3)

Number of gal = (5,606 ft^3)(7.48 gal/ft^3) = 41,932.88 gal, round to **41,900 gal**

186. **Given the following data, what is the total amount of sludge in gallons from the digester applied to drying beds 1 and 2?**

Drying bed 1 = 246 ft by 56.7 ft
Drying bed 2 = 245.8 ft by 56.75 ft
Digester diameter = 50.4 ft in diameter
Digester is drawn down 3.06 ft

Equation: **Volume, gal = (0.785)(Diameter, ft)^2(Depth, ft)(7.48 gal/ft^3)**

Volume, gal = (0.785)(50.4 ft)(50.4 ft)(3.06 ft)(7.48 gal/ft^3) = 45,641 gal, round to **45,600 gal**

187. **Given the following data, calculate the average gallons applied to a drying bed for each cycle period and the solids loading in lb/yr/ft²:**

Drying bed = 278 ft long and 47.1 ft wide.
Average sludge application per cycle = 4.01 inches
Average percent solids = 4.76%
Drying and removal cycle on average = 20.3 days

First, convert 4.01 inches to feet.

Number of feet = 4.01 in./12 in./ft = 0.334 ft

Next, determine the volume in ft³ sent to the drying bed.

Volume, ft³ = (278 ft)(47.1 ft)(0.334 ft) = 4,373.33 ft³

Next, calculate the average volume in gallons sent to the sand drying beds.

Number of gal = (4,373.33 ft³)(7.48 gal/ft³) = 32,712.51 gal, round to **32,700 gal**

Next, calculate the number of lb.

Number of lb = (4,373.33 ft³)(7.48 gal/ft³)(8.34 lb/gal) = 272,822 lb

Lastly, calculate the solids loading rate.

Equation: Solids loading rate, lb/yr/ft² = $\dfrac{\dfrac{(\text{lb})(365\text{ days})(\text{Percent solids})}{(\text{Drying cycle})(\text{yr})(100\%)}}{\text{Drying bed area, ft}^2}$

Substitute values and solve.

Solids loading rate, lb/yr/ft² = $\dfrac{\dfrac{(272,822\text{ lb})(365\text{ days})(4.76\%\text{ solids})}{(20.3\text{ days})(\text{yr})(100\%)}}{(278\text{ ft})(47.1\text{ ft})}$

Solids loading rate, lb/yr/ft² = $\dfrac{(13,439.51\text{ lb/day})(365\text{ days/yr})(0.0476)}{13,093.8\text{ ft}^2}$

Solids loading rate, lb/yr/ft² = **17.8 lb/yr/ft²**

DEWATERING CALCULATIONS

This section contains several types of dewatering problems. The more water that is removed from sludge, the less cost associated with further processing or disposal. The problems are important to the operator because they are helpful in evaluating process control or in informing the operator of process efficiency. See Figure E-12 in Appendix E for one type of sludge process using dewatering.

188. Given the following data, determine the surface area of a vacuum filter dewatering unit:

Applied digested sludge = 136 gpm
Solids concentration = 5.3%
Filter loading = 8.45 lb/hr/ft²
Specific gravity of the digested sludge = 1.04

Equation: **Filter loading, lb/hr/ft²** =

$$\frac{(\text{Sludge, gpm})(60 \text{ min/hr})(8.34 \text{ lb/gal})(\text{Sludge sp gr})(\% \text{ solids})}{\text{Vacuum surface filter area, ft}^2}$$

Rearrange equation to solve for vacuum surface filter area.

$$\text{Vacuum filter area, ft}^2 = \frac{(\text{Sludge, gpm})(60 \text{ min/hr})(8.34 \text{ lb/gal})(\text{Sludge sp gr})(\text{Percent solids})}{\text{Filter loading, lb/hr/ft}^2}$$

Substitute values and solve.

$$\text{Vacuum filter area, ft}^2 = \frac{(136 \text{ gpm})(60 \text{ min/hr})(8.34 \text{ lb/gal})(1.04 \text{ sp gr})(5.3\%/100\%)}{8.45 \text{ lb/hr/ft}^2}$$

Vacuum filter area, ft² = 443.92 ft², round to **440 ft²**

189. A vacuum filter has a wet cake flow of 4,375 lb/hr and a filter that is 12.1 ft by 25.3 ft. Calculate the filter yield in lb/hr/ft², if the percent solids are 13.6%.

First, determine the area of the filter.

Area of filter, ft² = (12.1 ft)(25.3 ft) = 306.13 ft²

Equation: **Filter yield, lb/hr/ft²** $= \dfrac{\text{(Wet cake flow, lb/hr)(Percent solids/100\%)}}{\text{Area, ft}^2}$

Substitute values and solve.

Filter yield, lb/hr/ft² $= \dfrac{(4,375 \text{ lb/hr})(13.6\%/100\%)}{306.13 \text{ ft}^2} = 1.944$ lb/hr/ft², round to **1.94 lb/hr/ft²**

SETTLEABLE SOLIDS CALCULATIONS

These tests are performed on samples from either the clarifier's influent or effluent or from a sedimentation tank. They are used to determine the percent and thus the efficiency of settleable solids. Calculations based on these tests follow.

190. A 2,000.0 mL of activated sludge was collected in a graduated cylinder. What is the percent of settleable solids, if after exactly 30 minutes the sludge solids that settled totaled 354 mL?

Equation: **Percent settleable solids** $= \dfrac{\text{(Settled sludge, mL)(100\%)}}{\text{Sample size, mL}}$

Substitute values and solve.

Percent settleable solids $= \dfrac{(354 \text{ mL})(100\%)}{2,000.0 \text{ mL}} = $ **17.7% Settled solids**

191. Given the following data, calculate the percent settleable solids.

Activated sludge sample = 0.427 gallon poured into a large graduated cylinder
Settling time is exactly = 30 minutes
Sludge solids in graduated cylinder = 319 mL

First, convert the sample in gallons to mL.

Number of mL = (0.427 gal)(3,785 mL/gal) = 1,616.195

Equation: **Percent settleable solids** $= \dfrac{(\text{Settled sludge, mL})(100\%)}{\text{Sample size, mL}}$

Substitute values and solve.

Percent settleable solids $= \dfrac{(319 \text{ mL})(100\%)}{1,616.195 \text{ mL}} = 19.74\%$, round to **19.7% Settled solids**

COMPOSTING CALCULATIONS

Composting is an aerobic biological process. This process decomposes organic matter to a stable end product. The optimum moisture content for composting ranges from 50 to 60 percent water. Several different composting calculations follow.

192. Given the following data for blending compost (BC) with wood chips, calculate the percent of the blended compost:

Bulk density of sludge $= 1,695$ lb/yd^3
Sludge volume $= 14.6$ yd^3
Sludge solids content $= 18.7\%$
Density of wood chips $= 595$ lb/yd^3
Wood chip solids $= 54.2\%$
Mix ratio (MR) of wood chips to sludge $= 3$ to 1

Equation: **Percent solids BC $=$**

$$\frac{[(\text{Sludge, yd}^3)(\text{lb/yd}^3)(\%\text{ solids}) + (\text{Sludge, yd}^3)(\text{MR})(\text{lb/yd}^3)(\%\text{ solids, chips})](100\%)}{(\text{Sludge, yd}^3)(\text{lb/yd}^3) + (\text{Sludge, yd}^3)(\text{Mix ratio})(\text{Wood chips, lb/yd}^3)}$$

Substitute values and solve.

Percent solids BC $=$

$$\frac{[(14.6\text{ yd}^3)(1,695\text{ lb/yd}^3)(18.7\%/100\%) + (14.6\text{ yd}^3)(3)(595\text{ lb/yd}^3)(54.2\%/100\%)](100\%)}{(14.6\text{ yd}^3)(1,695\text{ lb/yd}^3) + (14.6\text{ yd}^3)(3)(595\text{ lb/yd}^3)}$$

$$\text{Percent solids BC} = \frac{(4,627.689 + 14,125.062)(100\%)}{24,747 + 26,061} = \frac{(18,752.751)(100\%)}{50,808}$$

Percent solids BC $= 36.91\%$, round to **37.0% Solids in blended compost**

193. What is the percent moisture content of a composting blend, given the following data?

5,150 lb of sludge was added and mixed with 4,520 lb of compost
Compost $= 63.2\%$ solids
Added sludge $= 19.8\%$ solids

Because percent moisture needs to be solved, first determine the percent moisture of both the compost and sludge.

Compost percent moisture $= 100\% - 63.2\%$ solids $= 36.8\%$ moisture content

Sludge percent moisture $= 100\% - 19.8\%$ solids $= 80.2\%$ moisture content

Equation: **Mixture's % moisture =**

$$\frac{[(\text{Sludge, lb})(\% \text{ moisture}) + (\text{Compost, lb})(\% \text{ moisture})]100\%}{\text{Sludge, lb} + \text{Compost, lb}}$$

Substitute values and solve.

$$\text{Mixture's percent moisture} = \frac{[(5{,}150 \text{ lb})(80.2\%/100\%) + (4{,}520 \text{ lb})(36.8\%/100\%)]100\%}{5{,}150 \text{ lb} + 4{,}520 \text{ lb}}$$

$$\text{Mixture's percent moisture} = \frac{(4{,}130.3 \text{ lb} + 1{,}663.36 \text{ lb})100\%}{9{,}670 \text{ lb}} = \mathbf{59.9\% \text{ moisture content}}$$

194. **Given the following data, calculate the amount of compost in lb/day that needs to be blended with a dewatered sludge to make a mixture that has a moisture content of 50.0%.**

Dewatered digester primary sludge = 8,625 lb/day
Dewatered sludge solids = 33.6%
Compost solids = 67.8%

First, determine the moisture content of the dewatered sludge and the compost.

Dewatered sludge moisture = 100% − 33.6% = 66.4%

Compost percent moisture = 100% − 67.8% = 32.2%

Equation: **Mixture's % moisture =**

$$\frac{[(\text{Sludge, lb/day})(\% \text{ moisture}) + (\text{Compost, lb/day})(\% \text{ moisture})]100\%}{\text{Sludge, lb} + \text{Compost, lb}}$$

Substitute values and solve.

$$50.0\% \text{ moisture content} = \frac{[(8{,}625 \text{ lb/day})(66.4\%/100\%) + (x \text{ lb/day})(32.2\%/100\%)]100\%}{8{,}625 \text{ lb/day} + x \text{ lb/day}}$$

Solve for x.

Divide both sides of the equation by 100%.

$$0.500 = \frac{(8,625 \text{ lb/day})(66.4\%/100\%) + (x \text{ lb/day})(32.2\%/100\%)}{8,625 \text{ lb/day} + x \text{ lb/day}}$$

Simplify terms in the numerator.

$$0.500 = \frac{5,727 \text{ lb/day} + 0.322x}{8,625 \text{ lb/day} + x \text{ lb/day}}$$

Multiply both sides of the equation by (8,625 lb/day + x lb/day).

$$0.500(8,625 \text{ lb/day} + x \text{ lb/day}) = 5,727 \text{ lb/day} + 0.322x$$

Multiply terms on the left side of the equation.

$$4,312.5 \text{ lb/day} + 0.500x = 5,727 \text{ lb/day} + 0.322x$$

Subtract $0.322x$ and 4,312.5 lb/day from both sides of the equation.

$$0.500x - 0.322x = 5,727 \text{ lb/day} - 4,312.5 \text{ lb/day}$$

Simplify both sides of the equation.

$$0.178x = 1,414.5 \text{ lb/day}$$

x = 7,946.63 lb/day, round to **7,950 lb/day of Compost required**

195. **Calculate the amount of compost in lb/day that needs to be blended with a dewatered sludge to make a mixture that has a moisture content of 55.0%, if the dewatered digester primary sludge is 6,880 lb/day, the dewatered sludge solids is 30.9%, and the compost solids moisture content is 31.6%.**

First, determine the moisture content of the dewatered sludge.

Dewatered sludge moisture = 100% − 30.9% = 69.1%

Compost percent moisture in this case is given at 31.6%.

Equation: **Mixture's % moisture =**

$$\frac{[(\text{Sludge, lb/day})(\% \text{ moisture}) + (\text{Compost, lb/day})(\% \text{ moisture})]100\%}{\text{Sludge, lb/day} + \text{Compost, lb/day}}$$

Substitute values and solve.

$$55.0\% \text{ moisture content} = \frac{[(6{,}880 \text{ lb/day})(69.1\%/100\%) + (x \text{ lb/day})(31.6\%/100\%)]100\%}{6{,}880 \text{ lb/day} + x \text{ lb/day}}$$

Solve for x.

Divide both sides of the equation by 100%.

$$0.550 = \frac{(6{,}880 \text{ lb/day})(69.1\%/100\%) + (x \text{ lb/day})(31.6\%/100\%)}{6{,}880 \text{ lb/day} + x \text{ lb/day}}$$

Simplify terms in the numerator.

$$0.550 = \frac{4{,}754.08 \text{ lb/day} + 0.316x}{6{,}880 \text{ lb/day} + x \text{ lb/day}}$$

Multiply both sides of the equation by (6,880 lb/day + x lb/day).

$$0.550(6{,}880 \text{ lb/day} + x \text{ lb/day}) = 4{,}754.08 \text{ lb/day} + 0.316x \text{ lb/day}$$

Multiply terms on the left side of the equation.

$$3{,}784 \text{ lb/day} + 0.550x \text{ lb/day} = 4{,}754.08 \text{ lb/day} + 0.316x \text{ lb/day}$$

Subtract $0.316x$ and 3,784 lb/day from both sides of the equation.

$$0.550x \text{ lb/day} - 0.316x \text{ lb/day} = 4{,}754.08 \text{ lb/day} - 3{,}784 \text{ lb/day}$$

Simplify both sides of the equation.

$$0.234x \text{ lb/day} = 970.08 \text{ lb/day}$$

$x = 4{,}145.64$ lb/day, round to **4,150 lb/day of Compost required**

196. Calculate the percent moisture content when 4,320 lb of dewatered biosolids and 2,170 lb of compost are mixed each day. The dewatered biosolids (DB) have a solids content of 29.5%, and the compost has a moisture content of 29.5%.

First, calculate the moisture content in the biosolids.

DB percent moisture content = $100\% - 29.5\% = 70.5\%$

Next, calculate the percent moisture in mixture.

Equation: **Percent moisture in mixture** =

$$\frac{[(DB, lb/day)(\text{Percent moisture DB}) + (\text{Compost lb/day})(\text{Percent moisture compost})]100\%}{DB, lb/day + \text{Compost, lb/day}}$$

Percent moisture in mixture = $\dfrac{[(4,320 \text{ lb/day})(70.5\%/100\%) + (2,170 \text{ lb/day})(29.5\%)]100\%}{4,320 \text{ lb/day} + 2,170, \text{lb/day}}$

Percent moisture in mixture = $\dfrac{(3,045.6 \text{ lb/day} + 640.15 \text{ lb/day})(100\%)}{6,490 \text{ lb/day}}$

Percent moisture in mixture = 56.79%, round to **56.8% Moisture content of mixture**

197. Given the following, calculate this particular system's composting cycle time for this set of data:

Site available capacity = **9,070 yd³**
Bulk density of wet sludge = **1,690 lb/yd³**
Bulk density of wet compost = **1,050 lb/yd³**
Sludge solids content = **18.8%**
Bulk density of wet wood chips = **725 lb/yd³**
Dry solids = **28,560 lb/day**
Mix ratio (MR) of wood chips to sludge = **exactly 3 to 1**

Equation: **Cycle time, days** $=$

$$\frac{(\text{Capacity, yd}^3)(\text{Bulk density of wet compost lb/yd}^3)}{\dfrac{\text{Dry solids, lb/day}}{\text{Percent solids}} + \dfrac{(\text{Dry solids, lb/day})}{\text{Percent solids}}\ (\text{MR})\ \dfrac{(\text{Bulk density of wood chips, lb/yd}^3)}{(\text{Bulk density of wet sludge lb/yd}^3)}}$$

Substitute values and solve.

$$\text{Cycle time, days} = \frac{(9{,}070\ \text{yd}^3)(1{,}050\ \text{lb/yd}^3)}{\dfrac{28{,}560\ \text{lb/day}}{18.8\%/100\%} + \dfrac{(28{,}560\ \text{lb/day})}{18.8\%/100\%}\ (3)\ \dfrac{(725\ \text{lb/yd}^3)}{1{,}690\ \text{lb/yd}^3}}$$

Reduce and simplify.

$$\text{Cycle time, days} = \frac{9{,}523{,}500\ \text{lb}}{\dfrac{28{,}560\ \text{lb/day}}{0.188} + \dfrac{(28{,}560\ \text{lb/day})(1.287)}{0.188}}$$

Reduce again.

$$\text{Cycle time, days} = \frac{9{,}523{,}500\ \text{lb}}{151{,}914.89\ \text{lb/day} + 195{,}514.47\ \text{lb/day}}$$

$$\text{Cycle time, days} = \frac{9{,}523{,}500\ \text{lb}}{347{,}429.36\ \text{lb/day}}$$

Cycle time, days $= 27.41$ days, round to **27.4 days**

198. What is the capacity for a compost site in processing dry sludge in lb/day and tons/day, given the following data?

Site capacity = 12,250 yd³
Average compost cycle = 21.4 days
Bulk density of wet sludge = 1,700 lb/yd³
Bulk density of wet compost = 925 lb/yd³
Sludge solids content = 19.5%
Bulk density of wet wood chips = 685 lb/yd³
Mix ratio (MR) of wood chips to sludge = exactly 3.3 to 1

Equation: **Cycle time, days =**

$$\frac{(\text{Capacity, yd}^3)(\text{Bulk density of wet compost lb/yd}^3)}{\dfrac{x\,\text{Dry solids, lb/day}}{\text{Percent solids}} + \dfrac{(x\,\text{Dry solids, lb/day})}{\text{Percent solids}}\,(\text{MR})\,\dfrac{(\text{Bulk density of wood chips, lb/yd}^3)}{(\text{Bulk density of wet sludge lb/yd}^3)}}$$

Substitute values and solve.

$$21.4\,\text{days} = \frac{(12{,}250\,\text{yd}^3)(925\,\text{lb/yd}^3)}{\dfrac{x\,\text{Dry solids, lb/day}}{19.5\%/100\%} + \dfrac{(x\,\text{Dry solids, lb/day})}{19.5\%/100\%}\,(3.3)\,\dfrac{(685\,\text{lb/yd}^3)}{1{,}700\,\text{lb/yd}^3}}$$

Reduce and simplify.

$$21.4\,\text{days} = \frac{11{,}331{,}250\,\text{lb}}{\dfrac{x\,\text{Dry solids, lb/day}}{0.195} + \dfrac{(x\,\text{Dry solids, lb/day})(1.3297)}{0.195}}$$

Reduce again by dividing 0.195 into 1x dry solids and multiply right side of denominator.

$$21.4\,\text{days} = \frac{11{,}331{,}250\,\text{lb}}{5.128x\,\text{lb/day} + 6.819x,\,\text{lb/day}}$$

$$21.4\,\text{days} = \frac{11{,}331{,}250\,\text{lb}}{11.947x\,\text{lb/day}}$$

Solve for x.

$$11.947x\,\text{lb/day} = \frac{11{,}331{,}250\,\text{lb}}{21.4\,\text{days}}$$

$$x = \frac{11{,}331{,}250\,\text{lb}}{(21.4\,\text{days})(11.947\,\text{lb/day})}$$

$x = 44{,}320$ lb/day, round to **44,000 lb/day Dry sludge**

$$\text{Dry sludge, tons/day} = \frac{44{,}320\,\text{lb/day}}{2{,}000\,\text{lb/ton}} = 22.16\ \text{tons/day, round to \textbf{22 tons/day Dry sludge}}$$

BIOSOLIDS DISPOSAL CALCULATIONS

Biosolids disposal is a beneficial and an environmentally safe reuse of the stable waste products in the wastewater treatment process. Biosolids are increasingly being land applied because they no longer can be dumped into the ocean, and landfill space is increasingly in short supply. A number of factors must be considered when applying biosolids to land, such as the amount of nitrogen available for plant growth, nitrogen loading rate, phosphorus, or metals in the biosolids. See Figure E-12 in Appendix E for one type of sludge process using land application.

199. **Determine the number of acres used in the land application of biosolids, given the following data:**

Hydraulic loading rate = 0.48 in./day
Flow = 875,000 gpd

First, determine the number of gallons per acre-inch.

Number of gals/acre-in. $= (43{,}560\ \text{ft}^2/\text{acre})(1\ \text{ft}/12\ \text{in.})(7.48\ \text{gal/ft}^3) = 27{,}152\ \text{gal/acre-in.}$

Equation: **Hydraulic loading rate, in./day** $= \dfrac{\text{Flow, gpd}}{(27{,}152\ \text{gal/acre-in.})(\text{Area, acres})}$

Rearrange the equation and solve for area in acres.

$$\text{Area, acres} = \frac{\text{Flow, gpd}}{(27{,}152\ \text{gal/acre-in.})(\text{Hydraulic loading rate, in./day})}$$

Substitute values and solve.

$$\text{Area, acres} = \frac{875{,}000\ \text{gpd}}{(27{,}152\ \text{gal/acre-in.})(0.48\ \text{in./day})}$$

Area, acres = 67.14 acres, round to **67 acres**

200. **Calculate the flow in gpd for a 98-acre land application site, if the hydraulic loading rate is 0.52.**

Know: 27,152 gal/acre-in. (see previous problem)

Equation: **Hydraulic loading rate, in./day** $= \dfrac{\text{Flow, gpd}}{(27{,}152 \text{ gal/acre-in.})(\text{Area, acres})}$

Rearrange the equation and solve for flow in gpd.

Flow, gpd = (Area, acres)(27,152 gal/acre-in.)(Hydraulic loading rate, in./day)

Substitute values and solve.

Flow, gpd = (98 acres)(27,152 gal/acre-in.)(0.52 in./day)

Flow, gpd = 1,383,665.92 gal, round to **1,400,000 gpd**

201. **What is the plant available nitrogen (PAN) in lb per dry ton, given the following data?**

Organic nitrogen (N) in biosolids = 24,700 mg/kg
Ammonia nitrogen (N) = 11,100 mg/kg
Biosolids volatilization rate (VR) = 0.50
Mineralization rate (MR) from activated sludge system = 0.20

Equation: **PAN, lb/dry ton =**

[(Organic N, mg/kg)(MR) + (Ammonia N, mg/kg)(VR)](0.002 lb/dry ton)

PAN, lb/dry ton = [(24,700 mg/kg)(0.20) + (11,100 mg/kg)(0.50)](0.002 lb/dry ton)

PAN, lb/dry ton = (4,940 + 5,550)(0.002 lb/dry ton)

PAN, lb/dry ton = (10,490)(0.002 lb/dry ton) = 20.98 lb/dry ton, round to **21 lb/dry ton**

202. **If the plant available nitrogen (PAN) of the biosolids is 27.5 lb/dry ton, how many dry tons per acre of PAN should be applied for a crop that requires 185 lb of nitrogen per acre?**

Equation: **PAN, dry tons/acre** $= \dfrac{\text{Plant nitrogen required, lb/acre}}{\text{PAN, lb/dry ton}}$

PAN, dry tons/acre $= \dfrac{185 \text{ lb/acre}}{27.5 \text{ lb/dry ton}} = $ **6.73 dry tons/acre PAN**

203. How many lb/acre/year of plant available nitrogen (PAN) will be applied to a wastewater land application field, given the following parameters?

Nitrate (NO_3) = 11.7 mg/L
Nitrite (NO_2) = 0.65 mg/L
Total Kjeldahl nitrogen (TKN)* = 64.2 mg/L
Ammonia (NH_3) = 19.8 mg/L
Applying 12.5 in/acre/year
Mineralization rate (MR) = 0.30
Volatilization rate (VR) = 0.50

First, determine the PAN applied per year to each acre.

Equation: **PAN, mg/L = [MR(TKN − NH_3)] + [1 − VR(NH_3)] + (NO_3 + NO_2)**

PAN, mg/L = [0.30(64.2 mg/L − 19.8 mg/L)] + [1 − 0.50(19.8 mg/L)] + (11.7 mg/L + 0.65 mg/L)

PAN, mg/L = 0.30(44.4 mg/L) + 9.9 mg/L + 12.35 mg/L

PAN, mg/L = 13.32 mg/L + 9.9 mg/L + 12.35 mg/L = 35.57 mg/L

Next, convert the hydraulic loading rate from inches to mil gal/acre/yr.

Flow, mil gal/acre/yr $= \dfrac{(43,560 \text{ ft}^3/\text{acre})(1 \text{ ft/12 in.})(7.48 \text{ gal/ft}^3)(12.5 \text{ in./acre/yr})(1 \text{ acre})}{1,000,000/\text{mil}}$

Flow, mil gal/acre/yr = 0.339405 mil gal/acre/yr

Lastly, solve for PAN in lb/acre/yr.

PAN, lb/acre/yr = (35.57 mg/L)(0.339405 mil gal/acre/yr)(8.34 lb/gal)

PAN, lb/acre/yr = 100.686 lb/acre/yr, round to **100 lb/acre/yr PAN**

Note: TKN = organic nitrogen plus ammonia

204. Given the parameters for the following land application site, calculate the concentration of total nitrogen; nitrogen loading rate, and phosphorus loading rate in lb/day, lb/yr, and lb/acre/yr, and COD growing season loading rate in lb/day and lb/acre/day:

Land application site = 245 acres
Growing season = 178 days/yr
Average hydraulic loading rate = 0.238 mgd
Total Kjeldahl nitrogen (TKN) = 22.3 mg/L
Nitrate (NO_3) = 0.34 mg/L
Nitrite (NO_2) = 0.15 mg/L
Ammonia (NH_3) = 7.9 mg/L
Total phosphate = 4.76 mg/L
COD = 295 mg/L

First, calculate the total nitrogen.

Total nitrogen (N) = Nitrate, mg/L + Nitrite, mg/L + TKN, mg/L

Total nitrogen (N) = 0.34 mg/L + 0.15 mg/L + 22.3 mg/L = 22.79 mg/L, round to **23 mg/L N**

Now, determine the total nitrogen loading rate.

Nitrogen loading rate, lb/day = (Total Nitrogen, mg/L)(mgd)(8.34 lb/gal)

Nitrogen loading rate, lb/day = (22.79 mg/L)(0.238 mgd)(8.34 lb/gal)

Nitrogen loading rate, lb/day = 45.236 lb/day, round to **45 lb/day Nitrogen**

Nitrogen loading rate, lb/yr = (45.236 lb/day)(178 days/yr)

Nitrogen loading rate, lb/yr = 8,052 lb/yr, round to **8,100 lb/yr Nitrogen**

Nitrogen loading rate, lb/acre/yr $= \dfrac{8,052\ \text{lb/yr}}{245\ \text{acres}}$

Nitrogen loading rate, lb/acre/yr $= 32.835$ lb/acre/yr, round to **33 lb/acre/yr Nitrogen**

Phosphorus (P) loading rate, lb/day $=$ (P, mg/L)(mgd)(8.34 lb/gal)

Phosphorus loading rate, lb/day $=$ (4.76 mg/L)(0.238 mgd)(8.34 lb/gal)

Phosphorus loading rate, lb/day $= 9.448$ lb/day, round to **9.45 lb/day Phosphorus**

Phosphorus loading rate, lb/yr $=$ (9.448 lb/day)(178 days/yr)

Phosphorus loading rate, lb/yr $= 1,681.74$ lb/yr, round to **1,680 lb/yr Phosphorus**

Phosphorus loading rate, lb/acre/yr $= \dfrac{1,681.74\ \text{lb/yr}}{245\ \text{acres}}$

Phosphorus loading rate, lb/acre/yr $= 6.864$ lb/acre/yr, round to **6.86 lb/acre/yr Phosphorus**

COD loading rate, lb/day $=$ (COD, mg/L)(mgd)(8.34 lb/gal)

COD loading rate, lb/day $=$ (295 mg/L)(0.238 mgd)(8.34 lb/gal)

COD loading rate, lb/day $= 585.55$ lb/day, round to **586 lb/day COD**

COD loading rate, lb/acre/yr $= \dfrac{585.55\ \text{lb/day}}{245\ \text{acres}}$

COD loading rate, lb/acre/yr $=$ **2.39 lb/acre/day COD**

205. Given the following data, calculate the sodium absorption ratio (SAR) for a wastewater:

Sodium (Na$^+$) = 107 mg/L
Calcium (Ca^{2+}) = 41.5 mg/L
Magnesium (Mg^{2+}) = 26.7 mg/L
Equivalent weight of sodium = 22.99
Equivalent weight of calcium = 20.04
Equivalent weight of magnesium = 12.15

Note: See next section, Laboratory Calculations, on how to calculate equivalent weights.

First, determine the milliequivalents (meq) for sodium, calcium, and magnesium.

$$\text{Sodium, meq} = \frac{107 \text{ mg/L}}{22.99} = 4.654 \text{ meq of sodium}$$

$$\text{Calcium, meq} = \frac{41.5 \text{ mg/L}}{20.04} = 2.071 \text{ meq of calcium}$$

$$\text{Magnesium, meq} = \frac{26.7 \text{ mg/L}}{12.15} = 2.198 \text{ meq of magnesium}$$

Now, solve for SAR.

Equation: $$\textbf{SAR} = \frac{\textbf{Na}^+}{[(0.5)(\textbf{Ca}^{2+} + \textbf{Mg}^{2+})]^{1/2}}$$

Substitute values and solve.

$$\text{SAR} = \frac{4.654 \text{ meq Na}^+}{[(0.5)(2.071 \text{ meq Ca}^{2+} + 2.198 \text{ meq Mg}^{2+})]^{1/2}}$$

$$\text{SAR} = \frac{4.654 \text{ meq Na}^+}{[(0.5)(4.269)]^{1/2}}$$

$$\text{SAR} = \frac{4.654 \text{ meq Na}^+}{[2.1345]^{1/2}}$$

Calculate the square root of 2.1345:

$$\text{SAR} = \frac{4.654 \text{ meq Na}^+}{1.461} = \textbf{3.19 SAR}$$

CHEMISTRY AND LABORATORY PROBLEMS

Operators should have a thorough understanding of many laboratory calculations for they help in evaluating plant processes and efficiencies. Following are a few examples.

206. What is the molarity of a sodium hydroxide (NaOH) solution, if 62.8 grams (g) of NaOH is dissolved in exactly 2 liters of deionized water?

Given: The following grams/mole atomic weights from Table C-1, Appendix C:

Na = 22.99 g/mole
O = 15.999 g/mole
H = 1.008 g/mole

First, determine the number of g/mole for NaOH.

Equation: **NaOH, g/mole = Na, g/mole + O, g/mole + H, g/mole**

NaOH, g/mole = 22.99 g/mole + 15.999 g/mole + 1.008 g/mole = 39.997 g/mole

Next, determine the number of moles that are in 62.8 grams of NaOH.

$$\text{Moles, NaOH} = \frac{62.8 \text{ g}}{39.997 \text{ g/mole}} = 1.57 \text{ moles of NaOH}$$

Now, calculate the molarity.

Equation: $\textbf{Molarity} = \dfrac{\textbf{Moles solute}}{\textbf{Liters solution}}$

$$\text{Molarity of solution} = \frac{1.57 \text{ moles of solute}}{2 \text{ liters}} = \textbf{0.785 Molarity NaOH solution}$$

207. What is the normality (N) of a Mg(OH)$_2$ solution if 0.75 equivalents are dissolved in 1.5 liters of solution?

Equation: **Normality** $= \dfrac{\text{Number of equivalents of solute}}{\text{Liters of solution}}$

Normality $= \dfrac{0.75 \text{ equivalents}}{1.5 \text{ liters}} =$ **0.5 N Mg(OH)$_2$**

Note: See page 160 for an explanation of equivalent weights.

208. What is the equivalent weight in grams of Ca(OH)$_2$ in the following reaction?

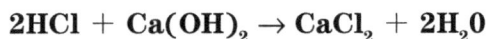

$2HCl + Ca(OH)_2 \rightarrow CaCl_2 + 2H_2O$

Because it takes 2 molecules HCl to neutralize 1 molecule of Ca(OH)$_2$, it follows that

2 eq of HCl will react with 1 eq of Ca(OH)$_2$ or written usually as 1 g-eq HCl to 1/2 g-eq Ca(OH)$_2$.

Looking up the formula weight of Ca(OH)$_2$ in Table C-3 shows it to be 74.095 g/fw.

Now, solve the problem.

Ca(OH)$_2$, g-eq $= \dfrac{74.095 \text{ g/fw}}{2 \text{ eq/fw}} =$ **37.048 g-eq of Ca(OH)$_2$**

209. What is the equivalent weight in grams of K$_3$PO$_4$ in its reaction with HCl?

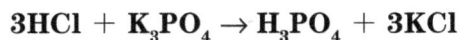

$3HCl + K_3PO_4 \rightarrow H_3PO_4 + 3KCl$

3 eq-wt of HCl will react with 1 of K$_3$PO$_4$, which is 3:1 or 1:⅓.

Next, look up in Table C-1, Appendix C, the atomic weights for potassium (K), phosphorus (P), and oxygen (O). Arrange the data in tabular form, as shown below. Multiply the atomic weight by the corresponding number of atoms for each element. Then, add the formula weights to find the number of g/fw of K_3PO_4.

Element	Number of Atoms	Atomic Weight, g	Formula Weight, g
K	3	39.0983	117.2949
P	1	30.97376	30.97376
O	4	15.9994	63.9976
			Total, g = 212.26626 g

Now, solve the problem.

$$K_3PO_4, \text{g-eq} = \frac{212.26626 \text{ g/fw}}{3 \text{ eq/fw}} = 70.75542 \text{ g-eq, round to } \mathbf{70.7554 \text{ g-eq of } K_3PO_4}$$

210. How many grams of HCl will react with 65.09 g of K_3PO_4?

Know: 1 g-eq of HCl reacts with 1/3 g-eq of K_3PO_4, which equals 70.7554 g-eq

First, determine the number of equivalents in 65.09 g of K_3PO_4.

$$\text{Number of g-eq in 65.09 g of } K_3PO_4 = \frac{65.09 \text{ g}}{70.7554 \text{ g/g-eq}} = 0.9199 \text{ g-eq } K_3PO_4$$

Thus, 0.9199 g-eq of HCl are required.

Know: eq-wt or g-eq of HCl equals 36.461 g/g-eq (Table C-3, Appendix C)

Number of g HCl = (0.9199 g-eq)(36.461 g/g-eq) = **33.54 g of HCl**

211. What is the percent Al in Alum $Al_2(SO_4)_3 \times 14(H_2O)$? Round atomic weights to nearest 0.01.

The equation for calculating the % Al in Alum $[Al_2(SO_4)_3 \times 14(H_2O)]$ is:

$$\textbf{Percent of element in compound} = \frac{(\text{Molecular Wt of the element})(100\%)}{\text{Molecular Wt of compound}}$$

First, determine the molecular weight of each of the elements in the compound. As in the above problem, find the atomic weights in Table C-1 in Appendix C.

Element	Number of Atoms		Atomic Wt		Molecular Wt
Al	2	×	26.98	=	53.96
S	3	×	32.06	=	96.18
O	26	×	16.00	=	416.00
H	28	×	1.01	=	28.28
		Molecular Wt of $Al_2(SO_4)_3 \times 14(H_2O)$		=	594.42

Molecular Wt of $Al_2(SO_4)_3 \times 14(H_2O) = 594.42$

The molecular wt of Al in $Al_2(SO_4)_3 \times 14(H_2O)$ is 53.96

Substituting in above formula.

$$\% \text{ Al} = \frac{(53.96)(100)}{594.42} = 9.078\%, \text{ round to } \textbf{9.08\% Al}$$

212. What is the concentration of alum in mg/L, if 10.3 mL of a 0.785 grams/liter alum solution is added to 1,000 mL of deionized water?

Equation: $\textbf{Alum, mg/L} = \dfrac{(\text{Stock, mL})(1,000 \text{ mg/gram})(\text{Concentration in grams/Liter})}{\text{Sample size, mL}}$

$$\text{Alum, mg/L} = \frac{(10.3 \text{ mL})(1,000 \text{ mg/gram})(0.785 \text{ grams/Liter})}{1,000 \text{ mL}} = \textbf{8.09 mg/L}$$

213. A 5,000-mg (0.5%) stock solution (5,000 ppm or 5,000 mg/L) is required for doing jar tests. If the alum has a specific gravity of 1.332 and is 48.0% aluminum sulfate, how many milliliters of alum are required to make exactly 1,000 mL of stock solution?

First, find the number of lb/gal of alum.

Alum, lb/gal = (sp gr)(8.34 lb/gal) = (1.332 sp gr)(8.34 lb/gal) = 11.1089 lb/gal

Next, determine the number of grams/mL.

$$\text{Polymer grams/mL, Alum} = \frac{(11.1089 \text{ lb/gal})(48.0\% \text{ Al}_2\text{SO}_4, \text{ Purity})(454 \text{ grams/lb})}{(3{,}785 \text{ mL/gal})(100\%)}$$

Polymer grams/mL, Alum = 0.6396 grams/mL

Convert grams/mL to milligrams/mL

Polymer mg/mL = (0.6396 grams/mL)(1,000 mg/g) = 639.6 mg/mL

Next, convert mL to liters by multiplying by 1,000.

Number mg/L = (639.6 mg/mL)(1,000 mL/liter) = 639,600 mg/liter

Next, determine the number of mL required.

Equation: $C_1V_1 = C_2V_2$

(639,600 mg/L)(x, mL) = (5,000 mg/Liter)(1,000 mL)

$$x, \text{mL} = \frac{(5{,}000 \text{ mg/liter})(1{,}000 \text{ mL})}{639{,}600 \text{ mg/liter}} = \textbf{7.82 mL, Alum}$$

Now, using a pipette, add 7.82 mL of the 48.0% alum solution to a clean, dry 1,000-mL flask. Dilute the alum to the 1,000-mL mark with deionized water. Add a magnetic stir bar and place the flask on a magnetic stirrer. Turn the magnetic stirrer on and mix this solution with the bar as vigorously as possible for at least 10 minutes.

Thus, every 1 mL of alum solution that is added to 1,000-mL sample of raw water will add a 5-mg/L dose (because of second dilution with raw water; 5,000 mg/L stock/1,000 mg/L raw water sample = 5 mg/L). If 5 mL of this stock solution were added to the 1,000-mL raw water sample, it would be a dose of 25 mg/L. If you are using the 2-liter square jars, simply double the milliliters added for each 5-mg/L dosage increase desired. Another way is to feed the alum neat by using a micropipette; pipette 0.00782 mL of alum into a 1,000 mL raw water sample to make a dose of 5 mg/L.

214. A 1.00% stock polymer solution (10,000 ppm or 10,000 mg/L) is desired for performing a jar test. If the polymer has a specific gravity of 1.36 and is 32.5% polymer, how many milliliters of polymer are required to make exactly 1,000-mL stock solution?

First, find the number of lb/gal of polymer.

Polymer, lb/gal = (sp gr)(8.34 lb/gal) = (1.36 sp gr)(8.34 lb/gal) = 11.3424 lb/gal

Next, determine the number of grams/mL.

$$\text{Number of grams/mL, Polymer} = \frac{(11.3424 \text{ lb/gal})(32.5\% \text{ Polymer})(454 \text{ grams/lb})}{(3,785 \text{ mL/gal})(100\%)}$$

Number of grams/mL, Polymer = 0.442 grams/mL

Convert grams/mL to milligrams/mL.

Number of mg/mL = (0.442 grams/mL)(1,000 mg/g) = 442 mg/mL

Next, convert mL to liters by multiplying by 1,000.

Number mg/L = (442 mg/mL)(1,000 mL/liter) = 442,000 mg/liter

Next, determine the number of mL required.

Equation: $C_1V_1 = C_2V_2$

(442,000 mg/L)(x, mL) = (10,000 mg/liter)(1,000 mL)

$$x, \text{mL} = \frac{(10,000 \text{ mg/liter})(1,000 \text{ mL})}{442,000 \text{ mg/liter}} = 22.62 \text{ mL, round to } \textbf{22.6 mL, Polymer}$$

Now, using a pipette, add 22.6 mL of the 32.5% polymer solution to a clean, dry 1,000-mL flask. Dilute the polymer to the 1,000-mL mark with deionized water. Add a magnetic stir bar and place the flask on a magnetic stirrer. Turn the magnetic stirrer on and mix this solution with the bar as vigorously as possible for at least 10 minutes.

Thus, every 1 mL of polymer solution that is added to 1,000 mL sample of raw water will add a 10 mg/L dose (because of second dilution with raw water; 10,000 mg/L/1,000 mg/L raw water sample = 10 mg/L). If 3 mL of this stock solution were added to the 1,000-mL raw water sample, it would be a dose of 30 mg/L. If you are using the 2-liter square jars, simply double the mL added for each 10 mg/L dosage increase desired. Another way is to feed the polymer neat by using a micropipette; pipette 0.0226 mL of polymer into a 1,000 mL raw water sample.

215. How many milliliters of 0.875 Normal (N) phosphoric acid (H_3PO_4) solution will neutralize 3.35 g NaOH?

Know: One gram-equivalent (g-eq) of H_3PO_4 will completely react with 1 g-eq of NaOH

Know: Equivalent weight of NaOH = Formula weight = 39.997 grams (Table C-3, Appendix C)

Gram-equivalents of NaOH = Gram-equivalents of H_3PO_4

Gram-equivalents of NaOH = (Liters H_3PO_4)(Normality H_3PO_4)

$$\frac{3.35 \text{ g NaOH}}{39.997 \text{ g/g eq}} = (\text{Liters } H_3PO_4)(0.875 \text{ N } H_3PO_4)$$

Solve for liters H_3PO_4.

$$\text{Liters } H_3PO_4 = \frac{3.35 \text{ g NaOH}}{(39.997 \text{ g/g eq})(0.875 \text{ N } H_3PO_4)} = 0.0957 \text{ L}$$

Finally, convert liters to milliliters.

$$\text{Number of mL} = (0.0957 \text{ L})(1,000 \text{ mL/L}) = \textbf{95.7 mL } H_3PO_4$$

216. Calculate the unseeded BOD$_5$ in mg/L, given the following data:

Start of test bottle dissolved oxygen (DO) = 8.4 mg/L
Bottle was incubated for 5 days in the dark at 20°C
After 5 days DO = 3.1 mg/L
Sample size = 140 mL
Total volume = 300 mL

Equation: $\text{\textbf{BOD}}_5 \text{ \textbf{unseeded, mg/L}} = \dfrac{(\text{Initial DO, mg/L} - \text{Final DO, mg/L})(\text{Total volume, mL})}{\text{Sample volume, mL}}$

$\text{BOD}_5 \text{ unseeded, mg/L} = \dfrac{(8.4 \text{ mg/L} - 3.1 \text{ mg/L})(300 \text{ mL})}{140 \text{ mL}} = 11.36$, round to **11 mg/L**

217. Calculate the seeded BOD$_5$ in mg/L, given the following data:

Sample size = 125 mL
Initial DO = 7.5 mg/L
Final DO = 2.6 mg/L
BOD$_5$ of seed stock = 82 mg/L
Seed stock = 4.0 mL
Total volume = 300 mL

First, calculate the seed correction in mg/L.

Equation: $\text{\textbf{Seed correction, mg/L}} = \dfrac{(\text{BOD}_5 \text{ of seed stock, mg/L})(\text{Seed stock, mg/L})}{\text{Total volume, mL}}$

$\text{Seed correction, mg/L} = \dfrac{(82 \text{ mg/L})(4.0 \text{ mg/L})}{300 \text{ mL}} = 1.09 \text{ mg/L}$

Next, calculate the BOD$_5$ seeded in mg/L.

Equation: $\text{\textbf{BOD}}_5 \text{ \textbf{seeded, mg/L}} =$

$\dfrac{(\text{Initial DO, mg/L} - \text{Final DO, mg/L} - \text{Seed correction, mg/L})(\text{Total volume, mL})}{\text{Sample volume, mL}}$

$\text{BOD}_5 \text{ seeded, mg/L} = \dfrac{(7.5 \text{ mg/L} - 2.6 \text{ mg/L} - 1.09 \text{ mg/L})(300 \text{ mL})}{125 \text{ mL}}$

$\text{BOD}_5 \text{ seeded, mg/L} = \dfrac{(3.81 \text{ mg/L})(300 \text{ mL})}{125 \text{ mL}} = 9.144 \text{ mg/L}$, round to **9.1 mg/L BOD$_5$**

218. Calculate the percent solids and percent volatile solids (VS) for a biosolids sample, given the following data:

	Biosolids Sample, g	Dried Sample, g	Burnt Sample (ash), g
Sample and dish wt.	128.47 g	29.89 g	25.26 g
Weight of dish	23.41 g	23.41 g	23.41 g

First, determine the original weight of the biosolids (sample).

Equation: Weight of biosolids, g = sample and dish wt, g – wt of dish, g

Weight of biosolids, g = 128.47 g – 23.41 g = 105.06 g

Next, determine the total solids by subtracting the weight of the dish from the weight of the dried sample.

Total solids, g = 29.89 g – 23.41 g = 6.48 g

Now, calculate the percent of total solids.

Equation: **Percent total solids** $= \dfrac{\text{(Weight of total solids, g)}(100\%)}{\text{Weight of biosolids sample, g}}$

Percent total solids $= \dfrac{(6.48 \text{ g})(100\%)}{105.06 \text{ g}} = 6.168\%$, round to **6.17%**

Lastly, calculate the percent VS.

First, find the solids lost in burning.

Solids (ash), g = Sample and dish dried, g – Burnt sample, g

Solids (ash), g = 29.89 g – 25.26 g = 4.63 g

Equation: **Percent VS** $= \dfrac{\text{(Solids lost, g)}(100\%)}{\text{Weight of total solids, g}}$

Percent VS $= \dfrac{(4.63 \text{ g})(100\%)}{6.48 \text{ g}} =$ **71.4%**

219. Calculate the percent solids and percent volatile solids (VS) for a biosolids sample, given the following data:

	Biosolids Sample, g	Dried Sample, g	Burnt Sample (ash), g
Sample and dish wt.	105.16 g	32.09 g	28.54 g
Weight of dish	27.38 g	27.38 g	27.38 g

First, determine the original weight of the biosolids (sample).

Equation: Weight of biosolids, g = sample and dish wt, g – wt of dish, g

Weight of biosolids, g = 105.16 g – 27.38 g = 77.78 g

Next, determine the total solids by subtracting weight of the dish from the weight of the dried sample.

Total solids, g = 32.09 g – 27.38 g = 4.71 g

Now, calculate the percent of total solids.

Equation: **Percent total solids** $= \dfrac{(\text{Weight of total solids, g})(100\%)}{\text{Weight of biosolids sample, g}}$

Percent total solids $= \dfrac{(4.71 \text{ g})(100\%)}{77.78 \text{ g}} = 6.0555\%$, round to **6.06%**

Lastly, calculate the percent VS.

First, find the solids lost in burning.

Solids (ash), g = Sample and dish dried, g – Burnt sample, g

Solids (ash), g = 32.09 g – 28.54 g = 3.55 g

Equation: **Percent VS** $= \dfrac{(\text{Solids (ash), g})(100\%)}{\text{Weight of total solids, g}}$

Percent VS $= \dfrac{(3.55 \text{ g})(100\%)}{4.71 \text{ g}} = $ **75.4%**

220. Calculate the number of mg/L of suspended solids (SS) and percent of volatile suspended solids (VSS) in a primary effluent sample, given the following data:

	Dried Sample, g	Burnt Sample (ash), g
Sample and dish wt.	27.0012 g	26.9982 g
Weight of dish	26.9940 g	26.9940 g
Volume of sample, mL	100.0 mL	

First, determine the amount of suspended solids in grams that were in the 50 mL sample.

$$SS, g = 27.0012 \text{ g} - 26.9940 \text{ g} = 0.0072 \text{ g SS}$$

Next, calculate the amount of SS in the primary effluent.

$$SS, mg/L = \frac{(0.0072 \text{ g})}{(100 \text{ mL})} \frac{(1,000 \text{ mg})}{(1 \text{ g})} \frac{(10)}{(10)} = \frac{72 \text{ mg}}{1,000 \text{ mL}}$$

Note: 10 is a multiplier to convert 100 mL to 1,000 mL.

$$SS, mg/L = \frac{(72 \text{ mg})(1,000 \text{ mL})}{(1,000 \text{ mL})(1 \text{ L})} = \textbf{72 mg/L}$$

Next, determine the weight of the VSS.

$$VSS, g = 26.9982 \text{ g} - 26.9940 \text{ g} = 0.0042 \text{ g VSS}$$

Now, calculate the percent VSS.

$$Percent \ VSS = \frac{(\text{Weight of VS, g})(100\%)}{\text{Weight of SS}}$$

$$Percent \ VSS = \frac{(0.0042 \text{ g VSS})(100\%)}{0.0072 \text{ g SS}} = \textbf{58.33\%}$$

221. A composite sample is being collected at a wastewater plant. If 12 samples totaling 6,000 mL are required and the average flow rate is 3.22 mgd, what will be the proportioning factor and what will be the number of mL for each sample time?

Time	Flow, mgd
0600 Hours	3.22
0800 Hours	3.29
1000 Hours	3.54
1200 Hours	3.60
1400 Hours	3.49
1600 Hours	3.42
1800 Hours	3.35
2000 Hours	3.48
2200 Hours	3.61
2400 Hours	3.37
0200 Hours	2.41
0400 Hours	1.92
Total Flow	38.70
Average Flow	3.225

First, calculate the proportioning factor.

Equation: $\text{Proportioning factor} = \dfrac{\text{Total mL required}}{(\text{Number of samples})(\text{Average flow, mgd})}$

Substitute values and solve.

$\text{Proportioning factor} = \dfrac{6,000 \text{ mL}}{(12 \text{ samples})(3.225 \text{ mgd})} = \textbf{155 mL}$

Next, calculate the volumes needed for each sample time.

$\text{Sample volume} = (\text{mgd})(\text{Proportioning factor})$

Substitute values and solve.

Sample volume 0600 Hours	=	(3.22)(155)	=	499.10 mL, round to **500 mL**
Sample volume 0600 Hours	=	(3.29)(155)	=	509.95 mL, round to **510 mL**
Sample volume 0600 Hours	=	(3.54)(155)	=	548.70 mL, round to **549 mL**
Sample volume 0600 Hours	=	(3.60)(155)	=	558.00 mL, round to **558 mL**
Sample volume 0600 Hours	=	(3.49)(155)	=	540.95 mL, round to **541 mL**
Sample volume 0600 Hours	=	(3.42)(155)	=	530.10 mL, round to **530 mL**
Sample volume 0600 Hours	=	(3.35)(155)	=	519.25 mL, round to **519 mL**
Sample volume 0600 Hours	=	(3.48)(155)	=	539.40 mL, round to **539 mL**
Sample volume 0600 Hours	=	(3.61)(155)	=	559.55 mL, round to **560 mL**
Sample volume 0600 Hours	=	(3.37)(155)	=	522.35 mL, round to **522 mL**
Sample volume 0600 Hours	=	(2.41)(155)	=	373.55 mL, round to **374 mL**
Sample volume 0600 Hours	=	(1.92)(155)	=	297.60 mL, round to **298 mL**

Total (cross check) = **6,000 mL**

BASIC ELECTRICITY PROBLEMS

Operators should have a basic understanding of electrical calculations, and they must always exercise safety in dealing with electricity at wastewater treatment plants or anywhere.

222. What is the voltage (E) on a circuit, if the current is 18.2 amperes (I or amps) and the resistance (R) is 26.1 ohms?

Equation: **Voltage = (Amps)(Resistance, ohms) or E = (I)(R)**

Substitute values and solve.

Voltage = (18.2 amps)(26.4 ohms) = 480.48 volts, round to **480 volts**

223. **What is the resistance on a wire, if the amperes are 15.8 and the voltage is 220.1 volts?**

Equation: **Voltage = (Amps)(Resistance, ohms)**

Rearrange the equation to solve for the resistance in ohms.

Resistance, ohms = Voltage/Amps

Substitute values and solve.

Resistance, ohms = 220.1 volts/15.8 amps = 13.93 ohms, round to **13.9 ohms**

224. **A process valve is controlled by a 4- to 20-mA signal from the SCADA system. If the valve is 38.4% open and has a 0% to 100% range, what must the signal be in mA from the SCADA system?**

Equation: **Current process reading** $= \dfrac{(\text{Live signal, mA} - 4 \text{ mA offset})(\text{Maximum capacity})}{16 \text{ mA span}}$

Substitute values and solve.

$38.4\% = \dfrac{(\text{Live signal, mA} - 4 \text{ mA offset})(100\% \text{ capacity})}{16 \text{ mA}}$

Rearrange the formula to solve for live signal in mA.

Live signal, mA $= \dfrac{(38.4\%)(16 \text{ mA})}{100\%} + 4 \text{ mA}$

Live signal, mA = 10.144 mA, round to **10.1 mA**

225. What would the SCADA reading be on the board in mA for a 4-mA to 20-mA signal, if a digester tank has a maximum level capacity of 30.4 ft and it currently has 19.3 ft of sludge water in the tank?

Equation: **Current process reading** $= \dfrac{(\text{Live signal, mA} - 4 \text{ mA offset})(\text{Maximum capacity})}{16 \text{ mA span}}$

Substitute values and solve.

$19.3 \text{ ft} = \dfrac{(\text{Live signal, mA} - 4 \text{ mA offset})(30.4 \text{ ft Maximum level})}{16 \text{ mA}}$

Rearrange the formula for solving live signal in mA as follows.

Live signal, mA $= \dfrac{(19.3 \text{ ft})(16 \text{ mA})}{30.4 \text{ ft}} + 4 \text{ mA}$

Live signal, mA $= 14.158$ mA, round to **14.2 mA**

226. The SCADA system at a wastewater plant uses a 4-mA to 20-mA signal to monitor a chemical polymer tank level. If the readout on a SCADA board reads 11.44 mA, what is the height in feet of the polymer in a tank with a capacity of 12.4 ft?

Know: 4 mA = 0 ft in the polymer tank and 20 mA = 12.4 ft in the tank

Equation: **Current process reading** $= \dfrac{(\text{Live signal, mA} - 4 \text{ mA offset})(\text{Maximum capacity})}{16 \text{ mA span}}$

Substitute values and solve.

Polymer level, ft $= \dfrac{(11.44 \text{ mA} - 4 \text{ mA offset})(12.4 \text{ ft Maximum level})}{16 \text{ mA}}$

Polymer level, ft = **5.77 ft**

KILOWATT DETERMINATIONS

As above, operators should have a basic understanding of kilowatt calculations, and they must always exercise safety in dealing with electricity at wastewater treatment plants or anywhere.

227. How many kilowatts will it take to operate a 198-hp pump, assuming the startup energy is 2.1 times?

kW = (Number of hp)(0.746 kW/hp)(Startup energy)

kW = (198 hp)(0.746 kW/hp)(2.1) = 310.19 kW, round to **310 kW**

228. Calculate the total kilowatts needed to operate the following small facility, if everything is operating:

Automatic screen	10 hp
Filter pump	15 hp
Chemical pumps	5 hp
Sludge pump	25 hp
Trickling filter	40 hp
Lighting	5.5 hp
Instrumentation	1.5 hp
First, add the total hp:	102.0 hp

Formula: **kW = (Number of hp)(0.746 kW/hp)**

kW = (102.0 hp)(0.746 kW/hp) = 76.092 kW, round to **76.1 kW**

229. What is the power cost in dollars and cents for a 200 mhp pump, if it operates an average of 6 hr and 20 min each day and the cost is $0.0734/kW-hr? Assume 30-day month.

First, determine the number of hours the pump operates each month.

Pump operating time, hr/month = 30 days/mo[(6 hr/day) + (20 min/day)(1 hr/60 min)]

Pump operating time, hr = 180 hr/mo + 10 hr/mo = 190 hr/month

Next, determine the kW for the pump.

Know: 746 watts per hp

$$\text{Number of kW} = \frac{(200 \text{ mhp})(746 \text{ watts/hp})}{1,000 \text{ watts/kW}} = 149.2 \text{ kW}$$

Now, calculate the cost to run the pump for one month.

Cost/month = (190 hr/mo)(149.2 kW)($0.0734/kW-hr) = **$2,080.74**

1. Given the following data, calculate the amount of volatile solids (VS) destroyed in lb/day/ft³ of digester capacity:

Digester radius = 34.9 ft
Average sludge height = 19.55 ft
Sludge flow (Flow) = 6,440 gpd
Sludge solids concentration (SSC) = 5.12%
Volatile solids content (VSC) = 61.7%
Volatile solids reduction (VSR) = 49.8%
Specific gravity of sludge = 1.03

2. How many gallons of a 32.75% solution must be mixed with a 10.2% solution to make exactly 650 gallons of an 18.0% solution? Solve by using the dilution triangle and give answer to nearest gallon.

3. What is the flow velocity in feet per second (ft/s) for a trapezoidal channel, given the following data?

Bottom width, w_1 = 10.8 ft
Water surface width, w_2 = 19.5 ft
Depth 4.63 ft
Flow = 78 ft³/sec

4. What is the gpm flow in a 36.0-inch sewage pipeline that is flowing at a velocity of 1.02 ft/s and the depth of the sewage averages 12.5 inches?

5. A trickling filter has a diameter of 179.9 ft. If the flow through the filter is 2.74 mgd and the recirculation rate is 24.0% of the flow rate, what is the hydraulic loading rate on a trickling filter in gallons per day per square foot (gpd/ft²)?

6. Determine the waste activated sludge (WAS) flow in gpm, given the following data.

Influent flow = 1.88 mgd
Clarifier radius = 40.2 ft
Clarifier depth = 15.8 ft
Aerator = 0.786 mil gal
Mixed liquor suspended solids (MLSS) = 2,695 mg/L
Return activated sludge (RAS) SS = 7,330 mg/L
Secondary effluent SS = 29.5 mg/L
Target solids retention time (SRT) = 14 days exactly

7. What are the Cl_2 and sulfur dioxide (SO_2) dosages in mg/L for a wastewater plant's effluent, given the following data?

Pounds of chlorine used = 315 lb/day
Flow = 2,280 gpm
Chlorine demand = 10.7 mg/L
Assume SO_2 is 3.00 mg/L higher than the chlorine residual

8. A wastewater plant is treating a flow of 1,375,000 gpd with an alum dose of 4.8 mg/L. If the alum is 48.5% pure and has a specific gravity of 1.315, what is the alum feed in mL/min?

9. What should the chemical feeder be set on in mL/min, given the following data?

    Polymer dosage = 9.50 mg/L
    Plant flow = 1,570 gpm
    Polymer, lb/gal = 12.5 lb/gal

10. What is the dosage in mg/L for magnesium hydroxide and the feed rate in grams/min
    (g/min), if the wastewater plant is treating 2,540 gpm and the magnesium feed rate is
    154 lb/day?

11. How many pounds of a dry polymer (43.1% active) are required to make exactly
    400 gallons of a solution that is exactly 5.25%, and what feed rate will be required in
    mL/min for a dosage of 3.25 mg/L, if the plant is treating 1.79 mgd?

12. Given the following data, estimate the dry sludge solids produced by a secondary
    clarifier in lb/day:

    Secondary influent flow = 1,190 gpm
    Influent BOD_5 = 238 mg/L
    Effluent BOD_5 = 22.7 mg/L
    Bacterial growth rate for this plant = 0.380 lb SS/lb BOD_5
    Suspended solids = 337 mg/L

13. Given the following data, determine the sludge age in days at an oxidation ditch
    wastewater treatment plant:

    Mixed liquor suspended solids (MLSS) = 3,935 mg/L
    Solids added = 488 lb/day
    Average top width of ditch at water surface = 16.1 ft
    Average depth = 5.11 ft
    Average bottom width = 9.47 ft
    Length of ditch = 212 ft
    Diameter of half circles = 114 ft

14. If the MCRT desired was 9.0 days, what would the waste rate be for the following system in lb/day?

Flow = 1,950 gpm
Aeration tank (AT) volume = 0.610 mil gal
Clarifier tank (CT) volume = 0.301 mil gal
MLSS = 2,640 mg/L
Effluent TSS = 18.0 mg/L

15. Calculate the mass balance for the following conventional biological system. Is there a problem with this system? If so, discuss.

Influent waste flow = 1.19 mgd
Influent BOD$_5$ = 228 mg/L
TSS = 274 mg/L
Effluent flow = 1.12 mgd
Effluent BOD$_5$ = 21.5 mg/L
Effluent TSS = 40.5 mg/L
Waste flow = 0.0231 mgd
Waste TSS = 7,965 mg/L

Given: This conventional activated biosolids system without primary has a lb solids/lb BOD$_5$ = 0.85 lb solids/lb BOD$_5$

16. How many pounds of lime will be required to neutralize a sour digester that is 49.9 ft in diameter, has a working sludge level of 21.3 ft, and has a volatile acid content of 2,925 mg/L?

17. Calculate the required waste rate from an aeration tank in mgd and gpm, given the following data:

Volume of aeration tanks = 1.205 mil gal
Desired COD lb/MLVSS lb = 0.175
Primary effluent flow = 3.82 mgd
Primary effluent COD = 136 mg/L
Mixed liquor volatile suspended solids (MLVSS) = 3,820 mg/L
Waste volatile solids (WVS) concentration = 4,290 mg/L

18. What are the solids loading for a dissolved air flotation (DAF) unit in lb/d/ft² that is 55.2 ft by 29.9 ft, where sludge flow averages 82 gpm, with a waste-activated sludge (WAS) concentration of 6,340 mg/L, and the sludge has a specific gravity of 1.04?

19. Given the following data, calculate the average gallons applied to a drying bed for each cycle period and the solids loading in lb/yr/ft²:

Drying bed = 261 feet long and 48.0 feet wide.
Average sludge application per cycle = 4.59 inches
Average percent solids = 4.88%
Drying and removal cycle on average = 21.6 days

20. Calculate the amount of compost in lb/day that needs to be blended with a dewatered sludge to make a mixture that has a moisture content of 50.0%, if the dewatered digester primary sludge is 5,350 lb/day, the dewatered sludge solids are 28.6%, and the compost solids moisture content is 32.0%.

21. What is the capacity for a compost site in processing dry sludge in lb/day and tons/day, given the following data?

Site capacity = 14,700 yd³
Average compost cycle = 23.7 days
Bulk density of wet sludge = 1,690 lb/yd³
Bulk density of wet compost = 1,050 lb/yd³
Sludge solids content = 19.1%
Bulk density of wet wood chips = 635 lb/yd³
Mix ratio (MR) of wood chips to sludge = exactly 3.25 to 1

22. How many lb/acre/year of plant available nitrogen (PAN) will be applied to a wastewater land application field, given the following parameters?

Nitrate (NO_3) = 13.1 mg/L
Nitrite (NO_2) = 0.73 mg/L
Total Kjeldahl nitrogen (TKN) = 61.4 mg/L
Ammonia (NH_3) = 20.7 mg/L
Applying 10.2 in/acre/year
Mineralization rate (MR) = 0.30
Volatilization rate = 0.50

23. Given the following data, calculate the sodium absorption ratio (SAR) for a wastewater:

Sodium (Na^+) = 118 mg/L
Calcium (Ca^{2+}) = 43.7 mg/L
Magnesium (Mg^{2+}) = 25.2 mg/L
Equivalent weight of sodium = 22.99
Equivalent weight of calcium = 20.04
Equivalent weight of magnesium = 12.15

24. A 20,000-mg (2.00%) stock solution (20,000 ppm or 20,000 mg/L) is required for doing jar tests. If the alum has a specific gravity of 1.296 and is 49.2% aluminum sulfate, how many milliliters of alum are required to make exactly 1,000 mL of stock solution?

25. The SCADA system at a wastewater plant uses a 4-mA to 20-mA signal to monitor a chemical polymer tank level. If the readout on a SCADA board reads 13.80 mA, what is the height of the polymer in a tank with a maximum level of 16.75 feet?

1. Given the following data, calculate the amount of volatile solids (VS) destroyed in lb/day/ft³ of digester capacity:

Digester radius = 34.9 ft
Average sludge height = 19.55 ft
Sludge flow (Flow) = 6,440 gpd
Sludge solids concentration (SSC) = 5.12%
Volatile solids content (VSC) = 61.7%
Volatile solids reduction (VSR) = 49.8%
Specific gravity of sludge = 1.03

First, determine the number of lb/gal for the sludge.

Sludge, lb/gal = (8.34 lb/gal)(1.03) = 8.59

Next, calculate the digester capacity in ft³.

Digester capacity, ft³ = π(radius)²(Height, ft)

Digester capacity, ft³ = 3.14(34.9 ft)(34.9 ft)(19.55 ft) = 74,769.98 ft³

Next, write the equation.

$$\text{VS destroyed, lb/day/ft}^3 = \frac{(\text{Flow, gpd})(\text{Sludge, lb/gal})(\text{SSC, \%})(\text{VSC, \%})(\text{VSR, \%})}{\text{Digester capacity, ft}^3}$$

Substitute values and solve.

$$\text{VS destroyed, lb/day/ft}^3 = \frac{(6,440 \text{ gpd}) (8.59 \text{ lb/gal}) (5.12\%/100\%) (61.7\%/100\%) (49.8\%/100\%)}{74,769.98 \text{ ft}^3}$$

VS destroyed, lb/day/ft³ = 0.012 lb/day/ft³ VS destroyed

2. **How many gallons of a 32.75% solution must be mixed with a 10.2% solution to make exactly 650 gallons of an 18.0% solution? Solve by using the dilution triangle and give answer to nearest gallon.**

32.75%   7.8   7.8 parts of the 32.75% solution are required for every 22.55 parts.

    18.0%

10.2% $\dfrac{14.75}{22.55 \text{ total parts}}$ 14.75 parts of the 10.2% solution are required for every 22.55 parts.

$$\frac{7.8 \text{ parts} (650 \text{ gal})}{22.55 \text{ parts}} = 224.83 \text{ gallons, round to } \textbf{225 gallons of the 32.75\% solution}$$

$$\frac{14.75 \text{ parts} (650 \text{ gal})}{22.55 \text{ parts}} = 425.17 \text{ gallons, round to } \underline{425} \textbf{ gallons of the 10.2\% solution}$$
$$\underline{650} \text{ gallons—added here to cross check math}$$

To make the 650 gallons of the 18.0% solution, mix 225 gallons of the 32.75% solution with 425 gallons of the 10.2% solution.

3. **What is the flow velocity in feet per second (ft/s) for a trapezoidal channel, given the following data?**

Bottom width, w_1 = 10.8 ft
Water surface width, w_2 = 19.5 ft
Depth 4.63 ft
Flow = 78 ft³/sec

Equation: Flow (Q), ft³/sec $= \dfrac{(w_1 + w_2)}{2}$ **(Depth, ft)(Velocity, ft/s)**

Rearrange the formula to solve for velocity in ft/s.

$$\text{Velocity, ft/s} = \frac{2 (Q, \text{ft}^3/\text{sec})}{(w_1 + w_2) (\text{Depth, ft})}$$

Substitute and solve.

$$\text{Velocity, ft/s} = \frac{2\,(78\ \text{ft}^3/\text{sec})}{(10.8\ \text{ft} + 19.5\ \text{ft})\,(4.63\ \text{ft})} = \frac{156\ \text{ft}^3/\text{sec}}{(30.3\ \text{ft})\,(4.63\ \text{ft})} = \textbf{1.11 ft/s}$$

4.   **What is the gpm flow in a 36.0-inch sewage pipeline that is flowing at a velocity of 1.02 ft/s and the depth of the sewage averages 12.5 inches?**

First, divide the depth of sewage flow by the diameter of the pipe.

Ratio = depth/Diameter = 12.5 in./36.0 in. = 0.3472, round to 0.35

Note: Do not use extrapolation in solving this problem.

Next, determine the factor that needs to be used.

In Appendix D, look up 0.35 under the column d/D. The number immediately to the right will be the factor that needs to be used. In this case it is 0.2450. This will be the number used rather than 0.785.

Next, convert the pipe's diameter from inches to feet.

$$\text{Number of feet} = \frac{36.0\ \text{in.}}{12\ \text{in./ft}} = 3.00$$

Equation: **Flow, ft³/sec = (Area, ft²)(Velocity, ft/s)**

Where the area = (Factor)(Diameter)²

Substitute values and solve.

Flow, ft³/s = (0.2450)(3.00 ft)(3.00 ft)(1.02 ft/s) = 2.249 ft³/s

Now, convert ft³/s to gpm.

Flow, gpm = (2.249 ft³/s)(60 s/min)(7.48 gal/ft³) = 1,009.35 gpm, round to **1,010 gpm**

5. **A trickling filter has a diameter of 179.9 ft. If the flow through the filter is 2.74 mgd and the recirculation rate is 24.0% of the flow rate, what is the hydraulic loading rate on a trickling filter in gallons per day per square foot (gpd/ft²)?**

First, determine the total flow in gallons per day (gpd) through the trickling filter.

Total flow, gal = [2.74 mgd + 2.74 mgd(24.0%/100%)](1,000,000/mil)

Total flow, gal = [2.74 mgd + 0.6576 mgd](1,000,000/mil) = 3,397,600 gpd

Next, determine the surface area in ft² for the clarifier.

Area = πr² where r = Diameter/2 = 179.9 ft/2 = 89.95 ft

Trickling filter surface area, ft² = (3.14)(89.95 ft)(89.95 ft) = 25,405.75 ft²

Lastly, calculate the hydraulic loading rate.

$$\text{Hydraulic loading rate} = \frac{\textbf{Total flow, gpd}}{\textbf{Surface area, ft}^2}$$

$$\text{Hydraulic loading rate} = \frac{3,397,600, \text{gpd}}{25,405.75 \text{ ft}^2} = 133.73 \text{ gpd/ft}^2, \text{round to } \textbf{134 gpd/ft}^2$$

6. **Determine the waste activated sludge (WAS) flow in gpm, given the following data.**

Influent flow = 1.88 mgd
Clarifier radius = 40.2 ft
Clarifier depth = 15.8 ft
Aerator = 0.786 mil gal
Mixed liquor suspended solids (MLSS) = 2,695 mg/L
Return activated sludge (RAS) SS = 7,330 mg/L
Secondary effluent SS = 29.5 mg/L

Target solids retention time (SRT) = 14 days exactly

First, determine the volume in mil gal for the clarifier and add to the aerator volume.

Equation: Clarifier, mil gal $= \dfrac{\pi\,(\text{radius})^2\,(\text{Depth, ft})\,(7.48\ \text{gal/ft}^3)}{1{,}000{,}000/\text{mil}}$

Clarifier, mil gal $= \dfrac{3.14\,(40.2\ \text{ft})\,(40.2\ \text{ft})\,(15.8\ \text{ft})\,(7.48\ \text{gal/ft}^3)}{1{,}000{,}000/\text{mil}} = 0.5997\ \text{mil gal}$

Total volume $= 0.5997\ \text{mil gal} + 0.786\ \text{mil gal} = 1.3857\ \text{mil gal}$

Equation: **Target SRT** $=$

$\dfrac{(\textbf{MLSS mg/L})\,(\textbf{Clarifier, Aerator Volume, mil gal})\,(\textbf{8.34 lb/gal})}{(\textbf{RAS SS mg/L})\,(x\ \textbf{mgd})\,(\textbf{8.34 lb/gal}) + (\textbf{Effluent SS, mg/L})\,(\textbf{Flow, mgd})\,(\textbf{8.34 lb/gal})}$

Substitute values and solve.

14 days SRT $= \dfrac{(2{,}695\ \text{mg/L})\,(1.3857\ \text{mil gal})\,(8.34\ \text{lb/gal})}{(7{,}330\ \text{mg/L})\,(x\ \text{mgd})\,(8.34\ \text{lb/gal}) + (29.5\ \text{mg/L})\,(1.88\ \text{mgd})\,(8.34\ \text{lb/gal})}$

14 days SRT $= \dfrac{31{,}145.41\ \text{lb MLSS}}{(7{,}330\ \text{mg/L})\,(x\ \text{mgd})\,(8.34\ \text{lb/gal}) + 462.536\ \text{lb/day}}$

Rearrange the equation so that x mgd is in the numerator.

$(7{,}330\ \text{mg/L})(x\ \text{mgd})(8.34\ \text{lb/gal}) + 462.536\ \text{lb/day} = \dfrac{31{,}145.41\ \text{lb MLSS}}{14\ \text{days SRT}}$

$(7{,}330\ \text{mg/L})(x\ \text{mgd})(8.34\ \text{lb/gal}) + 462.536\ \text{lb/day} = 2{,}224.67\ \text{lb/day}$

Subtract 462.536 lb/day from both sides of the equation.

$(7{,}330\ \text{mg/L})(x\ \text{mgd})(8.34\ \text{lb/gal}) = 2{,}224.67\ \text{lb/day} - 462.536\ \text{lb/day}$

$x\ \text{mgd} = \dfrac{1{,}762.13\ \text{lb/day}}{(7{,}330\ \text{mg/L})(8.34\ \text{lb/gal})} = 0.028825\ \text{mgd}$

Lastly, convert mgd to gpm.

WAS flow, gpm $= \dfrac{(0.028825\ \text{mgd})\,(1{,}000{,}000/\text{mil})}{1{,}440\ \text{min/day}} = 20.02\ \text{gpm, round to }\textbf{20.0 gpm WAS}$

7.   **What are the Cl$_2$ and sulfur dioxide (SO$_2$) dosages in mg/L for a wastewater plant's effluent, given the following data?**

Pounds of chlorine used = 315 lb/day
Flow = 2,280 gpm
Chlorine demand = 10.7 mg/L
Assume SO$_2$ is 3.00 mg/L higher than the chlorine residual

First, convert gpm to mgd.

$$\text{Number of mgd} = \frac{(2,280 \text{ gpm}) (1,440 \text{ min/day})}{1,000,000/\text{mil}} = 3.2832 \text{ mgd}$$

Next, determine the chlorine dosage in mg/L.

Equation: **Number of lb/day = (Dosage, mg/L)(Number of mgd)(8.34 lb/gal)**

Rearrange the equation to solve for the chlorine dosage.

$$\text{Chlorine dosage, mg/L} = \frac{\text{Number of lb/day}}{(\text{mgd}) (8.34 \text{ lb/gal})}$$

Substitute values and solve.

$$\text{Chlorine dosage, mg/L} = \frac{315 \text{ lb/day}}{(3.2832 \text{ mgd}) (8.34 \text{ lb/gal})} = 11.50 \text{ mg/L Cl}_2\text{, round to } \mathbf{11.5 \text{ mg/L Cl}_2}$$

In order to determine the sulfur dioxide dosage, the chlorine residual needs to be known.

Chlorine residual = Chlorine dosage − Chlorine demand

Chlorine residual = 11.50 mg/L − 10.7 mg/L = 0.80 mg/L Cl$_2$ residual

Now, calculate the SO$_2$ dosage.

SO$_2$ dosage, mg/L = Chlorine residual, mg/L + 3.5 mg/L

SO$_2$ dosage, mg/L = 0.80 mg/L + 3.00 mg/L = **3.80 mg/L SO$_2$**

8. A wastewater plant is treating a flow of 1,375,000 gpd with an alum dose of 4.8 mg/L. If the alum is 48.5% pure and has a specific gravity of 1.315, what is the alum feed in mL/min?

First, convert gpd to mgd.

$$\text{Number of mgd} = \frac{1,375,000 \text{ gpd}}{1,000,000/\text{mil}} = 1.375 \text{ mgd}$$

Next, calculate the number of lb/day of alum required.

Next, determine the lb/gal for the alum.

$$\text{Number of lb/gal} = (8.34 \text{ lb/gal})(1.315 \text{ sp gr}) = 10.9671$$

Equation: $\textbf{mL/min} = \dfrac{(\text{mg/L})(3,785 \text{ mL/gal})(\text{mgd})(8.34 \text{ lb/gal})}{(1,440 \text{ min/day})(\text{Percent purity}/100\%)(\text{Alum, lb/gal})}$

$$\text{Alum, mL/min} = \frac{(4.8 \text{ mg/L})(3,785 \text{ mL/gal})(1.375 \text{ mgd})(8.34 \text{ lb/gal})}{(1,440 \text{ min/day})(48.5\%/100\%)(10.9671 \text{ lb/gal})}$$

Alum, mL/min = 27.20 mL/min, round to **27 mL/min of Alum**

9.   **What should the chemical feeder be set on in mL/min, given the following data?**

Polymer dosage = 9.50 mg/L
Plant flow = 1,570 gpm
Polymer, lb/gal = 12.5 lb/gal

First, convert gpm flow to mgd.

$$\text{Number of mgd} = \frac{(1,570 \text{ gpm})(1,440 \text{ min/day})}{1,000,000/\text{mil}} = 2.2608 \text{ mgd}$$

Next, determine the lb/day of polymer using the pounds formula.

Polymer, lb/day = (Dosage, mg/L)(mgd)(8.34 lb/gal)

Polymer, lb/day = (9.50 mg/L)(2.2608 mgd)(8.34 lb/gal) = 179.12 lb/day

Next, calculate the number of gallons polymer used.

$$\text{Polymer, gal} = \frac{(179.12 \text{ lb/day})}{(12.5 \text{ lb/gal})} = 14.33 \text{ gal}$$

Now, using the following equation, calculate the mL/min of polymer being used.

$$\text{Equation: } \textbf{Number of mL/min} = \frac{(\text{Number of gallons used})(3,785 \text{ mL/gal})}{1,440 \text{ min/day}}$$

Substitute values and solve.

$$\text{Polymer, mL/min} = \frac{(14.33 \text{ gal})(3,785 \text{ mL/gal})}{1,440 \text{ min/day}} = \textbf{37.7 mL/min Polymer}$$

10. **What is the dosage in mg/L for magnesium hydroxide and the feed rate in grams/min (g/min), if the wastewater plant is treating 2,540 gpm and the magnesium feed rate is 154 lb/day?**

First, calculate the feed rate in g/min by converting lb/day to g/min.

$$\text{Feed rate, g/min} = \frac{(154 \text{ lb/day})(454 \text{ g/lb})}{1,440 \text{ min/day}} = 48.553 \text{ g/min, round to } \textbf{48.6 g/min}$$

Next, convert gpm to mgd.

$$\text{Number of mgd} = \frac{(2,540 \text{ gpm})(1,440 \text{ min/day})}{1,000,000/\text{mi}} = 3.6576 \text{ mgd}$$

Next, find the dosage in mg/L by using the "pounds" equation.

Equation: **Chemical, lb/day = (Dosage, mg/L)(mgd)(8.34 lb/gal)**

Rearrange the "pounds" equation and solve for dosage in mg/L.

$$\text{Dosage, mg/L} = \frac{\text{Chemical, lb/day}}{(\text{mgd})(8.34 \text{ lb/gal})}$$

Substitute values and solve.

$$\text{Dosage, mg/L} = \frac{154 \text{ lb/day}}{(3.6576 \text{ mgd})(8.34 \text{ lb/gal})} = \textbf{5.05 mg/L of Magnesium hydroxide}$$

11. **How many pounds of a dry polymer (43.1% active) are required to make exactly 400 gallons of a solution that is exactly 5.25%, and what feed rate will be required in mL/min for a dosage of 3.25 mg/L, if the plant is treating 1.79 mgd?**

First, convert 250 gallons to be mixed to mil gal.

$$\text{Number of mil gal} = \frac{400 \text{ gallons}}{1,000,000/\text{mil}} = 0.000400 \text{ mil gal}$$

Know: 1% = 10,000 mg/L

Therefore, 5.25% = 52,500 mg/L

Next, calculate the number of lb of dry polymer required.

Equation: **Dry polymer, lb** $= \dfrac{(\text{Dose, mg/L})\,(\text{mil gal})\,(8.34 \text{ lb/gal})}{\textbf{Percent purity}}$

Substitute values and solve.

$$\text{Dry polymer, lb} = \frac{(52,500 \text{ mg/L})(0.000400 \text{ mil gal})(8.34 \text{ lb/gal})}{43.1\%/100\%}$$

Dry polymer, lb = 406.36 lb, round to **406 lb of Dry polymer**

Next, determine the number of pounds of polymer per gallon in this solution.

$$\text{Dry polymer, lb/gal} = \frac{406.36 \text{ lb}}{400 \text{ gal}} = 1.0159 \text{ lb/gal}$$

Now, calculate the mL/min required for a dosage of 3.25 mg/L.

Equation: **Dosage desired, mg/L** $= \dfrac{(\text{mL/min})\,(1,440 \text{ min/day})\,(\text{lb/gal})}{(3,785 \text{ mL/gal})\,(8.34 \text{ lb/gal})\,(\text{mgd})}$

Rearrange the formula to solve for mL/min.

$$\text{Dry polymer feed, mL/min} = \frac{(\text{Dosage, mg/L})\,(3,785 \text{ mL/gal})\,(8.34 \text{ lb/gal})\,(\text{mgd})}{(1,440 \text{ min/day})\,(\text{lb/gal})}$$

Substitute values and solve.

$$\text{Dry polymer feed, mL/min} = \frac{(3.25 \text{ mg/L})(3,785 \text{ mL/gal})(8.34 \text{ lb/gal})(1.79 \text{ mgd})}{(1,440 \text{ min/day})(1.0159 \text{ lb/gal})}$$

Dry Polymer feed = 125.53 mL/min, round to **126 mL/min of Dry polymer**

12. Given the following data, estimate the dry sludge solids produced by a secondary clarifier in lb/day:

Secondary influent flow = 1,190 gpm
Influent BOD$_5$ = 238 mg/L
Effluent BOD$_5$ = 22.7 mg/L
Bacterial growth rate for this plant = 0.380 lb SS/lb BOD$_5$
Suspended solids = 337 mg/L

First, determine the BOD$_5$ removal.

BOD$_5$ removal, mg/L = 238 mg/L − 22.7 mg/L = 215.3 mg/L

Next, convert gpm to mgd.

$$\text{Number of mgd} = \frac{(1,190 \text{ gpm})(1,440 \text{ min/day})}{1,000,000/\text{mil}} = 1.7136 \text{ mgd}$$

Next, determine the BOD$_5$ removal in lb/day.

Equation: **BOD$_5$ removal, lb/day = (BOD$_5$, mg/L)(Flow, mgd)(8.34 lb/gal)**

BOD$_5$ removal, lb/day = (215.3 mg/L)(1.7136 mgd)(8.34 lb/gal) = 3,076.94 lb/day

Lastly, use the bacterial growth rate to calculate the estimated lb/day of solids produced.

$$\frac{x \text{ lb/day Solids produced}}{3,076.94 \text{ lb/day BOD}_5 \text{ removed}} = \frac{0.380 \text{ lb SS/lb BOD}_5}{1 \text{ lb BOD}_5 \text{ removed}}$$

$$x \text{ lb/day Solids produced} = \frac{(3,076.94 \text{ lb/day BOD}_5 \text{ removed})(0.380 \text{ lb SS/lb BOD}_5)}{1 \text{ lb BOD}_5 \text{ removed}}$$

x lb/day Solids produced = 1,169.24 lb/day, round to **1,170 lb/day Solids produced**

13. Given the following data, determine the sludge age in days at an oxidation ditch wastewater treatment plant:

Mixed liquor suspended solids (MLSS) = 3,935 mg/L
Solids added = 488 lb/day
Average top width of ditch at water surface = 16.1 ft
Average depth = 5.11 ft
Average bottom width = 9.47 ft
Length of ditch = 212 ft
Diameter of half circles = 114 ft

First, determine the number of mil gal in the ditch.

Equation: $\dfrac{[(b_1 + b_2)\,\text{Depth}]}{2}$ (Length of 2 sides + Length of 2 half circles)(7.48 gal/ft³)

Where b_1 = bottom width of ditch and b_2 = top width of ditch at water surface, and the length of the two half circles is π(Diameter)

Substitute and solve.

Volume, gal $= \dfrac{[(9.47\ \text{ft} + 16.1\ \text{ft})\,(5.11\ \text{ft})]}{2}$ [(2)(212 ft) + (3.14)(114 ft)](7.48 gal/ft³)

Volume, gal = (65.331 ft²)(424 ft + 357.96 ft)(7.48 gal/ft³)

Volume, gal = (65.331 ft²)(781.96 ft)(7.48 gal/ft³)

Volume, gal = 382,125 gal

Next, convert gallons to mil gal.

Number of mil gal $= \dfrac{382,125}{1,000,000/\text{mil}} = 0.382125$ mil gal

Next, calculate the amount of solids under aeration.

Equation: **Solids under aeration, lb = (Ditch volume, mil gal)(MLSS. Mg/L)(8.34 lb/gal)**

Solids under aeration, lb = (0.382125 MD)(3,935 mg/L)(8.34 lb/gal) = 12,541 lb

Next, determine the sludge age in days.

Equation: **Sludge age, days** $= \dfrac{\text{Solids under aeration, lb}}{\text{Solids added, lb/day}}$

Sludge age, days $= \dfrac{12,541 \text{ lb}}{488 \text{ lb/day}} = 25.699$ days, round to **25.7 days**

14. **If the MCRT desired was 9.0 days, what would the waste rate be for the following system in lb/day?**

Flow = 1,950 gpm
Aeration tank (AT) volume = 0.610 mil gal
Clarifier tank (CT) volume = 0.301 mil gal
Mixed liquor suspended solids (MLSS) = 2,640 mg/L
Effluent TSS = 18.0 mg/L

First, convert the flow in gpm to mgd.

Flow, mgd $= \dfrac{(1,950)\,(1,440 \text{ min/day})}{1,000,000/\text{mil}} = 2.808$ mgd

Equation: **Waste rate, lb/day =**

$$\dfrac{\textbf{MLSS, mg/L}\,[\textbf{AT, mil gal} + \textbf{CT, mil gal}]\,(\textbf{8.34 lb/gal})}{\textbf{Desired MCRT}} - (\textbf{TSS, mg/L})(\textbf{Flow, mgd})(\textbf{8.34 lb/gal})$$

Waste rate, lb/day =

$$\dfrac{2,640 \text{ mg/L}\,[0.610 \text{ mil gal} + 0.301 \text{ mil gal}]\,(8.34 \text{ lb/gal})}{9.0 \text{ days, Desired MCRT}} - (18.0 \text{ mg/L TSS})(2.808 \text{ mgd})(8.34 \text{ lb/gal})$$

Waste rate, lb/day =

$$\dfrac{(2,640 \text{ mg/L})\,(0.911 \text{ mil gal})\,(8.34 \text{ lb/gal})}{9.0 \text{ days, Desired MCRT}} - (18.0 \text{ mg/L TSS})(2.808 \text{ mgd})(8.34 \text{ lb/gal})$$

Waste rate, lb/day $= 2,228.67$ lb/day $- 421.537$ lb/day

Waste rate, lb/day $= 1,807.13$ lb/day, round to **1,810 lb/day**

15. Calculate the mass balance for the following conventional biological system. Is there a problem with this system and if so discuss?

Influent waste flow = 1.19 mgd
Influent BOD_5 = 228 mg/L
Total suspended solids (TSS) = 274 mg/L
Effluent flow = 1.12 mgd
Effluent BOD_5 = 21.5 mg/L
Effluent TSS = 40.5 mg/L
Waste flow = 0.0231 mgd
Waste TSS = 7,965 mg/L

Given: This conventional activated biosolids system without primary has a lb solids/lb BOD_5 = 0.85 lb solids/lb BOD_5

First, calculate the BOD_5 influent and then the effluent in lb/day.

Influent BOD_5, lb/day = (228 mg/L)(1.19 mgd)(8.34 lb/gal) = 2,262.81 lb/day

Effluent BOD_5, lb/day = (21.5 mg/L)(1.12 mgd)(8.34 lb/gal) = 200.83 lb/day

BOD_5 removed, lb/day = 2,262.81 lb/day − 200.83 lb/day = 2,061.98 lb/day

Next, determine the solids produced in lb/day.

Equation: **Solids produced, lb/day = (BOD_5 removed, lb/day)(0.85 lb solids/lb BOD_5)**

Solids produced, lb/day = (2,061.98 lb/day)(0.85 lb/lb BOD_5) = 1,752.68 lb/day

Next, calculate the solids and sludge removed.

Effluent solids, lb/day = (40.5 mg/L)(1.19 mgd)(8.34 lb/gal) = 401.95 lb/day

Effluent sludge, lb/day = (7,965 mg/L)(0.0231 mgd)(8.34 lb/gal) = 1,534.49 lb/day

Next, calculate the total solids removed.

Total solids removed, lb/day = 401.95 lb/day + 1,534.49 lb/day = 1,936.44 lb/day

Now, calculate the percent mass balance of the system.

Equation: **Percent mass balance** $= \dfrac{(\text{Solids produced, lb/day} + \text{Solids removed, lb/day})\,(100\%)}{\text{Solids produced, lb/day}}$

Percent mass balance $= \dfrac{(1{,}752.68\ \text{lb/day} - 1{,}936.44\ \text{lb/day})\,(100\%)}{1{,}752.68\ \text{lb/day}} = \mathbf{10.5\%}$

The negative sign is not needed. This system is not in balance. Problems: Process problems; too many solids being removed; laboratory or sampling error(s).

16. **How many pounds of lime will be required to neutralize a sour digester that is 49.9 ft in diameter, has a working sludge level of 21.3 ft, and has a volatile acid content of 2,925 mg/L?**

Know: 1 mg/L of lime will neutralize 1 mg/L of volatile acids

First, calculate the volume of the digester in gallons.

Number of gallons = (0.785)(49.9 ft)(49.9 ft)(21.3 ft)(7.48 gal/ft³) = 311,424 gal

Next, convert gallons to mil gal.

Number of mil gal $= \dfrac{311{,}424\ \text{gal}}{1{,}000{,}000/\text{mil}} = 0.311424$ mil gal

Equation: **Number of lb = (Volatile acids, mg/L)(mil gal)(8.34 lb/gal)**

Substitute and solve.

Lime, lb = (2,925 mg/L)(0.311424 mil gal)(8.34 lb/gal) = 7,597 lb, round to **7,600 lb of Lime**

17. **Calculate the required waste rate from an aeration tank in mgd and gpm, given the following data:**

Volume of aeration tanks = 1.205 mil gal
Desired COD lb/MLVSS lb = 0.175
Primary effluent flow = 3.82 mgd
Primary effluent COD = 136 mg/L
Mixed liquor volatile suspended solids (MLVSS) = 3,820 mg/L
Waste volatile solids (WVS) concentration = 4,290 mg/L

First, find the existing MLVSS in pounds.

Equation: **Existing MLVSS, lb = (MLVSS, mg/L)(Aeration Tank, mil gal)(8.34 lb/gal)**

Existing MLVSS, lb = (3,820 mg/L)(1.205 mil gal)(8.34 lb/gal) = 38,390 lb MLVSS

Next, determine the desired MLVSS in pounds.

Equation: $$\textbf{Desired MLVSS, lb} = \frac{(\text{Primary effluent COD, mg/L})(\text{mgd})(8.34\text{ lb/gal})}{\text{Desired COD lb/MLVSS lb}}$$

$$\text{Desired MLVSS, lb} = \frac{(136\text{ mg/L})(3.82\text{ mgd})(8.34\text{ lb/gal})}{0.175\text{ COD lb/MLVSS lb}} = 24{,}759\text{ lb MLVSS}$$

Next, subtract the existing MLVSS from the desired MLVSS to find the waste in pounds.

Waste, lb = 38,390 lb − 24,759 lb = 13,631 lb

Next, calculate the waste rate in mgd.

Equation: $$\textbf{Waste rate, mgd} = \frac{\text{Waste, lb}}{(\text{WVS concentration, mg/L})(8.34\text{ lb/gal})}$$

$$\text{Waste rate, mgd} = \frac{13{,}631\text{ lb}}{(4{,}290\text{ mg/L})(8.34\text{ lb/gal})} = 0.38098\text{ mgd, round to } \textbf{0.381 mgd}$$

Lastly, calculate the waste rate in gpm.

$$\text{Waste rate, gpm} = \frac{(0.38098\text{ mgd})(1{,}000{,}000\text{ gpd/mgd})}{1{,}440\text{ min/day}} = 264.57\text{ gpm, round to } \textbf{265 gpm}$$

18. **What are the solids loading for a dissolved air flotation (DAF) unit in lb/d/ft² that is 55.2 ft by 29.9 ft, where sludge flow averages 82 gpm, with a waste-activated sludge (WAS) concentration of 6,340 mg/L, and the sludge has a specific gravity of 1.04?**

First, determine the area of the DAF unit in ft².

DAF area, ft² = (55.2 ft)(29.9 ft) = 1,650.48 ft²

Next, convert gpm to mgd.

$$\text{Number of mgd} = \frac{(82\text{ gpm})(1,440\text{ min}/\text{day})}{1,000,000/\text{mil}} = 0.11808\text{ mgd}$$

Next, calculate the weight of the sludge in lb/gal.

Sludge, lb/gal = (8.34 lb/gal)(1.04 sp gr) = 8.6736 lb/gal

Next, calculate the solids loading.

Equation: **Solids loading, lb/d/ft²** $= \dfrac{(\text{WAS, mg/L})(\text{mgd})(\text{lb/gal, Sludge})}{\text{DAF area, ft}^2}$

$$\text{Solids loading, lb/d/ft}^2 = \frac{(6,340\text{ mg/L, WAS})(0.11808\text{ mgd})(8.6736\text{ lb/gal, Sludge})}{1,650.48\text{ ft}^2\text{ DAF}}$$

Solids loading, lb/d/ft² = 3.934 lb/d/ft², round to **3.9 lb/d/ft²**

19. **Given the following data, calculate the average gallons applied to a drying bed for each cycle period and the solids loading in lb/yr/ft²:**

Drying bed = 261 feet long and 48.0 feet wide
Average sludge application per cycle = 4.59 inches
Average percent solids = 4.88%
Drying and removal cycle on average = 21.6 days

First, convert 4.59 inches to feet.

Number of feet = 4.59 in./12 in./ft = 0.3825 ft

Next, determine the volume in ft^3 sent to the drying bed.

Volume, ft^3 = (261 ft)(48.0 ft)(0.3825 ft) = 4,791.96 ft^3

Next, calculate the volume in gallons sent to the sand drying beds.

Number of gal = (4,791.96 ft^3)(7.48 gal/ft^3) = 35,843.86 gal, round to **35,800 gal**

Next, calculate the number of lb. Assume 8.34/gal.

Number of lb = (35,843.86 gal)(8.34 lb/gal) = 298,938 lb

Lastly, calculate the solids loading rate.

Equation: **Solids loading rate, lb/yr/ft^2** = $\dfrac{\dfrac{\textbf{(lb)(365 days)(Percent solids)}}{\textbf{(Drying cycle)(yr)(100\%)}}}{\textbf{Drying bed area, ft}^2}$

Substitute values and solve.

Solids loading rate, lb/yr/ft^2 = $\dfrac{\dfrac{(298{,}938 \text{ lb})(365 \text{ days})(4.88\% \text{ solids})}{(21.6 \text{ days})(\text{yr})(100\%)}}{(261 \text{ ft})(48.0 \text{ ft})}$

Solids loading rate, lb/yr/ft^2 = $\dfrac{(13{,}839.72 \text{ lb/day})(365 \text{ days/yr})(0.0488)}{12{,}528 \text{ ft}^2}$

Solids loading rate, lb/yr/ft^2 = **19.7 lb/yr/ft^2**

20. **Calculate the amount of compost in lb/day that needs to be blended with a dewatered sludge to make a mixture that has a moisture content of 50.0%, if the dewatered digester primary sludge is 5,350 lb/day, the dewatered sludge solids is 28.6%, and the compost solids moisture content is 32.0%.**

First, determine the moisture content of the dewatered sludge.

Dewatered sludge moisture = 100% − 28.6% = 71.4%

Compost percent moisture in this case is given at 32.0%.

Equation: **Mixture's percent moisture =**

$$\frac{[(\text{Sludge, lb/day})(\% \text{ moisture}) + (\text{Compost, lb/day})(\% \text{ moisture})]100\%}{\text{Sludge, lb} + \text{Compost, lb}}$$

Substitute values and solve.

$$50.0\% \text{ moisture content} = \frac{[(5,350 \text{ lb/day})(71.4\%/100\%) + (x \text{ lb/day})(32.0\%/100\%)]100\%}{5,350 \text{ lb} + x \text{ lb/day}}$$

Solve for x.

Divide both sides of the equation by 100%.

$$0.500 = \frac{(5,350 \text{ lb/day})(71.4\%/100\%) + (x \text{ lb/day})(32.0\%/100\%)}{5,350 \text{ lb} + x \text{ lb/day}}$$

Simplify terms in the numerator.

$$0.500 = \frac{3,819.9 \text{ lb/day} + 0.320x}{5,350 \text{ lb/day} + x \text{ lb}}$$

Multiply both sides of the equation by (5,350 lb/day + x lb)

$$0.500(5,350 \text{ lb/day} + x \text{ lb/day}) = 3,819.9 \text{ lb/day} + 0.320x$$

Multiply terms on the left side of the equation.

$$2,675 \text{ lb/day} + 0.500x = 3,819.9 \text{ lb/day} + 0.320x$$

Subtract $0.320x$ and 2,675 lb from both sides of the equation.

$$0.500x - 0.320x = 3,819.9 \text{ lb/day} - 2,675 \text{ lb/day}$$

Simplify both sides of the equation.

$$0.180x = 1,144.9 \text{ lb/day}$$

$x = 6,360.56$ lb/day, round to **6,360 lb/day of Compost required**

21. **What is the capacity for a compost site in processing dry sludge in lb/day and tons/ day, given the following data?**

Site capacity = 14,700 yd³
Average compost cycle = 23.7 days
Bulk density of wet sludge = 1,690 lb/yd³
Bulk density of wet compost = 1,050 lb/yd³
Sludge solids content = 19.1%
Bulk density of wet wood chips = 635 lb/yd³
Mix ratio (MR) of wood chips to sludge = exactly 3.25 to 1

Equation: **Cycle time, days** $=$

$$\frac{(\text{Capacity, yd}^3)(\text{Bulk density of wet compost lb/yd}^3)}{\dfrac{x \text{ Dry solids, lb/day}}{\text{Percent solids}} + \dfrac{(x \text{ Dry solids, lb/day})}{\text{Percent solids}}(\text{MR})\dfrac{(\text{Bulk density of wood chips, lb/yd}^3)}{(\text{Bulk density of wet sludge lb/yd}^3)}}$$

Substitute values and solve.

$$23.7 \text{ days} = \frac{(14,700 \text{ yd}^3)(1,050 \text{ lb/yd}^3)}{\dfrac{x \text{ Dry solids, lb/day}}{19.1\%/100\%} + \dfrac{(x \text{ Dry solids, lb/day})}{19.1\%/100\%}(3.25)\dfrac{(635 \text{ lb/yd}^3)}{1,690 \text{ lb/yd}^3}}$$

Reduce and simplify.

$$23.7 \text{ days} = \frac{15,435,000 \text{ lb}}{\dfrac{x \text{ Dry solids, lb/day}}{0.191} + \dfrac{(x \text{ Dry solids, lb/day})(1.2212)}{0.191}}$$

Reduce again:

$$23.7 \text{ days} = \frac{15,435,000 \text{ lb}}{5.236x \text{ lb/day} + 6.394x, \text{ lb/day}}$$

$$23.7 \text{ days} = \frac{15,435,000 \text{ lb}}{11.629x}$$

Solve for x.

$$11.629x = \frac{15,435,000 \text{ lb}}{23.7 \text{ days}}$$

$$x = \frac{15,435,000 \text{ lb}}{(23.7 \text{ days})(11.629)}$$

$x = 56,004$ lb/day, round to **56,000 lb/day Dry sludge**

$$\text{Dry sludge, tons/day} = \frac{56,0004\,\text{lb/day}}{2,000\,\text{lb/ton}} = 28.002\ \text{tons/day, round to }\mathbf{28.0\ tons/day\ Dry\ sludge}$$

22. How many lb/acre/year of plant available nitrogen (PAN) will be applied to a wastewater land application field, given the following parameters?

Nitrate (NO_3) = 13.1 mg/L
Nitrite (NO_2) = 0.73 mg/L
Total Kjeldahl nitrogen (TKN) = 61.4 mg/L
Ammonia (NH_3) = 20.7 mg/L
Applying 10.2 in./acre/year
Mineralization rate (MR) = 0.30
Volatilization rate = 0.50

First, determine the PAN applied per year to each acre.

Equation: **PAN, mg/L = [MR(TKN − NH$_3$)] + [1 − VR(NH$_3$)] + (NO$_3$ + NO$_2$)**

PAN, mg/L = [0.30(61.4 mg/L − 20.7 mg/L)] + [1 − 0.50(20.7 mg/L)] + (13.1 mg/L + 0.73 mg/L)

PAN, mg/L = [0.30(40.7 mg/L)] + 10.35 mg/L + 13.83 mg/L

PAN, mg/L = 12.21 mg/L + 10.35 mg/L + 13.83 mg/L = 36.39 mg/L

Next, convert the hydraulic loading rate from inches to mil gal/acre/yr.

$$\text{Flow, mil gal/acre/yr} = \frac{(43,560\,\text{ft}^3/\text{acre})(1\,\text{ft}/12\,\text{in.})(7.48\,\text{gal/ft}^3)(10.2\,\text{in./acre/yr})(1\,\text{acre})}{1,000,000/\text{mil}}$$

Flow, mil gal/acre/yr = 0.27695 mil gal/acre/yr

Lastly, solve for PAN in lb/acre/yr.

PAN, lb/acre/yr = (36.39 mg/L)(0.27695 mil gal/acre/yr)(8.34 lb/gal)

PAN, lb/acre/yr = 84.05 lb/acre/yr, round to **84 lb/acre/yr PAN**

23. Given the following data, calculate the sodium absorption ratio (SAR) for a wastewater:

Sodium (Na^+) = 118 mg/L
Calcium (Ca^{2+}) = 43.7 mg/L
Magnesium (Mg^{2+}) = 25.2 mg/L
Equivalent weight of sodium = 22.99
Equivalent weight of calcium = 20.04
Equivalent weight of magnesium = 12.15

First, determine the milliequivalents (meq) for sodium, calcium, and magnesium.

$$\text{Sodium, meq} = \frac{118\,\text{mg/L}}{22.99} = 5.1327\,\text{meq of sodium}$$

$$\text{Calcium, meq} = \frac{43.7\,\text{mg/L}}{20.04} = 2.1806\,\text{meq of calcium}$$

$$\text{Magnesium, meq} = \frac{25.2\,\text{mg/L}}{12.15} = 2.074\,\text{meq of magnesium}$$

Now, solve for SAR.

Equation: $$\textbf{SAR} = \frac{Na^+}{[(0.5)(Ca^{2+} + Mg^{2+})]^{1/2}}$$

Substitute known values and solve.

$$SAR = \frac{5.1327\,\text{meq}\,Na^+}{[(0.5)(2.1806\,\text{meq}\,Ca^{2+} + 2.074\,\text{meq}\,Mg^{2+})]^{1/2}}$$

$$SAR = \frac{5.1327\,\text{meq}\,Na^+}{[(0.5)(4.2546)]^{1/2}}$$

$$SAR = \frac{5.1327\,\text{meq}\,Na^+}{[2.1273]^{1/2}}$$

Calculate the square root of 2.1273:

$$SAR = \frac{5.1327\,\text{meq}\,Na^+}{1.4585} = \textbf{3.52 SAR}$$

24. **A 20,000-mg (2.00%) stock solution (20,000 ppm or 20,000 mg/L) is required for doing jar tests. If the alum has a specific gravity of 1.296 and is 49.2% aluminum sulfate, how many milliliters of alum are required to make exactly 1,000 mL of stock solution?**

First, find the number of lb/gal of alum.

Alum, lb/gal = (sp gr)(8.34 lb/gal) = (1.296)(8.34 lb/gal) = 10.8086 lb/gal

Next, determine the number of grams/mL.

$$\text{Polymer grams/mL, Alum} = \frac{(10.8086 \text{ lb/gal})(49.2\% \text{ Al}_2\text{SO}_4, \text{Purity})(454 \text{ grams/lb})}{(3,785 \text{ mL/gal})(100\%)}$$

Polymer grams/mL, Alum = 0.6379 grams/mL

Convert grams/mL to milligrams/mL.

Polymer mg/mL = (0.6379 grams/mL)(1,000 mg/g) = 637.9 mg/mL

Next, convert mL to liters by multiplying by 1,000.

Number mg/L = (637.9 mg/mL)(1,000 mL/liter) = 637,900 mg/Liter

Next, determine the number of mL required.

Equation: $\mathbf{C_1V_1 = C_2V_2}$

(637,900 mg/L)(x, mL) = (20,000 mg/liter)(1,000 mL)

$$x, \text{mL} = \frac{(20,000 \text{ mg/Liter})(1,000 \text{ mL})}{637,900 \text{ mg/liter}} = \textbf{31.35 mL, Alum}$$

Now, using a pipette, add 31.35 mL of the 49.2% alum solution to a clean, dry 1,000-mL flask. Dilute the alum to the 1,000-mL mark with deionized water. Add a magnetic stir bar and place the flask on a magnetic stirrer. Turn the magnetic stirrer on and mix this solution with the bar as vigorously as possible for at least 10 minutes.

Thus, every 1 mL of alum solution that is added to 1,000-mL sample of raw water will add a 20 mg/L dose (because of second dilution with raw water; 20,000 mg/L/1,000 mg/L raw water sample = 20 mg/L). If 2 mL of this stock solution were added to the 1,000-mL raw water sample, it would be a dose of 40 mg/L. If you are using the 2-liter square jars, simply double the milliliters added for each 20 mg/L dosage increase desired. Another way is to feed the alum neat by using a micropipette; pipette 0.03135 mL of alum into a 1,000-mL raw water sample to make a dose of 20 mg/L.

25. **The SCADA system at a wastewater plant uses a 4-mA to 20-mA signal to monitor a chemical polymer tank level. If the readout on a SCADA board reads 13.80 mA, what is the height of the polymer in a tank with a maximum level of 16.75 feet?**

Know: 4 mA = 0 feet in the polymer tank and that 20 mA = 16.75 feet in the tank

Equation: **Current process reading** $= \dfrac{(\text{Live signal, mA} - 4\,\text{mA offset})(\text{Maximum capacity})}{16\,\text{milliamp span}}$

Substitute known values and solve.

Polymer level, ft $= \dfrac{(13.80\,\text{mA} - 4\,\text{mA offset})(16.75\,\text{ft Maximum level})}{16\,\text{mA}}$

Polymer level, ft = **10.26 ft**

APPENDIXES

APPENDIX

A

COMMON CONVERSION FACTORS

AREA

1 acre (ac) = 43,560 square feet (ft^2)
1 acre-ft = 43,560 cubic feet (ft^3)
2.4711 ac = 1 hectare
1 hectare = 0.4047 acre

1 hectare = 10,000 square meters
1 square mile = 640 acres

CONCENTRATION

1% solution = 1 part in 100 parts 1 ppm = 1 milligram per liter (mg/L)
1% solution = 10,000 parts per million (ppm) 1 grain per gal (gpg) = 17.12 ppm

DENSITY

Water has a density of 1 gram per mL (1g/mL) or 8.34 lb/gal or 62.4 lb/ft^3

FLOW

1 miner's inch = 1.5 ft^3/min
1 ft^3/s = 448.8 gal/min (gpm)

1 ft^3/s = 0.6463 million gallons per day (mgd)
1 mgd = 1.547 ft^3/s

LENGTH

1 inch = 2.54 centimeter (cm)
100 cm = 1 meter (m)
1 m = 39.37 inches
1 m = 3.281 feet (ft

1 yard = 0.9144 m
1,000 m = 1 kilometer (km)
1 km = 1.609 miles

POWER

1 horsepower (hp) = 0.746 kilowatts (kw)
1 kw = 1.341 hp)

PRESSURE

1 lb per sq in (psi) = 2.307 ft. of water
1 foot of water = 0.4335 psi
1 atmosphere (atm) = 14.7 psi

1 atmosphere = 29.92 inches of mercury
1 atm = 33.90 ft of water
1 atm = 760 mm of mercury

TEMPERATURE

Degrees Fahrenheit (°F) = (9 °F/5 °C)(°C) + 32°F
Degrees Celsius (°C) = (°F - 32 °F)(5 °C /9 °F)

VOLUME

2 pints = 1 quart
8 pints = 1 gallon (gal)
4 quarts = 1 gallon
1 quart = 32 fluid ounces
1 gallon = 128 fluid ounces
1 mL = 1 cubic centimeter
1 cubic foot = 7.48 gallons
1 acre-foot = 325,829 gal

1 gallon = 3.785 liters
1 liter = 1.0567 quarts
1 liter = 1,000 milliliters (mL)
3,785 ml = 1 gallon
1,000 liters = 1 cubic meter
1 cubic meter = 35.3 ft^3
1 MG = 3.07 acre-feet

WEIGHT

1 gram (g) = 1,000 milligrams
1,000 gm = 1 kilogram (kg)
1 lb = 454 g
1 lb = 7,000 grains (gr)

1 kg = 2.205 pounds (lb)
2,000 lb = 1 ton
1 mg/L = 1 part per million (ppm)
1 grain per gal (gpg) = 17.1 ppm

APPENDIX B

SUMMARY OF WASTEWATER TREATMENT EQUATIONS

AREA EQUATIONS

Area of a rectangle = (Length)(Width)

Area of a circle(tank) = $(0.785)(\text{Diameter})^2$ or πr^2

Area of a parallelogram = (Base)(Height)

$$\text{Area of a trapezoid} = \frac{(\text{Altitude})(\text{Base}_1 + \text{Base}_2)}{2}$$

BASIC ELECTRICITY FORMULAS

Voltage = (Amps)(Resistance, ohms)

Resistance, ohms = Voltage/Amps

$$\text{Current process reading} = \frac{(\text{Live signal, mA} - 4\,\text{mA offset})(\text{Maximum capacity})}{16\,\text{mA span}}$$

BIOCHEMICAL OXYGEN DEMAND LOADING EQUATION

BOD_5, lb/day = (BOD_5, mg/L)(Number of mgd)(8.34 lb/gal)

BIOCHEMICAL OXYGEN DEMAND UNSEEDED AND SEEDED EQUATIONS

$$BOD_5 \text{ unseeded, mg/L} = \frac{(\text{Initial DO, mg/L} - \text{Final DO, mg/L})(\text{Total volume, mL})}{\text{Sample volume, mL}}$$

$$\text{Seed correction, mg/L} = \frac{(BOD_5 \text{ of seed stock, mg/L})(\text{Seed stock, mg/L})}{\text{Total volume, mL}}$$

BOD_5 seeded, mg/L =

$$\frac{(\text{Initial DO, mg/L} - \text{Final DO, mg/L} - \text{Seed correction, mg/L})(\text{Total volume, mL})}{\text{Sample volume, mL}}$$

BIOSOLIDS CONCENTRATION FACTOR

$$CF = \frac{\text{Sample volume, mL}}{\text{Percent influent biosolids}}$$

BIOSOLIDS PUMPING AND PRODUCTION FORMULAS

Estimated pumping rate =

$$\frac{(\text{Influent TSS, mg/L} - \text{Effluent TSS, mg/L})(\text{Flow, mgd})(8.34 \text{ lb/gal})}{(\text{Percent solids in sludge})(\text{Sludge, lb/gal})(1,440 \text{ min/day})}$$

$$\text{Biosolids, lb/mil gal} = \frac{(\text{Biosolids, gal})(8.34 \text{ lb/gal})}{(\text{Flow, mgd})(\text{Number of days})}$$

$$\text{Biosolids, lb/mil gal} = \frac{(\text{Settled sludge, mL})(100\%)}{\text{Total sample vol., mL}}$$

$$\text{Biosolids, wet tons/yr} = \frac{(\text{Biosolids, lb/mil gal})(\text{mgd})(365 \text{ days/yr})}{2,000 \text{ lb/ton}}$$

Estimated pumping rate =

$$\frac{(\text{Influent TSS, mg/L} - \text{Effluent TSS, mg/L}) (\text{Flow, mgd}) (8.34 \text{ lb/gal})}{(\text{Percent solids in sludge}) (\text{Sludge, lb/gal}) (1,440 \text{ min/day})}$$

BIOSOLIDS RETENTION TIME EQUATION

$$\text{BRT, days} = \frac{\text{Digester working volume, gal}}{\text{Influent flow, gpd}}$$

BIOSOLIDS VOLUME INDEX AND BIOSOLIDS DENSITY INDEX EQUATIONS

$$\text{BVI} = \frac{(\text{Settled biosolids, mL/L}) (1,000 \text{ mg/g})}{\text{MLSS, mg/L}}$$

$$\text{BDI} = \frac{(\text{Capacity of digester}) (\text{Percent seed sl}\text{u}}{100\%}$$

CENTRIFUGE THICKENING EQUATIONS

Hydraulic loading, gal/day = (Sludge flow, gpm)(1,440 min/day)

$$\text{Feed time, min} = \frac{(\text{Capacity, ft}^3) (\text{Solids, }\%/100\%) (7.48 \text{ gal/ft}^3) (8.34 \text{ lb/gal})}{(\text{Flow, gpm}) (\text{Solids concentration, }\%/100\%) (8.34 \text{ lb/gal})}$$

$$\text{Simplified: Feed time, min} = \frac{(\text{Capacity, ft}^3) (\text{Solids, }\%) (7.48 \text{ gal/ft}^3)}{(\text{Flow, gpm}) (\text{Solids concentration, }\%)}$$

CHEMICAL FEED SOLUTION SETTINGS

$$\text{Feed rate, mL/min} = \frac{(\text{gpd}) (3,785 \text{ mL/gal})}{1,440 \text{ min/day}}$$

or

$$\text{Number of mL/min} = \frac{(\text{Number of gallons used}) (3,785 \text{ mL/gal})}{1,440 \text{ min/day}}$$

CHEMICAL OXYGEN DEMAND LOADING FORMULA

COD, lb/day = (COD, mg/L)(Number of mgd)(8.34 lb/gal)

CHEMISTRY AND LABORATORY EQUATIONS

$$\text{Moles} = \frac{\text{Grams of chemical}}{\text{Gram formula weight}}$$

$$\text{Percent of element in compound} = \frac{(\text{Molecular Wt of the element})(100\%)}{\text{Molecular Wt of compound}}$$

$$\text{BOD}_5 \text{ unseeded, mg/L} = \frac{(\text{Initial DO, mg/L} - \text{Final DO, mg/L})(\text{Total volume, mL})}{\text{Sample volume, mL}}$$

$$\text{Seed correction, mg/L} = \frac{(\text{BOD}_5 \text{ of seed stock, mg/L})(\text{Seed stock, mg/L})}{\text{Total volume, mL}}$$

BOD$_5$ seeded, mg/L =

$$\frac{(\text{Initial DO, mg/L} - \text{Final DO, mg/L} - \text{Seed correction, mg/L})(\text{Total volume, mL})}{\text{Sample volume, mL}}$$

$$\text{Percent of an element in a compound} = \frac{\text{Flow, gpm}}{(\text{Width, ft})(\text{Depth, ft})(60 \text{ sec/min})(7.}$$

$$\text{Molarity} = \frac{\text{Moles solute}}{\text{Liters solution}}$$

$$\text{Normality (N)} = \frac{\text{Number of grams} - \text{equivalents of solute}}{\text{Number of liters of solution}}$$

$$\text{Dosage, mg/L} = \frac{(\text{Stock, mL})(1{,}000 \text{ mg/gram})(\text{Concentration in grams/liter})}{\text{Sample size, mL}}$$

$$\text{Percent VS} = \frac{(\text{Solids lost, g})(100\%)}{\text{Weight of total solids, g}}$$

$$\text{Alum, mg/L} = \frac{(\text{Stock, mL})(1{,}000 \text{ mg/gram})(\text{Concentration in grams/Liter})}{\text{Sample size, mL}}$$

Solids (ash), g = Sample and dish dried, g - Burnt sample, g

CIRCUMFERENCE FORMULAS

Circumference = π(Diameter)

Circumference = 2π(radius) or $2\pi r$

COMMON CONVERSION FACTORS

Gallons to pounds: Number of pounds (lb) = (Number of gal)(8.34 lb/gal)

Gallons to cubic feet: Number of ft^3 = $\dfrac{(\text{Number of gal})(1 \text{ ft}^3)}{7.48 \text{ gal}}$

Acre-feet to cubic feet: Number of ft^3 = (Number of acre-ft)(43,560 ft^3/acre-ft)

mgd to ft^3/s: Number ft^3/s = (Number of mgd)$\dfrac{(1,000,000 \text{ gal})}{(1 \text{ mil gal})} \dfrac{(1 \text{ ft}^3)}{(7.48 \text{ gal})} \dfrac{(1 \text{ day})}{(1,440 \text{ min})} \dfrac{(1 \text{ min})}{(60 \text{ sec})}$

ft^3/s to mgd: Number of mgd = $\dfrac{(\text{Number of ft}^3)}{\text{sec}} \dfrac{(60 \text{ sec})}{\text{min}} \dfrac{(1,440 \text{ min})}{\text{day}} \dfrac{(7.48 \text{ gal})}{\text{ft}^3} \dfrac{(1 \text{ mil gal})}{1,000,000 \text{ gal}}$

ft^3/s to gpd = Number of gpd = (Number of ft^3/s)(86,400 s/day)(7.48 gal/ft^3)

gpm to ft^3/s: Number of ft^3/s = $\dfrac{\text{Number of gpm}}{(60 \text{ s/min})(7.48 \text{ gal/ft}^3)}$

ppm to Percent: Percent solution = (Known ppm) $\dfrac{1\%}{10,000 \text{ ppm}}$

Gallons to liters: Number of liters = (Number of gal)(3.785 L/1 gal)

COMPOSTING EQUATIONS

Mixture's % moisture = $\dfrac{[(\text{Sludge, lb})(\% \text{ moisture}) + (\text{Compost, lb})(\% \text{ moisture})]100\%}{\text{Sludge, lb} + \text{Compost, lb}}$

Percent solids BC =

$$\frac{(\text{Sludge, yd}^3)(\text{lb/yd}^3)(\% \text{ solids, sludge}) + (\text{Sludge, yd}^3)(\text{MR})(\text{lb/yd}^3)(\% \text{ solids, chips})(100\%)}{(\text{Sludge, yd}^3)(\text{lb/yd}^3) + (\text{Sludge, yd}^3)(\text{Mix ratio})(\text{lb/yd}^3)}$$

$$\text{Compost cycle, days} = \frac{(\text{Site capacity, yd}^3)(\text{Density of compost, lb/yd}^3)}{x \text{ Wet compost, lb/day}}$$

Cycle time, days =

$$\frac{(\text{Capacity, yd}^3)(\text{Bulk density of wet compost lb/yd}^3)}{\dfrac{x \text{ Dry solids, lb/day}}{\text{Percent solids}} + \dfrac{(x \text{ Dry solids, lb/day})}{\text{Percent solids}}(\text{MR})\dfrac{(\text{Bulk density of wood chips, lb/yd}^3)}{(\text{Bulk density of wet sludge lb/yd}^3)}}$$

Percent moisture in mixture =

$$\frac{[(\text{DB, lb/day})(\text{Percent moisture DB}) + (\text{Compost lb/day})(\text{Percent moisture compost})]100\%}{\text{DB, lb/day} + \text{Compost, lb/day}}$$

DENSITY EQUATIONS

Density = Mass/Volume

Number of g/cm^3 = (Number of lb/gal)(454 g/1 lb)(1 gal/3,785 cm^3)

DETENTION TIME EQUATIONS

$$\text{Detention time, hr} = \frac{\text{Volume, gal}}{\text{Flow rate, gal/ hour}}$$

$$\text{Detention time, hr} = \frac{(\text{Volume, gal})(24 \text{ hr/day})}{\text{Flow, gpd}}$$

DEWATERING FORMULAS

Total nonfilterable residue, mg/L = Total residue, mg/L · Total filterable residue, mg/L

$$\text{Filter yield, lb/hr/ft}^2 = \frac{(\text{Wet cake flow, lb/hr})(\text{Percent solids}/100\%)}{\text{Area, ft}^2}$$

$$\text{Vacuum filter loading, lb/day/ft}^2 = \frac{(\text{Biosolids, gpd})(8.34 \text{ lb/gal})(\text{Percent solids})}{\text{Vacuum filter area, ft}^2}$$

DIGESTER GAS PRODUCTION FORMULAS

$$\text{Gas produced, ft}^3/\text{lb VS destroyed} = \frac{\text{Gas production, ft}^3/\text{day}}{\text{VS destroyed, lb/day}}$$

$$\text{Gas produced, m}^3/\text{lb VS destroyed} = \frac{(\text{Gas production, ft}^3/\text{day})}{(\text{VS destroyed, lb/day})(35.3 \text{ m}^3/\text{ft}^3)}$$

DIGESTER LOADING RATE EQUATIONS

$$\text{Digester loading rate, lb VSA/d/ft}^3 = \frac{\text{VSA, lb}}{\text{Volume of digester, ft}^3}$$

Where VSA is volatile solids added

Digester loading, lb VS/day/ft³ =

$$\frac{(\text{Flow, gpd})(8.34 \text{ lb/gal})(\text{sp gr})(\text{Percent sludge})(\text{Percent volatile solids})}{(0.785)(\text{Diameter})^2(\text{Sludge level})}$$

Digester loading, lb VS/day/1,000 ft³ =

$$\frac{(\text{Flow, gpd})(8.34 \text{ lb/gal})(\text{sp gr})(\text{Percent sludge})(\text{Percent volatile solids})}{(0.785)(\text{Diameter})^2(\text{Sludge level})}$$

DIGESTER VOLATILE SOLIDS RATIO FORMULAS

$$\text{Digester VS ratio} = \frac{\text{VS added lb/day}}{\text{lb VS in digester}}$$

$$\text{Digester VS ratio} = \frac{\text{VS added lb/day}}{(\text{lb VS in digester})(\text{TS}\%/100\%)(\text{VS}\%/100\%)}$$

DILUTION TRIANGLE

Concentration (Conc.)$_1$ Conc.$_2$ $-$ Conc. Desired $=$ Number of parts

       Conc. Desired

     Conc.$_2$ Conc.$_1$ $-$ Conc. Desired $= \dfrac{\text{Number of parts}}{\text{Total number of parts}}$

$$\frac{(\text{Number of gallons})\,(\text{Number of parts of Conc.}_1)}{\text{Total number of parts}} = \text{Number of gallons of Conc.}_1 \text{ required}$$

$$\frac{(\text{Number of gallons})\,(\text{Number of parts of Conc.}_2)}{\text{Total number of parts}} = \text{Number of gallons of Conc.}_2 \text{ required}$$

DISSOLVED AIR FLOTATION: AIR RATE FLOW EQUATIONS

Air, lb/day $=$ (Air flow, ft^3/min)(1,440 min/day)(0.0807 lb/ft^3, Air)

Air, lb/day $=$ (Air flow, ft^3/min)(60 min/hr)(0.0807 lb/ft^3, Air)

DISSOLVED AIR FLOTATION: AIR-TO-SOLIDS RATIO EQUATION

$$\text{Air-to-solids ratio} = \frac{(\text{Air flow, ft}^3/\min)\,(\text{Air, lb/ft}^3)}{(\text{gpm})\,(\text{Percent solids}/100\%)\,(8.34 \text{ lb/gal})}$$

DISSOLVED AIR FLOTATION: THICKENER SOLIDS LOADING EQUATIONS

$$\text{Solids loading, lb/d/ft}^2 = \frac{(\text{WAS, mg/L})\,(\text{mgd})\,(\text{Sludge, lb/gal})}{\text{DAF area, ft}^2}$$

$$\text{Solids loading, lb/hr/ft}^2 = \frac{(\text{WAS, mg/L})\,(\text{mgd})\,(\text{Sludge, lb/gal})}{(\text{DAF area, ft}^2)\,(24 \text{ hr/day})}$$

DOSAGE FORMULAS

Chlorine Dose = Chlorine demand + Chlorine residual

Chemical feed, lb/day = (Flow, mgd)(Dosage, mg/L)(8.34 lb/gal)

or rearranging to solve for dosage:

$$\text{Dosage, mg/L} = \frac{\text{lb/day}}{(\text{mgd})(8.34 \text{ lb/gal})}$$

$$\text{lb/day} = \frac{(\text{mgd})(\text{Dosage, mg/L})(8.34 \text{ lb/gal})}{(\text{Percent purity}/100\%)}$$

Above formula used when the purity of a substance or solution is less than 100%.

$$(\text{mgd})(x, \text{mg/L})(8.34 \text{ lb/gal}) = (\text{mgd})(\text{Dosage, mg/L})(8.34 \text{ lb/gal})$$

$$\text{Chlorine, lb/day} = \frac{(\text{Dosage, mg/L})(\text{mgd})(8.34 \text{ lb/gal})}{\text{Percent available chlorine}/100\%}$$

$$\text{Chemical dosage, mg/L} = \frac{(\text{mL/min})(1{,}440 \text{ min/day})(\text{Chemical, lb/gal})}{(3{,}785 \text{ mL/gal})(\text{mgd})(8.34 \text{ lb/gal})}$$

$$\text{Chemical dosage, mg/L} = \frac{(\text{mL/min})(1{,}440 \text{ min/day})(\text{Chemical, lb/gal})}{(3{,}785 \text{ mL/gal})(\text{mgd})(8.34 \text{ lb/gal})(\text{Percent Polymer})}$$

DRY CHEMICAL FEED SETTING EQUATION

$$\text{Chemical, lb/day} = \frac{(\text{Number of g/min})(1{,}440 \text{ min/day})}{454 \text{ g/lb}}$$

EXTRAPOLATION: USED FOR PIPES NOT FLOWING FULL

$$\text{Division Factor} = \frac{(\text{High } d/D - \text{Low } d/D)}{\text{Ratio } d/D}$$

FOOD/MICROORGANISM RATIO FORMULAS

$$F/M = \frac{(BOD_5, mg/L)(Flow, mgd)(8.34\ lb/gal)}{(mg/L\ MLVSS)(Volume\ in\ tank, mil\ gal)(8.34\ lb/gal)}$$

or as follows since 8.34 lb/gal in the numerator and denominator cancel each other out.

$$F/M = \frac{(BOD_5, mg/L)(Flow, mgd)}{(mg/L\ MLVSS)(Volume\ of\ tank, mil\ gal)}$$

FLOW RATE EQUATIONS

Flow = Volume/Time

Q (Flow) = (Area)(Velocity)

Example: Q, flow in ft^3/s = (Area, ft^2)(Velocity in feet per sec)

Flow in a pipe that changes size: (Area 1, sq ft)(Velocity 1, ft/s) = (Area 2, sq ft)(Velocity 2, ft/s)

Flow (Q), $ft^3/sec = \frac{(w_1 + w_2)}{2}$ (Depth, ft)(Velocity, ft/s)

GRAVITY THICKENER SOLIDS LOADING FORMULAS

$$Solids\ loading,\ lb/d/ft^2 = \frac{(Flow, gpm)(1,440\ min/day)(Percent\ solids)}{(Gravity\ thickener\ area)(100\%)}$$

$$Solids\ loading,\ lb/d/ft^2 = \frac{(Flow, gpd)(Percent\ solids)}{(Gravity\ thickener\ area)(100\%)}$$

GRIT REMOVAL FORMULA

$$Grit\ removal,\ ft^3/mil\ gal = \frac{Number\ of\ gallons\ removed}{(7.48\ gal/ft^3)(mil\ gal\ treated)}$$

HYDRAULIC DIGESTION TIME EQUATIONS

$$\text{Digestion time, days} = \frac{\text{Number of gallons}}{\text{Influent sludge flow, gal/day}}$$

$$\text{Digestion time, days} = \frac{(0.785)(\text{Diameter})^2(\text{Depth, ft})(7.48 \text{ gal/ft}^3)}{\text{Influent sludge flow, gal/day}}$$

HYDRAULIC LOADING RATE EQUATIONS

$$\text{Hydraulic loading rate} = \frac{\text{Total flow, gpd}}{\text{Surface area, ft}^2}$$

$$\text{Hydraulic loading rate, in/day} = \frac{\text{Flow, gpd}}{(27,152 \text{ gal/acre in.})(\text{Area, acres})}$$

KILOWATT FORMULAS

$$kW = (\text{Number of hp})(0.746 \text{ kW/hp})$$

$$\text{Kilowatt-hr/day} = (\text{hp})(0.746 \text{ kW/hp})(\text{Operating time, hr/day})$$

$$kW = (\text{Number of hp})(0.746 \text{ kW/hp})(\text{Startup energy})$$

LIME NEUTRALIZATION FORMULA

$$\text{Lime, lb} = (\text{Volatile acids, mg/L})(\text{mil gal})(8.34 \text{ lb/gal})$$

MASS BALANCE (PERCENT) EQUATION

$$\text{Percent mass balance} = \frac{(\text{Solids produced, lb/day} + \text{Solids removed, lb/day})(100\%)}{\text{Solids produced, lb/day}}$$

MEAN CELL RESIDENCE TIME EQUATIONS

$$\text{MCRT, days} = \frac{(\text{MLSS, mg/L})\,(\text{mil gal})\,(8.34\ \text{lb/day})}{\text{SS wasted, lb/day} + \text{SS lb/day}}$$

MCRT, days =

$$\frac{(\text{MLSS, mg/L})(\text{Aeration tank mil gal} + \text{Clarifier tank mil gal})(8.34\ \text{lb/gal})}{(\text{WAS, mg/L})(\text{Waste rate, mgd})(8.34\ \text{lb/gal}) + (\text{TSS, mg/L})(\text{Flow, mgd})(8.34\ \text{lb/gal})}$$

MIXTURE FORMULA

Percent mixture strength =

$$\frac{\text{Solution}_1\ \text{gal (Available \%/100\%)} + \text{Solution}_2\ \text{gal (Available \%/100\%)}(100\%)}{\text{gal of Solution}_1 + \text{gal of Solution}_2}$$

NITROGEN LOADING RATE EQUATION

Nitrogen loading rate, lb/day = (Total Nitrogen, mg/L)(mgd)(8.34 lb/gal)

NITROGEN (TOTAL) EQUATION

Total nitrogen (N) = Nitrate, mg/L + Nitrite, mg/L + TKN, mg/L

OPERATING TIME FORMULAS

$$\text{Operating time} = \frac{\text{Treated water}}{\text{Flow rate}}$$

$$\text{Operating time, min/hr} = \frac{(\text{Flow, mgd})(\text{Influent SS, mg/L} - \text{Effluent SS, mg/L})(100\%)}{(\text{Sludge pump, gpm})(\text{Percent solids})(24\ \text{hr/day})}$$

ORGANIC LOADING RATE EQUATIONS

$$\text{Organic loading rate, lb BOD}_5/\text{d/acre} = \frac{(\text{BOD}_5, \text{mg/L}) \, (\text{Flow, mgd}) \, (8.34 \, \text{lb/gal})}{\text{Surface area of pond, acre ft}}$$

$$\text{Organic loading rate, lb BOD}_5/\text{d/1,000 ft}^3 = \frac{(\text{BOD}_5, \text{mg/L}) \, (\text{Flow, mgd}) \, (8.34 \, \text{lb/gal})}{\text{Volume of trickling filter, ft}^3/1,000 \, \text{ft}^3}$$

PARTICULATE AND SOLUBLE BIOCHEMICAL OXYGEN DEMAND FORMULAS

Particulate BOD_5, mg/L = (SS, mg/L)(K value)

Soluble BOD_5 = Total BOD_5 - (K factor)(Total SS)

Total BOD_5 = (Particulate BOD_5)(K factor) + Soluble BOD_5

PERIMETER FORMULAS

Circumference = π(Diameter) = (3.14)(Diameter)

Rectangle = 2(Length) + 2(Width)

PLANT AVAILABLE NITROGEN FORMULAS

PAN, mg/L = [MR(TKN - NH_3)] + [1 – VR(NH_3)] + (NO_3 + NO_2)

$$\text{PAN, dry tons/acre} = \frac{\text{Plant nitrogen required, lb/acre}}{\text{PAN, lb/dry ton}}$$

PAN, lb/dry ton = [(Organic N, mg/kg)(MR) + (Ammonia N, mg/kg)(VR)](0.002 lb/dry ton)

POPULATION EQUIVALENT EQUATION

$$\text{Number of people} = \frac{(\text{BOD}_5, \text{mg/L})\,(\text{mgd})\,(8.34\,\text{lb/gal})}{\text{lb/day of BOD}_5/\text{person}}$$

POPULATION LOADING EQUATION

$$\text{Population loading, people/acre} = \frac{\text{Number of people served}}{\text{Area of pond(s), acres}}$$

PRESSURE FORMULAS

$$\text{psi} = \frac{\text{Depth, ft}}{2.31\,\text{ft/psi}}$$

$$\text{Height, ft} = (\text{psi})(2.31\,\text{ft/psi})$$

$$\text{psi} = (\text{Depth, ft})(0.433\,\text{psi/ft})$$

$$\text{Pressure} = \frac{\text{Force, lb}}{\text{Area, ft}^2}$$

$$\text{Pressure, lb/ft}^2 = (\text{Height or Depth, ft})(\text{Density, } 62.4\,\text{lb/ft}^3)$$

$$\frac{\text{Pressure}_A}{w} + \frac{\text{Velocity}_A^2}{2g} = \frac{\text{Pressure}_B}{w} + \frac{\text{Velocity}_B^2}{2g}$$

PERCENT MIXTURE FORMULAS

Percent mixture strength=

$$\frac{(\text{Solution}_1\,\text{lb})(\text{Available \%}/100\%) + (\text{Solution}_2, \text{lb})(\text{Available \%}/100\%)(100\%)}{\text{Solution}_1, \text{lb} + \text{Solution}_2, \text{lb}}$$

Percent mixture strength =

$$\frac{[(\text{Solution}_1\,\text{gal})(\text{lb/gal})(\text{Avail \%}/100\%) + (\text{Solution}_2\,\text{gal})(\text{lb/gal})(\text{Avail \%}/100\%)]100\%}{(\text{Solution}_1, \text{gal})(\text{lb/gal}) + (\text{Solution}_2, \text{gal})(\text{lb/gal})}$$

PERCENT RECOVERY EQUATION

$$\text{Percent recovery} = \frac{\text{Cake TS, \% (Feed sludge TSS, \% } - \text{ Return flow TSS, \%) (100\%)}}{\text{Feed sludge TSS, \% (Cake TS, \% } - \text{ Return flow TSS, \%)}}$$

PERCENT REDUCTION EQUATIONS

$$\text{Percent VS reduction} = \frac{(\text{Influent} - \text{Effluent}) (100\%)}{\text{Effluent} - (\text{Effluent}) (\text{Influent})}$$

$$\text{Percent VM reduction} = \frac{(\text{Percent influent VM} - \text{Percent effluent VM}) (100\%)}{[\text{Percent influent VM} - (\text{Percent influent VM}) (\text{Percent effluent VM})]}$$

Percent moisture reduction =

$$\frac{(\text{Percent influent moisture} - \text{Percent moisture, after digestion}) (100\%)}{[\text{Percent influent moisture} - (\text{Percent influent moisture}) (\text{Percent moisture, after digestion})]}$$

PERCENT REMOVAL FORMULAS

$$\text{Percent ntu removal} = \frac{(\text{Influent ntu} - \text{Effluent ntu}) (100\%)}{\text{Influent ntu}} \quad \text{or} \quad \frac{(\text{In} - \text{Out}) (100\%)}{\text{In}}$$

$$\text{Percent BOD}_5 \text{ removal} = \frac{(\text{Influent BOD}_5 - \text{Effluent BOD}_5) (100\%)}{\text{Influent BOD}_5} \quad \text{or} \quad \frac{(\text{In} - \text{Out}) (100\%)}{\text{In}}$$

$$\text{Percent removal efficiency} = \frac{(\text{Solids removed, mg/L}) (100\%)}{\text{Influent solids, mg/L}}$$

PERCENT STRENGTH OF SOLUTIONS AND SOLIDS

$$\text{Percent strength} = \frac{(\text{Number of lb of chemical}) (100\%)}{\text{Number of lb, Water} + \text{lb chemical}}$$

$$\text{Percent strength} = \frac{(\text{Number of grams of chemical}) (100\%)}{\text{Number of grams, water} + \text{grams chemical}}$$

$$\text{Percent total solids} = \frac{(\text{Dry sample in grams}) (100\%)}{\text{Sludge sample in grams}}$$

PERCENT SETTLED SLUDGE AND SOLIDS FORMULAS

$$\text{Percent settled sludge} = \frac{(\text{Settled sludge, mL})\,(100\%)}{\text{Total sample vol., mL}}$$

or similarly:

$$\text{Percent settleable solids} = \frac{(\text{Settled sludge, mL})\,(100\%)}{\text{Sample size, mL}}$$

$$\text{Percent inorganic solids} = \frac{(\text{Dry sample in grams})\,(100\%)}{\text{Sludge sample in grams}}$$

$(2° \text{ gpd})(2° \text{ sludge lb/gal})(x\% \ 2° \text{ sludge}) = (1° \text{ sludge, gpd})(1° \text{ sludge lb/gal})(\% \ 1° \text{ sludge})$

PHOSPHATE LOADING RATE EQUATION

Phosphorus (P) loading rate, lb/day = (P, mg/L)(mgd)(8.34 lb/gal)

PIT VOLUME AND DAYS TO FILL FORMULAS

$$\text{Screenings, ft}^3/\text{mil gal} = \frac{\text{Number of ft}^3/\text{day}}{\text{Number of mgd}}$$

$$\text{Number of days to fill} = \frac{\text{Pit volume, ft}^3}{\text{Screenings removed, ft}^3/\text{day}}$$

PUMPING AND PUMPING COST FORMULAS

$$\text{mhp} = \frac{(\text{whp})}{(\text{Motor efficiency})\,(\text{Pump efficiency})}$$

Where mhp = motor horsepower and whp = water horsepower

$$\text{mhp} = \frac{(\text{Flow, gpm})\,(\text{TH, ft})}{(3,960)\,(\text{Motor efficiency})\,(\text{Pump efficiency})}$$

Where TH = Total head

$$bhp = \frac{(Flow,\ gpm)\ (Differential\ pressure,\ psi)}{(1,714)\ (Pump\ efficiency)}$$

Brake hp = (hp)(Motor efficiency)

Where hp = horsepower

Water hp = (mhp)(Motor efficiency)(Pump efficiency)

Water hp = (bhp)(Pump efficiency)

Motor hp = bhp/Motor efficiency

Brake hp = whp/Pump efficiency

Cost, $/day = (Motor hp)(24 hr/day)(0.746 kW/hp)(Cost/kW-hr)

PUMPING RATE EQUATIONS

Pumping rate = Flow, gal/Time, min

$$Pump's\ discharge\ rate,\ gpm = \frac{Discharge,\ gal}{Time,\ min}$$

Discharge rate, gpm = Influent flow, gpm + Level drop, gpm

Number of gal per stroke = $(0.785)(Bore\ diameter,\ ft)^2(Stroke,\ ft)(7.48\ gal/ft^3)$

RATIOS

$$\text{Ratio} = \frac{\text{Re circulated flow}}{\text{Plant influent flow}}$$

$$\frac{\text{Speed setting}_1, \text{Percent}}{\text{Polymer dosage}_1, \text{mL}} = \frac{\text{Speed setting}_2, \text{Percent}}{\text{Polymer dosage}_2, \text{mL}}$$

$$\frac{\text{Chlorine dosage}_1, \text{mg/L}}{\text{Flow}_1, \text{mgd}} = \frac{\text{Chlorine dosage}_2, \text{mg/L}}{\text{Flow}_2, \text{mgd}}$$

$$\frac{\text{Digester solids}_1, \text{lb/day}}{\text{Flow}_1, \text{gpm}} = \frac{\text{Digester solids}_2, \text{lb/day}}{\text{Flow}_2, \text{gpm}}$$

$$\frac{\text{Alum dosage}_1, \text{mL}}{\text{Speed setting}_1, \%} = \frac{\text{Alum dosage}_2, \text{mL}}{\text{Speed setting}_2, \%}$$

SCREENINGS FORMULA

$$\text{Screenings, ft}^3/\text{mil gal} = \frac{\text{Number of ft}^3/\text{day}}{\text{Number of mgd}}$$

SEED SLUDGE EQUATION

$$\text{Seed sludge, gal} = \frac{(\text{Capacity of digester})(\text{Percent seed sludge required})}{100\%}$$

SLUDGE AGE (GOULD) EQUATIONS

$$\text{Sludge age, days} = \frac{\text{Solids under aeration, lb}}{\text{Solids added, lb/day}}$$

$$\text{Sludge age, days} = \frac{(\text{MLSS, mg/L})(\text{Volume of aeration tank})(8.34 \text{ lb/gal})}{(\text{SS, mg/L})(\text{Flow, mgd})(8.34 \text{ lb/gal})}$$

SLUDGE PUMPING FORMULAS

Sludge, lb/day = (Pumping, min/day)(24 hr/day)(Pump rate, gpm)(8.34 lb/gal)(sp gr of sludge) (% solids)

(Primary sludge, gal)(Primary sludge, lb/gal)(Percent PSS) =

(x Thickened sludge, gal)(Thickened sludge, lb/gal)(Percent TSS)

SLUDGE REMOVED EQUATION

SS removed, lb/day = (SS removed, mg/L)(Number of mgd)(8.34 lb/gal)

SLUDGE VOLUME INDEX AND SLUDGE VOLUME DENSITY EQUATIONS

$$SVI = \frac{(SS, mL)}{MLSS, g/L}$$

$$SDI = \frac{(MLSS, g)(100\%)}{SS, mL}$$

SODIUM ABSORPTION RATIO FORMULA

$$\text{Sodium absorption ratio} = \frac{Na^+}{[(0.5)(Ca^{2+} + Mg^{2+})]^{1/2}}$$

SOLIDS BALANCE (DIGESTER) EQUATIONS

Total solids, lb/day = (Raw sludge, lb/day)(Percent solids)

Fixed solids, lb/day = Total solids, lb/day - VS, lb/day

Water in sludge, lb/day = Sludge, lb/day - Total solids, lb/day

Percent VSR = $\dfrac{(\text{In} - \text{Out})(100\%)}{\text{In} - (\text{In})(\text{Out})}$

Gas produced, lb/day = (Effluent VS, lb/day)(Percent VSR)

VS in digested sludge, lb/day = Influent VS, lb/day - Destroyed VS, lb/day

Total digested solids, lb/day = $\dfrac{\text{VS digested, lb/day}}{\text{Percent digested VS}}$

Fixed solids, lb/day = Total digested solids, lb/day - VS digested, lb/day

Digested sludge, lb/day = $\dfrac{\text{Total digested solids, lb/day}}{\text{Digested sludge percent solids}}$

Water in digested sludge, lb/day = Sludge, lb/day - Total solids, lb/day

SOLIDS LOADING RATE EQUATIONS

Solids loading rate = $\dfrac{(\text{MLSS, mg/L})(\text{mgd})(8.34\ \text{lb/gal})}{\text{Area, ft}^2}$

Solids loading rate, lb/d/ft^2 = $\dfrac{(\text{Percent solids})(\text{Biosolids added, gpd})(8.34\ \text{lb/gal})}{(\text{Surface area, ft}^2)}$

SOLIDS PRODUCED FORMULA

Solids produced, lb/day = (BOD$_5$ removed, lb/day)(0.85 lb solids/lb BOD$_5$)

SOLIDS PUMPING EQUATIONS

$$\text{Solids, lb/day} = \frac{(\text{Sludge, lb/day})(\text{Percent solids})}{100\%}$$

Solids, lb/day = (Pumping, min/day)(24 hr/day)(Pump rate, gpm)(8.34 lb/gal)(sp gr of sludge)

Solids, lb/day = (Time, min/cycle)(cycles/day)(Pump rate, gpm)(8.34 lb/gal)(Percent solids)

SOLIDS RETENTION TIME EQUATION

Target SRT =

$$\frac{(\text{MLSS mg/L})(\text{Clarifier, Aerator Volume, mil gal})(8.34\ \text{lb/gal})}{(\text{RAS SS mg/L})(x\ \text{mgd})(8.34\ \text{lb/gal}) + (\text{Effluent SS, mg/L})(\text{Flow, mgd})(8.34\ \text{lb/gal})}$$

SOLIDS UNDER AERATION

Number of lb, solids = (SS, mg/L)(Number of mil gal)(8.34 lb/gal)

SOLUTION FORMULAS

(Concentration$_1$)(Volume$_1$) + (Concentration$_2$)(Volume$_2$) = (Concentration$_3$)(Volume$_3$)

or abbreviating the above equation: $C_1V_1 + C_2V_2 = C_3V_3$

$$\text{Percent HTH solution} = \frac{(\text{lb HTH})(100\%)}{(\text{Number of gal})(8.34\ \text{lb/gal})}$$

(Solution$_1$ percent)(x gal, Solution$_1$)(Solution$_1$, lb/gal) = (Solution$_2$ percent)(Solution$_2$, gal)(Solution$_2$, lb/gal)

SPECIFIC GRAVITY EQUATIONS

Specific gravity (sp gr) = Density of substance/Density of water

$$\text{Sp gr} = \frac{\text{Solute, lb/gal}}{8.34 \text{ lb/gal}}$$

$$\text{Sp gr} = \frac{\text{Number of lb/ft}^3}{62.4 \text{ lb/ft}^3}$$

STATISTIC FORMULAS

$$\text{Average} = \frac{\text{Sum of all measurements}}{\text{Number of measurements}}$$

or

$$\text{Arithmetic mean} = \frac{\text{Sum of all measurements}}{\text{Number of measurements}}$$

Median is the middle value.

Range = Largest value - Smallest value

Mode is the valve that occurs most frequently.

Note: There also can be two or more modes within a set of numbers.

Geometric mean = $[(x_1)(x_2)(x_3)(x_4).....(x_n)]^{1/n}$

Standard deviation = $[\Sigma f(x - X^-)^2/n - 1]^{1/2}$

SURFACE LOADING FORMULA

$$\text{Surface loading rate} = \frac{\text{gallons per day (gpd)}}{\text{Number of ft}^2}$$

SUSPENDED SOLIDS LOADING EQUATION

Suspended solids, lb/day = (SS, mg/L)(Number of mgd)(8.34 lb/gal)

TEMPERATURE FORMULAS

$$°C = 5/9 \ (°F - 32)$$

$$°F = 9/5 \ (°C) + 32$$

TOTAL FORCE AND HYDRAULIC PRESS EQUATIONS

Total force, pounds = (Area, in.²)(psig) or Total force = (Pressure)(Area)

$$\text{Pressure} = \frac{\text{Total force, lb}}{\text{Area, ft}^2}$$

TOTAL HEAD FORMULAS

Total head, ft = Total static head, ft + Head losses, ft

$$\text{TDH} = \frac{(\text{Differential pressure})\,(2.31 \text{ ft/psi})}{\text{Specific gravity}}$$

VACUUM FILTER OPERATING TIME EQUATION

$$\text{Filter yield, lb/hr/ft}^2 = \frac{\dfrac{(\text{Solids, lb/day})(\text{Percent recovery})}{(\text{Filter operation, lb/day})(100\% \ )}}{\text{Filter area ft}^2}$$

VELOCITY EQUATIONS

$$\text{Velocity, ft/s} = \frac{\text{Flow, gpm}}{(\text{Width, ft})(\text{Depth, ft})(60 \text{ sec}/\text{min})(7.48 \text{ gal/ft}^3)}$$

or

$$\text{Velocity, ft/s} = \frac{\text{Flow, gpm}}{(\text{Area, ft}^2)(60 \text{ sec}/\text{min})(7.48 \text{ gal/ft}^3)}$$

VOLATILE-ACIDS-TO-ALKALINITY RATIO FORMULAS

Ratio = Volatile acids/Alkalinity

Volatile acids = (Alkalinity)(Ratio)

$$\text{Digester VS ratio} = \frac{\text{VS added lb/day}}{\text{lb VS in digester}}$$

VOLATILE SOLIDS DESTROYED EQUATION

$$\text{VS destroyed, lb/day/ft}^3 = \frac{(\text{Flow, gpd})(\text{Sludge, lb/gal})(\text{SSC, }\%)(\text{VSC, }\%)(\text{VSR, }\%)}{\text{Digester capacity, }1{,}000 \text{ ft}^3}$$

VOLATILE SOLIDS: LB/DAY (PERCENT)

VS, lb/day = (Number of lb/day, sent to digester)(Percent VS/100%)

$$\text{VS, lb/day} = (\text{Number of gpd to digester})\frac{(\text{Percent solids})}{100\%}\frac{(\text{Percent VS})}{100\%}(8.34 \text{ lb/gal})$$

VOLATILE SOLIDS PUMPING FORMULA

VS, lb/day =

(Time, min/cycle)(cycles/day)(Pump rate, gpm)(8.34 lb/gal)(Percent, solids)(Percent VM)

VOLUME EQUATIONS

Volume of a basin in ft^3 or m^3 = (Length)(Width)(Depth)

Volume of a basin in gallons = (Length)(Width)(Depth)(7.48 gal/ft^3)

Volume of a cone in ft^3 or m^3 = $1/3\pi r^2$(Height or Depth)

Volume of a circular tank in ft^3 or m^3 = πr^2(Height) or Volume of a pipe = πr^2(Length) or

Volume of a cylindrical tank in ft^3 or m^3 = (0.785)(Diameter)^2(Height)

Volume of a trough in gallons = $\dfrac{(b_1 - b_2)}{2}$(Depth of water)(Length)(7.48 gal/ft^3)

Volume, gal = $\dfrac{(Length_1 + Length_2)}{2}\ \dfrac{(Width_1 + Width_2)}{2}$(Depth, ft)(7.48 gal/ft^3)

Volume of sphere, ft^3 = $\dfrac{4\pi r^3}{3}$

Digester capacity, ft^3 = π(radius)^2(Height, ft)

Equation below is for a partially filled pipe division factor for determining flow:

Division factor = $\dfrac{(High\ d/D - Low\ d/D)}{Ratio\ d/D}$

WASTE ACTIVATED SLUDGE LOADING RATE EQUATION

WAS, lb/day = (WAS, mg/L)(Number of mgd)(8.34 lb/gal)

WASTE ACTIVATED SLUDGE PUMPING RATE FORMULA

$$\text{Number of mgd} = \frac{\text{WAS, lb/day}}{(\text{Number of mg/L WAS})(8.34\,\text{lb/gal})}$$

WASTE RATE EQUATIONS

$$\text{Waste rate, mgd} = \frac{\text{Waste, lb/day}}{(\text{WAS, mg/L})(8.34\,\text{lb/gal})}$$

(WVS concentration, mg/L)(8.34 lb/gal)

The following two equations are used to find the waste rate (above equation).

$$\text{Desired MLVSS, lb} = \frac{(\text{Primary effluent COD, mg/L})(\text{mgd})(8.34\,\text{lb/gal})}{\text{Desired COD lb/MLVSS lb}}$$

Existing MLVSS, lb = (MLVSS, mg/L)(Aeration tank, mil gal)(8.34 lb/gal)

$$\text{Waste rate, gpd} = \frac{(\text{Solids produced, lb/day})(1,000,000/\text{mil})}{(\text{Waste TSS, mg/L})(8.34\,\text{lb/gal})}$$

Waste rate, lb/day =

$$\frac{\text{MLSS, mg/L}\,[\text{AT, mil gal} + \text{CT, mil gal}]\,(8.34\,\text{lb/gal})}{\text{Desired MCRT}} - (\text{TSS,mg/L})(\text{Flow, mgd})(8.34\,\text{lb/gal})$$

WEIR LENGTH FORMULAS

Weir length, ft = π(Diameter, ft)

Weir length, ft = 2π(radius, ft)

WEIR AND SURFACE OVERFLOW RATE EQUATIONS

Weir overflow rate $= \dfrac{\text{Flow, gpd}}{\text{Weir Length, ft}}$

Surface overflow rate $= \dfrac{\text{Flow, gpd}}{\text{Area, ft}^2}$

APPENDIX

C

TABLE C-1 **International Atomic Weights (Based on Carbon-12)**							
Element	**Symbol**	**Atomic Number**	**Atomic Weight**	**Element**	**Symbol**	**Atomic Number**	**Atomic Weight**
Actinium	Ac	89	227.0278	Chromium	Cr	24	51.996
Aluminum	Al	13	26.98154	Cobalt	Co	27	58.9332
Americium	Am	95	(243)*	Copper	Cu	29	63.546
Antimony	Sb	51	121.75	Curium	Cm	96	(247)*
Argon	Ar	18	39.948	Curium	Cm	96	(247)*
Arsenic	As	33	74.9216	Dubnium	Db	105	(262.0)
Astatine	At	85	(210)	Dysprosium	Dy	66	162.50
Barium	Ba	56	137.33	Einsteinium	Es	99	(252)
Berkelium	Bk	97	(247)	Erbium	Er	68	167.26
Beryllium	Be	4	9.01218	Europium	Eu	63	151.96
Bismuth	Bi	83	208.9804	Fermium	Fm	100	(257)
Bohrium	Bh	107	(262.0)	Fluorine	F	9	18.998403
Boron	B	5	10.81	Francium	Fr	87	(223)
Bromine	Br	35	79.904	Gadolinium	Gd	64	157.25
Cadmium	Cd	48	112.41	Gallium	Ga	31	69.72
Calcium	Ca	20	40.08	Germanium	Ge	32	72.59
Californium	Cf	98	(251)	Gold	Au	79	196.9665
Carbon	C	6	12.011	Hafnium	Hf	72	178.49
Cerium	Ce	58	140.12	Hassium	Hs	108	(265.0)
Cesium	Cs	55	132.9054	Helium	He	2	4.00260
Chlorine	Cl	17	35.453	Holmium	Ho	67	164.9304

*The numbers in parentheses indicate mass number of most stable known isotope.

Element	Symbol	Atomic Number	Atomic Weight	Element	Symbol	Atomic Number	Atomic Weight
Hydrogen	H	1	1.00794	Radon	Rn	86	(222)
Indium	In	49	114.82	Rhenium	Re	75	186.207
Iodine	I	53	126.9045	Rhodium	Rh	45	102.9055
Iridium	Ir	77	192.22	Rubidium	Rb	37	85.4678
Iron	Fe	26	55.847	Ruthenium	Ru	44	101.07
Krypton	Kr	36	83.80	Rutherfordium	Rf	104	(261.0)
Lanthanum	La	57	138.9055	Samarium	Sm	62	150.36
Lawrencium	Lw	103	(260)	Scandium	Sc	21	44.9559
Lead	Pb	82	207.2	Seaborgium	Sg	106	(263.0)
Lithium	Li	3	6.941	Selenium	Se	34	78.96
Lutetium	Lu	71	174.967	Silicon	Si	14	28.0855
Magnesium	Mg	12	24.305	Silver	Ag	47	107.8682
Manganese	Mn	25	54.9380	Sodium	Na	11	22.98977
Meitnerium	Mt	109	(266.0)	Strontium	Sr	38	87.62
Mendelevium	Mv	101	(258)	Sulfur	S	16	32.06
Mercury	Hg	80	200.59	Tantalum	Ta	73	180.9479
Molybdenum	Mo	42	95.94	Technetium	Tc	43	(98)
Neodymium	Nd	60	144.24	Tellurium	Te	52	127.60
Neon	Ne	10	20.179	Terbium	Tb	65	158.9254
Neptunium	Np	93	237.0482	Thallium	Tl	81	204.383
Nickel	Ni	28	58.69	Thorium	Th	90	232.0381
Niobium	Nb	41	92.9064	Thulium	Tm	69	168.9342
Nitrogen	N	7	14.0067	Tin	Sn	50	118.69
Nobelium	No	102	(259)	Titanium	Ti	22	47.88
Osmium	Os	76	190.2	Tungsten	W	74	183.85
Oxygen	O	8	15.9994	Unumbium	Uub	112	(277.0)
Palladium	Pd	46	106.42	Ununnilium	Uun	110	(269.0)
Phosphorus	P	15	30.97376	Unununium	Uuu	111	(272.0)
Platinum	Pt	78	195.08	Uranium	U	92	238.0289
Plutonium	Pu	94	(244)	Vanadium	V	23	50.9415
Polonium	Po	84	(209)	Xenon	Xe	54	131.29
Potassium	K	19	39.0983	Ytterbium	Yb	70	173.04
Praseodymium	Pr	59	140.9077	Yttrium	Y	39	88.9059
Promethium	Pm	61	(145)	Zinc	Zn	30	65.38
Protactinium	Pa	91	231.0359	Zirconium	Zr	40	91.22
Radium	Ra	88	226.0254				

Source: US Government Printing Office
*The numbers in parentheses indicate mass number of most stable known isotope.

TABLE C-2			
Common Chemicals Used and Encountered in the Water Industry			
Chemical	**Chemical Formula**	**Use**	**Miscellaneous**
Alum	$Al_2(SO_4)_3 \cdot 14(H_2O)$	Coagulant	
Ammonia	NH_3	Check for chlorine leaks	Make chloramine
Bicarbonate	HCO_3^-		Carbonate hardness
Calcium carbonate	$CaCO_3$	Primary hardness chemical	
Calcium hypochlorite	$Ca(OCl)_2$	Disinfectant	
Carbon dioxide	CO_2	Re-carbonation	
Caustic soda	$NaOH$	Adjust pH	
Chlorine	Cl_2	Disinfectant	
Chlorine dioxide	ClO_2	Disinfectant	
Copper sulfate	$CuSO_4$	Algae control	
Dichloramine	$NHCl_2$	Disinfectant	Formed when ammonia is and to water containing Cl
Ferric chloride	$FeCl_3$	Coagulant	
Ferric sulfate	$Fe_2(SO_4)_3$	Coagulant	
Ferrous sulfate	$Fe_2(SO_4)_3 \cdot 7(H_2O)$	Coagulant	
Fluorosilicic acid*	H_2SiF_6	Fluoridation	
Hydrated lime	$Ca(OH)_2$	Increase pH and alkalinity	
Hydrochloric acid	HCl	Laboratory, cleaning	
Hydroxide ion	OH^-		Naturally found in water
Hypochlorite ion	OCl^-	Disinfectant	
Magnesium hydroxide	$Mg(OH)_2$		Formed in the lime-soda softening process
Monochloramine	NH_2Cl	Disinfectant	Formed when ammonia is and to water containing Cl
Nitrate	NO_3^-		Fertilizer, sewage, natural deposits that are eroded
Quicklime	CaO	Water stabilization, increase pH and alkalinity	Water softening
Ozone	O_3	Disinfectant	
Potassium permanganate	$KMnO_4$	Control tastes-and-odors substances	Oxidize Fe and Mn
Sulfuric acid	H_2SO_4	Decrease pH and alkalinity	Water stabilization
Soda ash	Na_2CO_3	Adjust pH	
Sodium aluminate	$Na_2Al_2O_4$	Coagulant	
Sodium bicarbonate	$NaHCO_3$	Increase pH and alkalinity	Water stabilization
* Formerly known as hydrofluosilicic acid or "silly acid."			
Sodium fluoride	NaF	Fluoridation	

Sodium hexametaphosphate	$(NaPO_3)_n \cdot Na_2O*$	Sequestering agent	
Sodium hydroxide	$NaOH$	Adjust pH	
Sodium silicate	$Na_2O \cdot (SiO_2)_x$	Coagulant	
Sodium fluorosilicate[†]	Na_2SiF_6	Fluoridation	
Trichloramine	NCl_3		Formed when ammonia is and to water containing Cl
Zinc orthophosphate	$Zn_3(PO_4)_2$	Forms protective coating	
* Typically n = 14 † Formerly known as sodium silicofluoride.			

TABLE C-3
Common Formula Weights*

Compound	Weight in Grams	Compound	Weight in Grams
$AgCl$	143.32	$KHC_8H_4O_4$ (phthalate)	204.224
Ag_2CrO_4	331.73	$KH(IO_3)_2$	389.912
$Al_2(SO_4)_3 \bullet 14(H_2O)$	594.35	K_2HPO_4	174.176
$BaSO_4$	233.39	KH_2PO_4	136.086
$CaCO_3$	100.089	$KHSO_4$	136.16
CaC_2O_4	128.100	KI	166.003
CaF_2	78.077	KIO_3	214.001
CaO	56.079	KIO_4	230.000
$Ca(OCl)_2$	142.985	$KMnO_4$	158.034
$Ca(OH)_2$	74.09468	KNO_3	101.101
CO_2	44.010	$Mg(OH)_2$	58.320
Cl_2	70.906	$MgSO_4$	120.36
ClO_2	67.452	MnO_2	86.937
Cr_2O_3	151.990	Mn_2O_3	157.874
CuO	79.545	Mn_3O_4	228.812
$CuSO_4$	159.60	$NaBr$	102.894
$NHCl_2$	85.92064	$NaCl$	58.443
$FeCl_3$	162.206	$NaCN$	49.007
FeO	71.846	Na_2CO_3	105.989
Fe_2O_3	159.692	$Na_2Al_2O_4$	163.940
Fe_3O_4	231.539	$NaHCO_3$	84.007
$Fe_2(SO_4)_3$	399.87	NaF	41.988
$Fe_2(SO_4)_3 \bullet 7(H_2O)$	525.97	$NaOH$	39.997
HBr	80.912	$Na_2O \bullet (SiO_2)_x$	Variable
$HC_2H_3O_2$ (acetic acid)	60.052	Na_2SiF_6	188.055
HCO_3^-	61.017	NCl_3	120.366
HCl	36.461	NH_3	17.03052
$HClO_4$	100.458	NH_2Cl	51.476
HNO_3	63.013	NH_4Cl	53.491
H_2O	18.015	NH_4NO_3	80.043
H_2O_2	34.015	$(NH_4)_2SO_4$	132.13
H_3PO_4	97.995	NO_3^-	62.005
H_2S	34.08	O_3 (ozone)	47.9982
H_2SO_3	82.07	OH^-	17.007
*Based on US Government Printing Office of atomic weights from Table C-1 above.			
H_2SO_4	98.07348	OCl^-	51.452
H_2SiF_6	144.092	$PbCrO_4$	323.2

Compound	Weight in Grams	Compound	Weight in Grams
HgO	216.59	$Pb(NO_3)_2$	331.2098
Hg_2Cl_2	472.09	PbO	223.2
$HgCl_2$	271.50	PbO_2	239.2
KBr	119.002	$PbSO_4$	303.3
$KBrO_3$	167.0005	P_2O_5	141.944
KCl	74.551	Sb_2S_3	339.68
$KClO_3$	122.550	SiO_2	60.084
KCN	65.116	$SnCl_2$	189.596
K_2CrO_4	194.190	SnO_2	150.689
$K_2Cr_2O_7$	294.184	SO_2	64.06
$K_3Fe(CN)_6$	329.248	SO_3	80.06
$K_4Fe(CN)_6$	368.346	$Zn_3(PO_4)_2$	304.703

LESS THAN FULL PIPELINE FLOW

	TABLE D-1 Depth/Diameter (d/D) Table						
d/D	**Factor**	**d/D**	**Factor**	**d/D**	**Factor**	**d/D**	**Factor**
0.01	0.0013	0.26	0.1623	0.51	0.4027	0.76	0.6405
0.02	0.0037	0.27	0.1711	0.52	0.4127	0.77	0.6489
0.03	0.0069	0.28	0.1800	0.53	0.4227	0.78	0.6573
0.04	0.0105	0.29	0.1890	0.54	0.4327	0.79	0.6655
0.05	0.0147	0.30	0.1982	0.55	0.4426	0.80	0.6736
0.06	0.0192	0.31	0.2074	0.56	0.4526	0.81	0.6815
0.07	0.0242	0.32	0.2167	0.57	0.4625	0.82	0.6893
0.08	0.0294	0.33	0.2260	0.58	0.4724	0.83	0.6969
0.09	0.0350	0.34	0.2355	0.59	0.4822	0.84	0.7043
0.10	0.0409	0.35	0.2450	0.60	0.4920	0.85	0.7115
0.11	0.0470	0.36	0.2545	0.61	0.5018	0.86	0.7186
0.12	0.0534	0.37	0.2642	0.62	0.5115	0.87	0.7254
0.13	0.0600	0.38	0.2739	0.63	0.5212	0.88	0.7320
0.14	0.0668	0.39	0.2836	0.64	0.5308	0.89	0.7384
0.15	0.0739	0.40	0.2934	0.65	0.5404	0.90	0.7445
0.16	0.0811	0.41	0.3032	0.66	0.5499	0.91	0.7504
0.17	0.0885	0.42	0.3130	0.67	0.5594	0.92	0.7560
0.18	0.0961	0.43	0.3229	0.68	0.5687	0.93	0.7612
0.19	0.1039	0.44	0.3328	0.69	0.5780	0.94	0.7662
0.20	0.1118	0.45	0.3428	0.70	0.5872	0.95	0.7707
0.21	0.1199	0.46	0.3527	0.71	0.5964	0.96	0.7749
0.22	0.1281	0.47	0.3627	0.72	0.6054	0.97	0.7785
0.23	0.1365	0.48	0.3727	0.73	0.6143	0.98	0.7816
0.24	0.1449	0.49	0.3827	0.74	0.6231	0.99	0.7841
0.25	0.1535	0.50	0.3927	0.75	0.6319	1.00	0.7854

APPENDIX

E

WASTEWATER FLOW CHART DIAGRAMS

The following 12 diagrams show some of the most common types of wastewater treatment plants with the last diagram showing sludge processing. Wastewater plants shown in figures E9 and E10 are newer technologies using membranes, nanofiltration, and electrodialysis.

Please note that not all the processes in any diagram are necessarily used. Also, some steps are not depicted. There are of course many more wastewater treatment plant types and process arrangements that are not shown. The number of different wastewater treatment plant arrangements is beyond the scope of this book. For further study of other wastewater treatment plants, please see the references.

WASTEWATER TREATMENT AND SLUDGE PROCESSING FLOW CHARTS

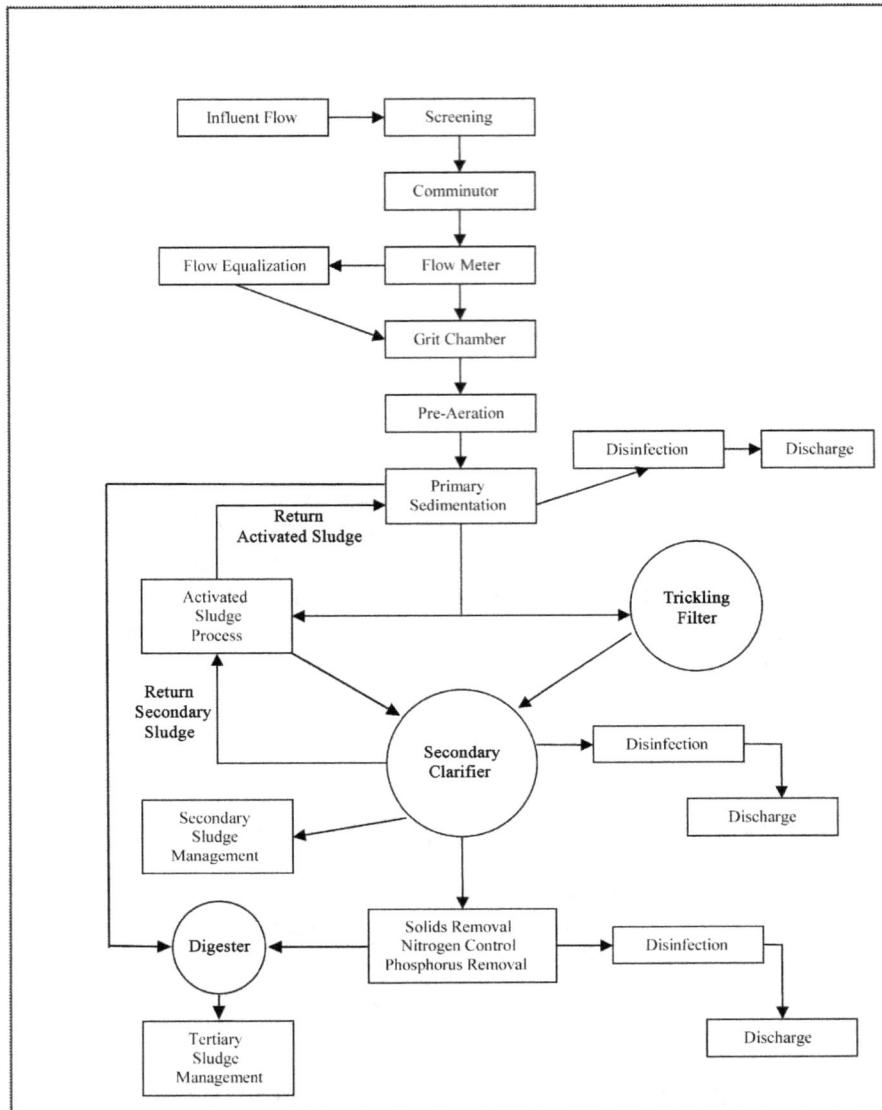

Figure E1 Flow chart of typical wastewater treatment processes

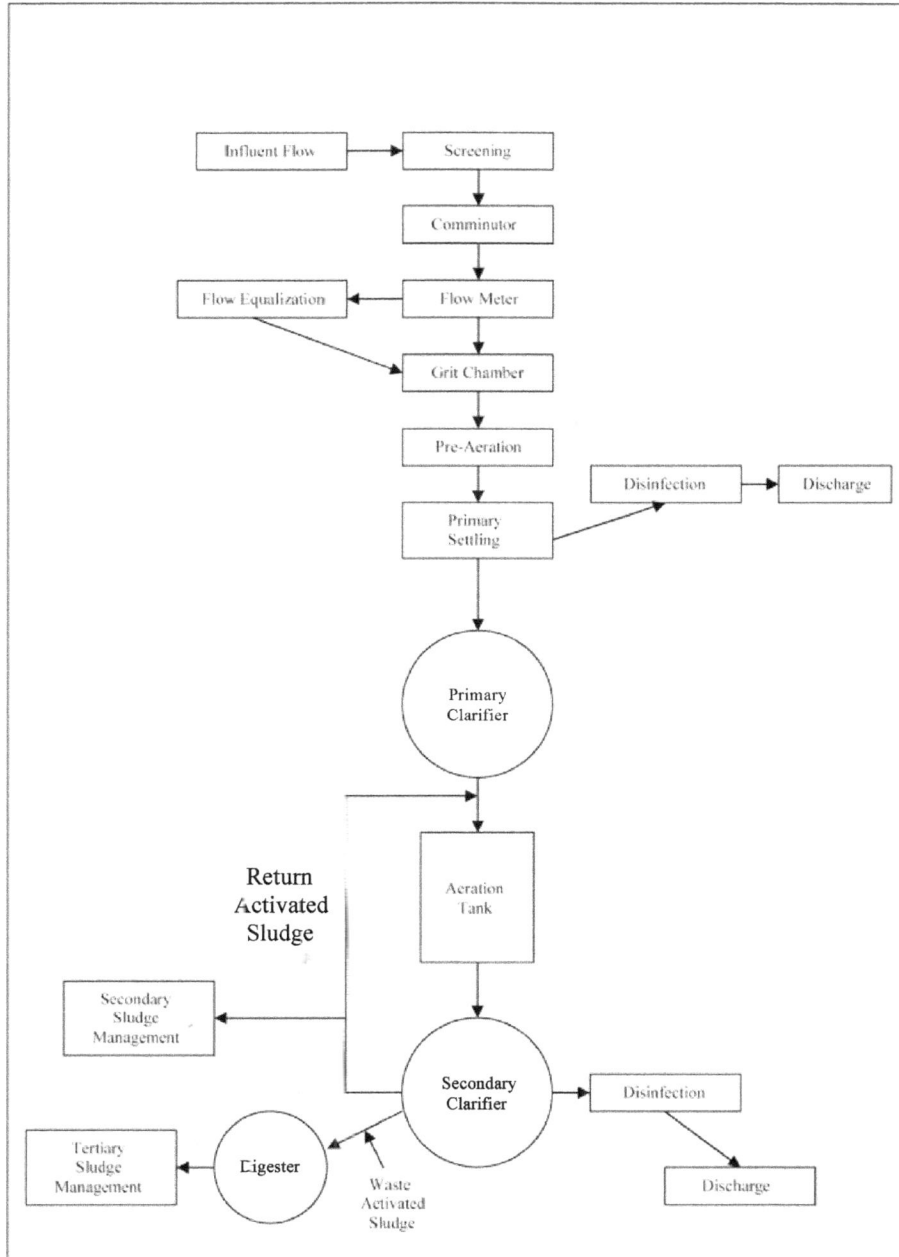

Figure E2 Flow chart of conventional activated sludge process

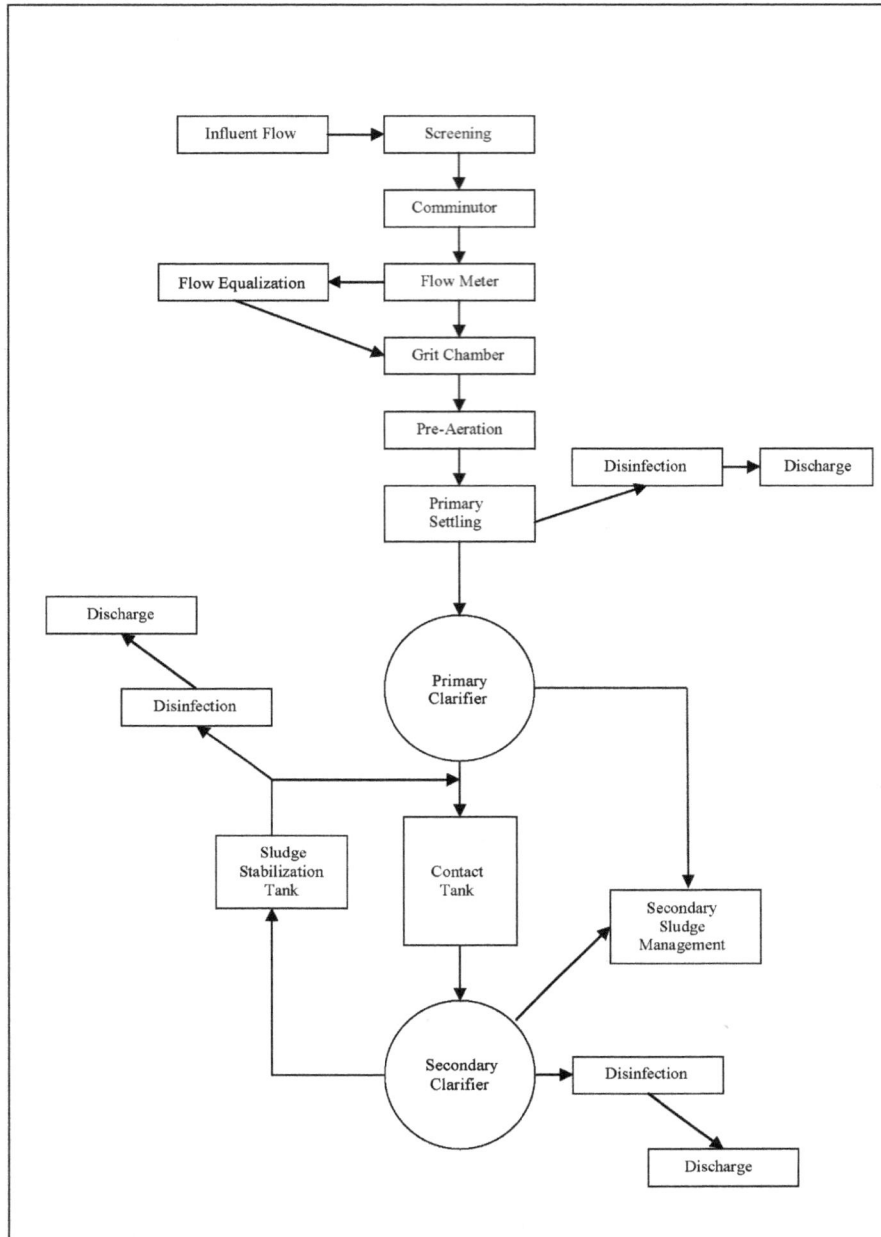

Figure E3 Flow chart of contact stabilization process

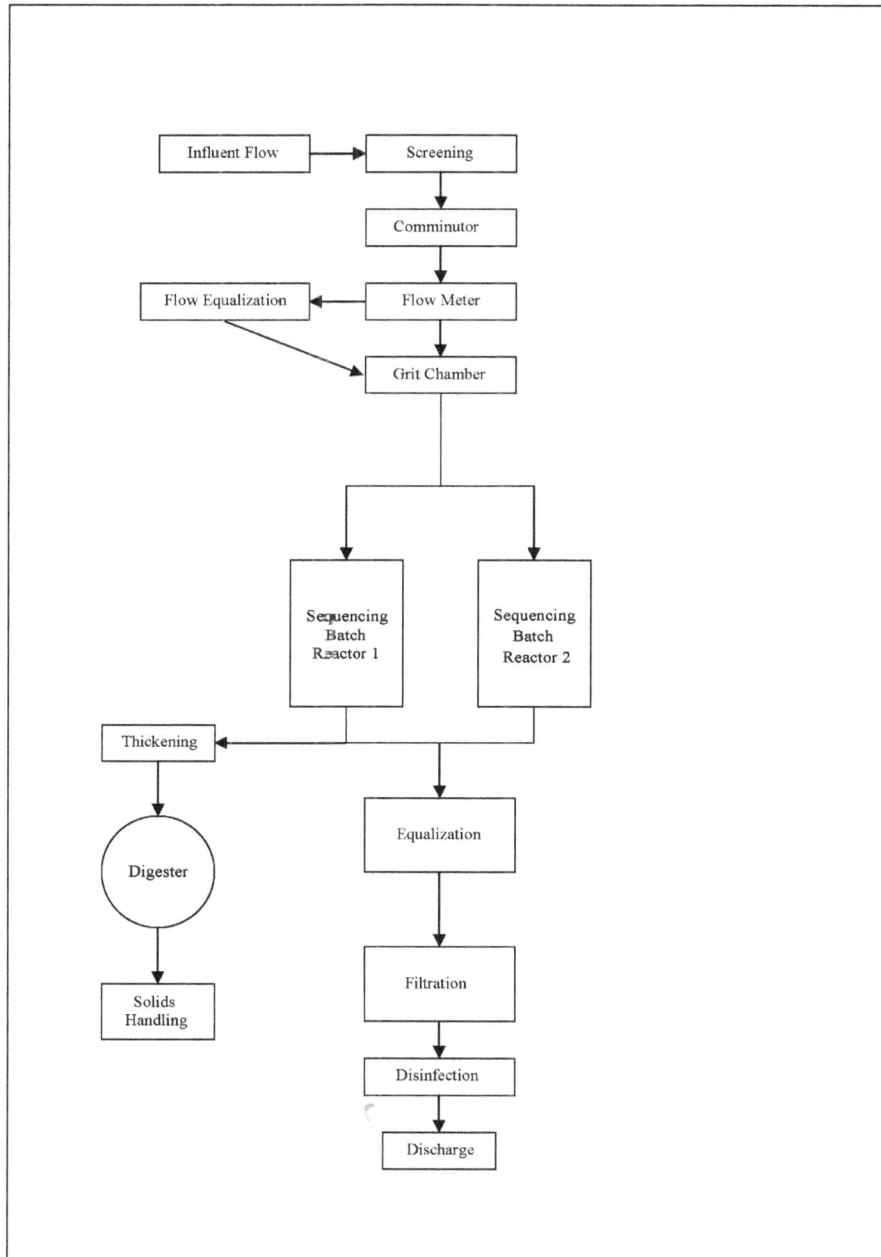

Figure E4 Flow chart of activated sequencing batch reactor

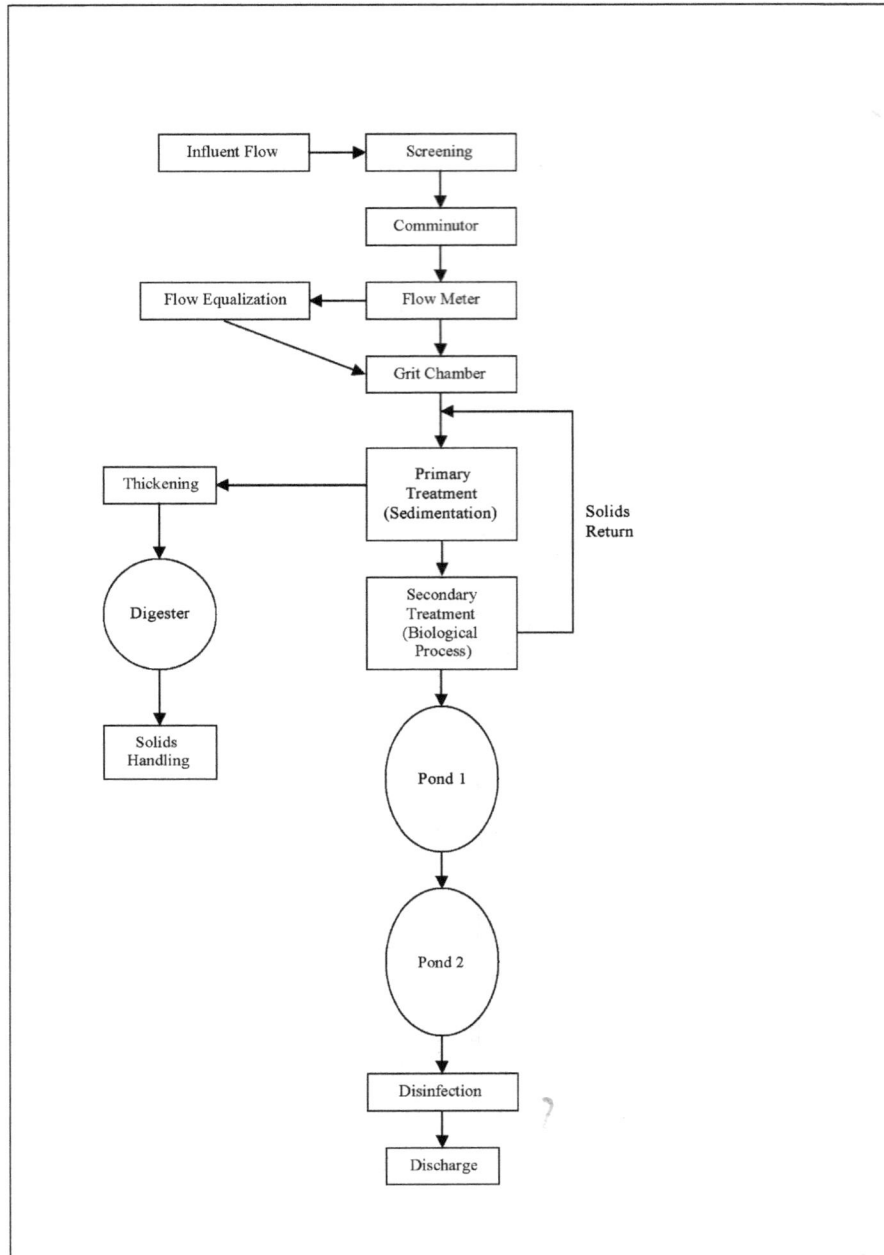

Figure E5 Flow chart of typical wastewater treatment using ponds after secondary treatment

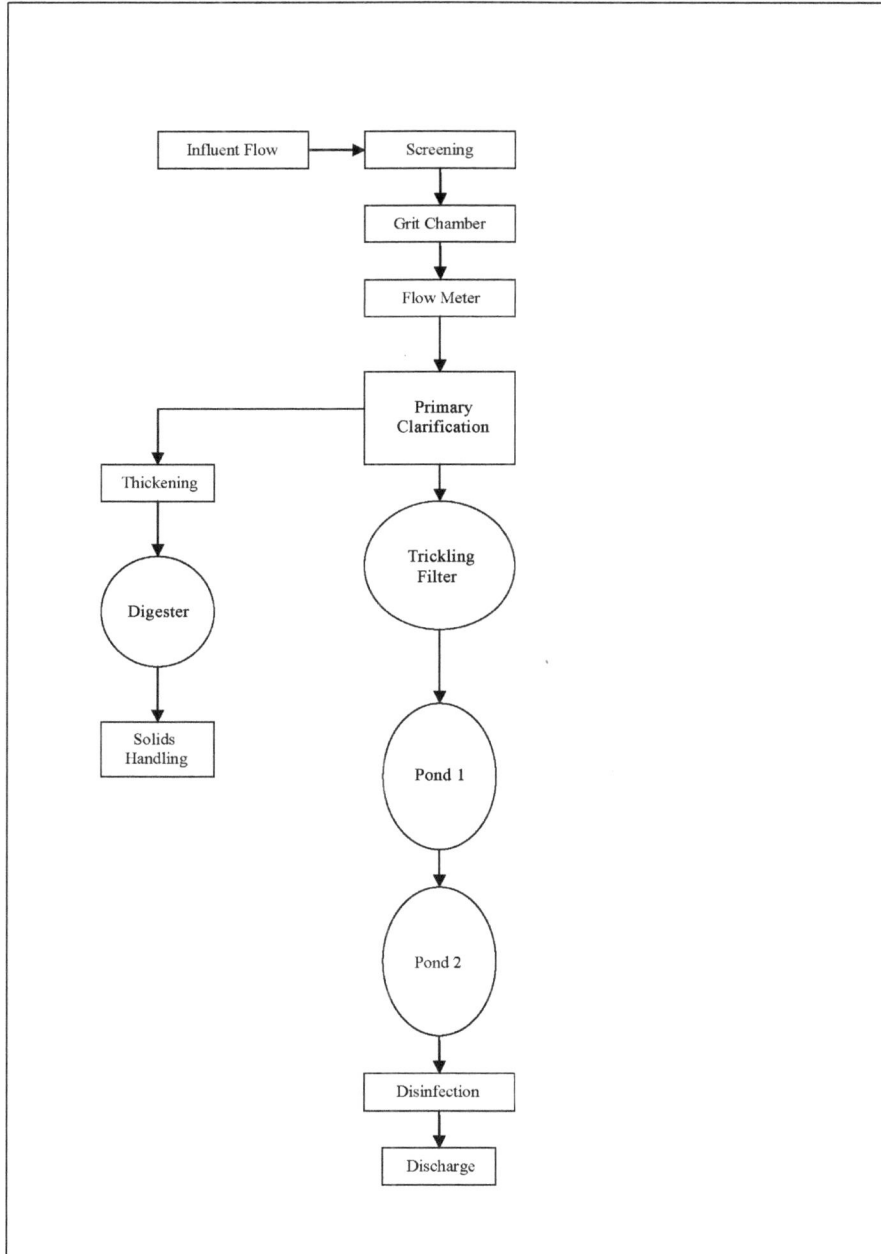

Figure E6 Flow chart of typical wastewater treatment using polishing ponds in series treatment

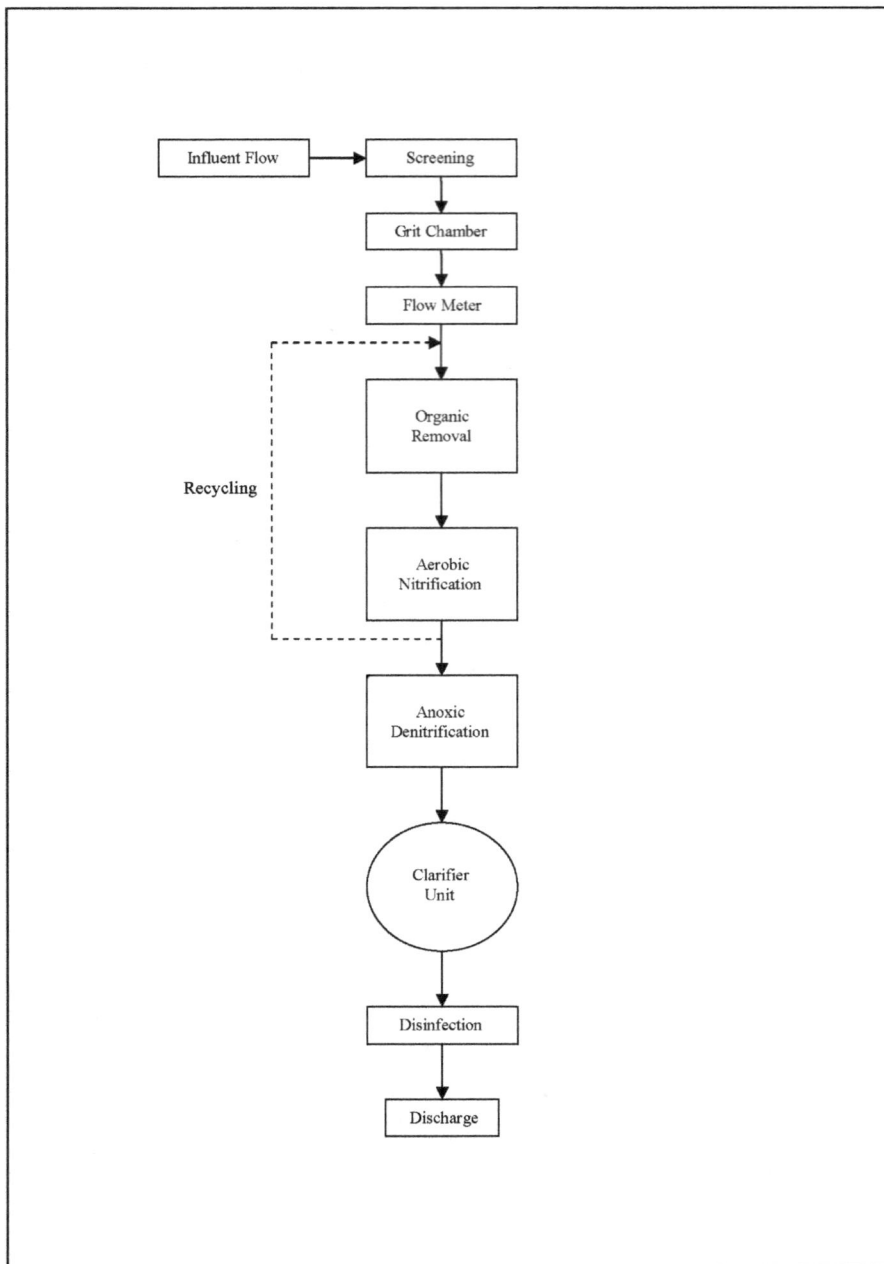

Figure E7 Flow chart of wastewater treatment using rotating biological contactor process

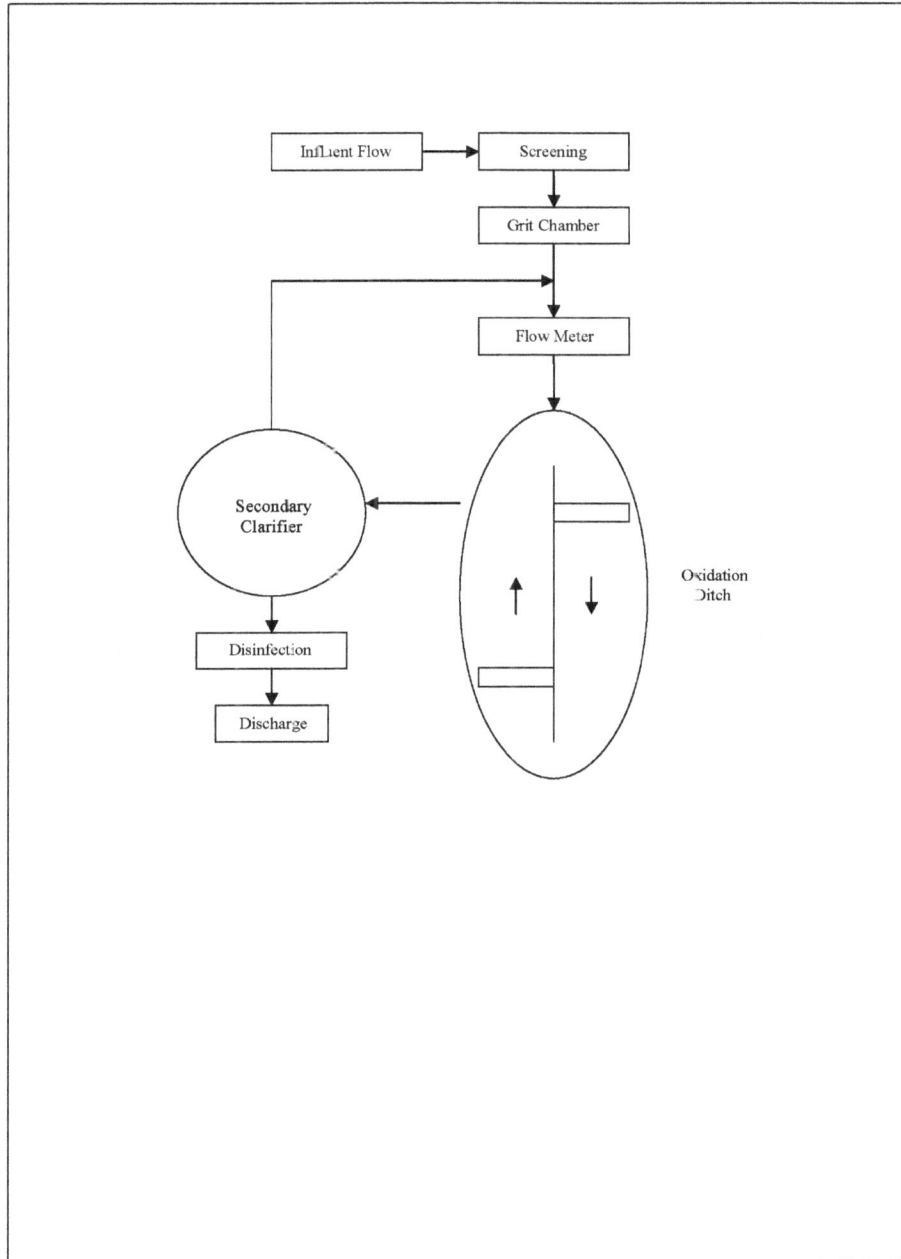

Figure E8 Flow chart of typical wastewater treatment using an oxidation ditch

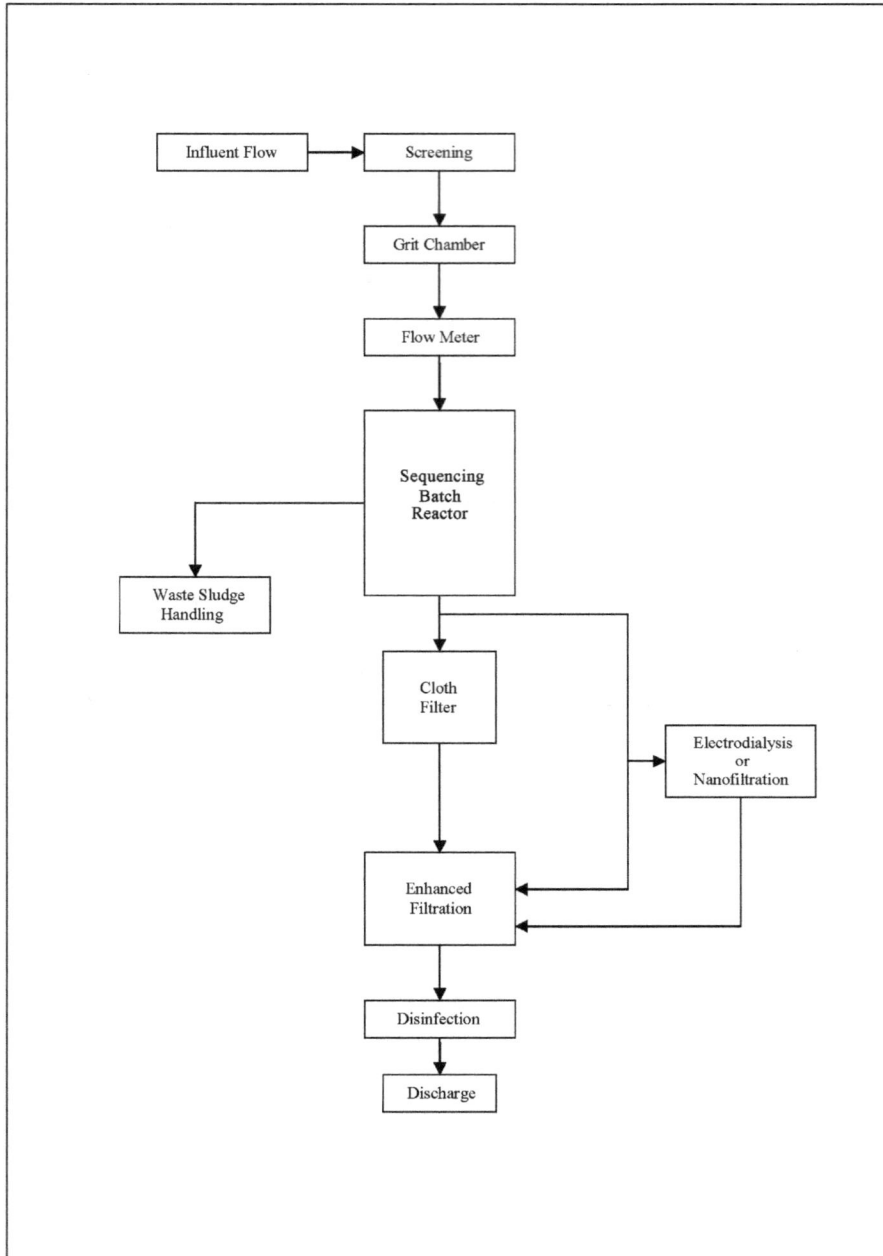

Figure E9 Flow chart of typical wastewater treatment using electrodialysis or nanofiltration

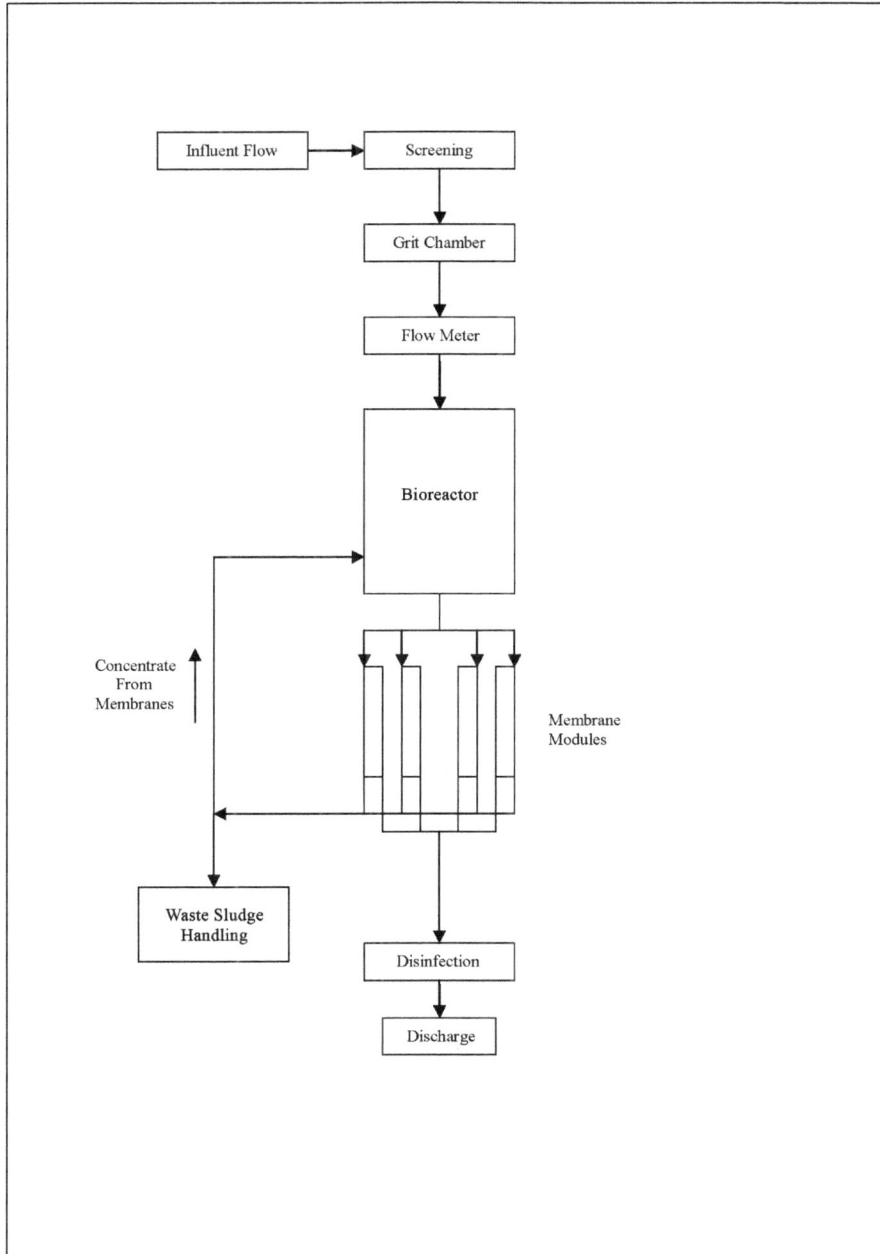

Figure E10 Flow chart of typical wastewater treatment using a membrane bioreactor

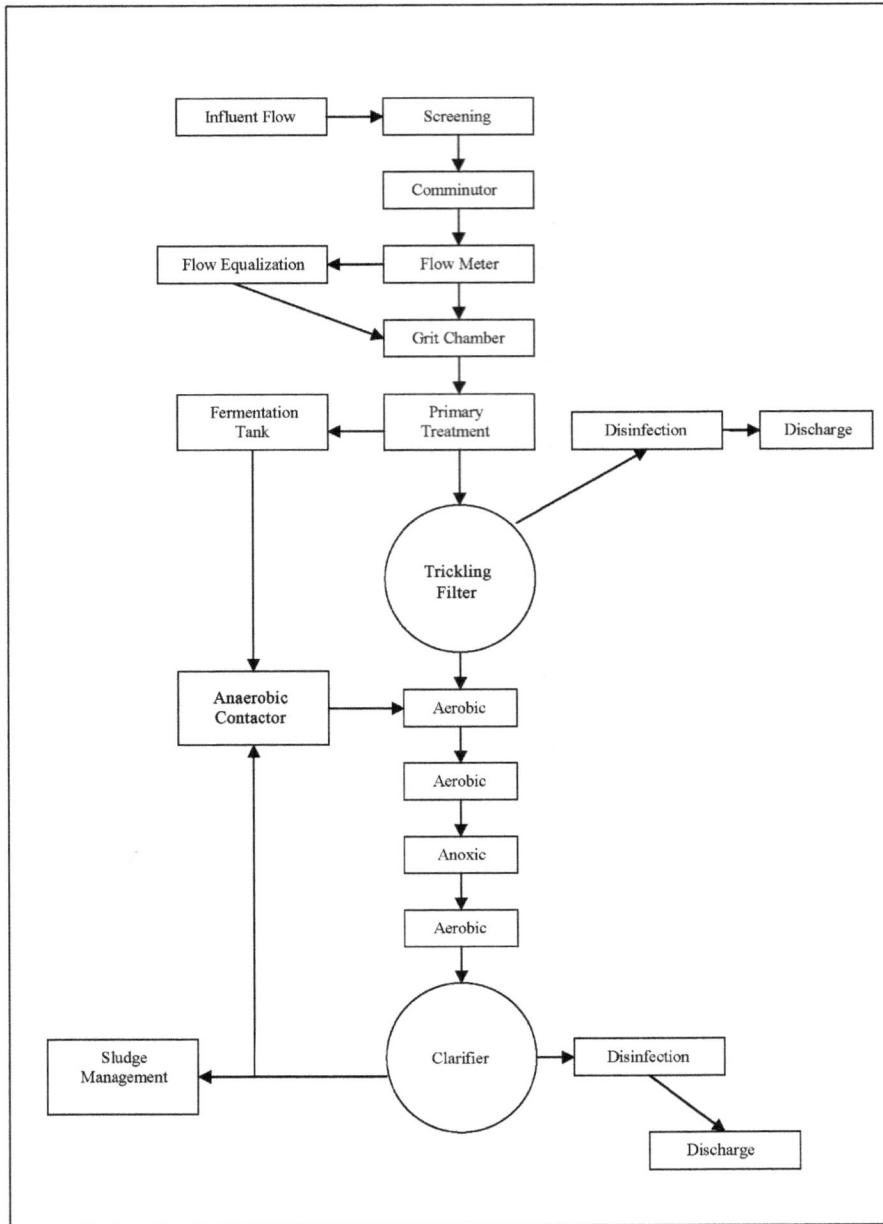

Figure E11 Flow chart of wastewater treatment nitrification process

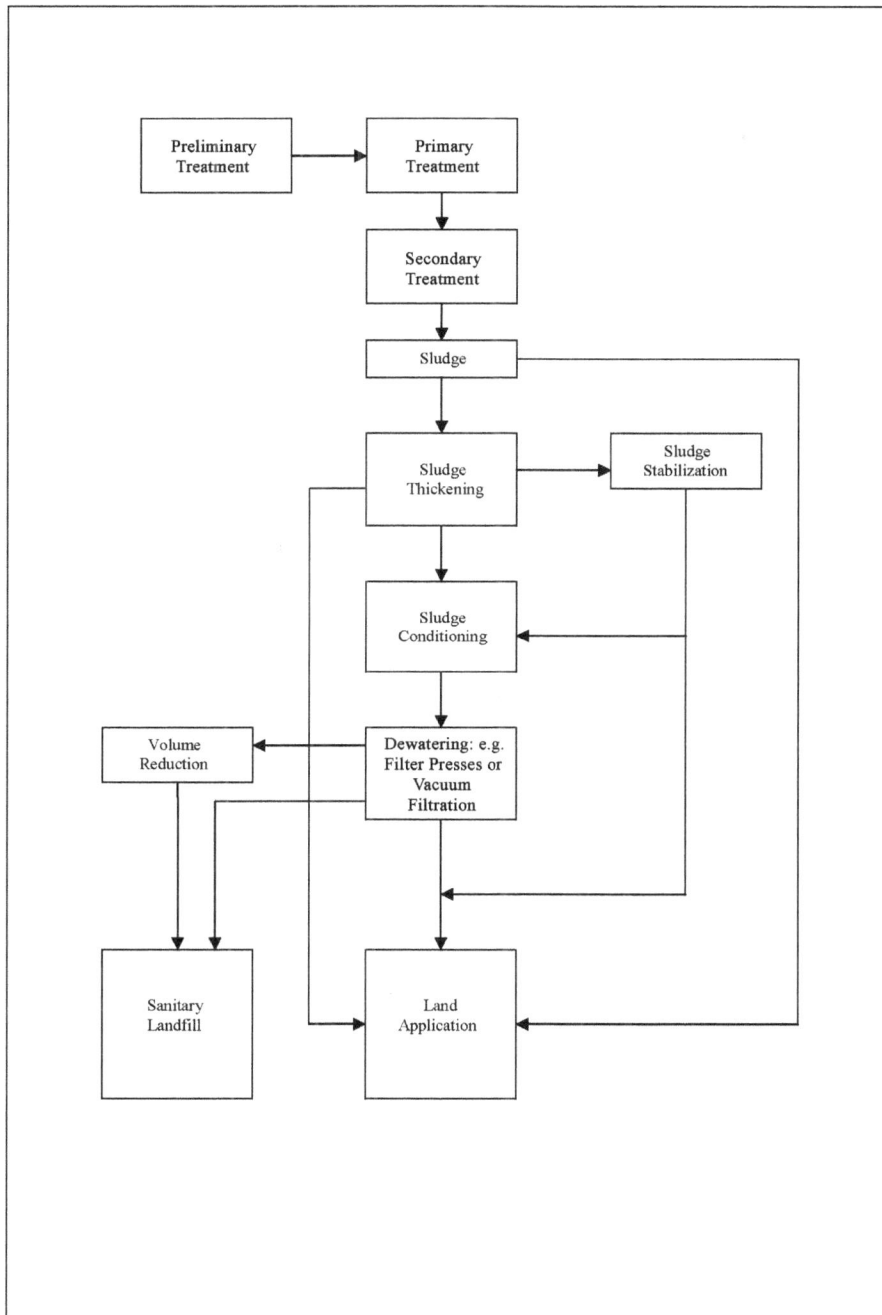

Figure E12 Flow chart of wastewater treatment plant processing sludge

APPENDIX F

ABBREVIATIONS

acre-ft	acre-feet
amps	amperes
AT	Aeration Tank
avail.	available
avg.	average
BC	Blending Compost
bhp	brake horsepower
BOD	Biochemical Oxygen Demand
°C	Degrees Centigrade
COD	Chemical Oxygen Demand
cm	centimeter(s)
CT	Clarifier Tank
D	Diameter or Depth (Note context)
d	day
DAF	Dissolved Air Flotation
DB	Dewatered Biosolids
DO	Dissolved Oxygen
effic.	efficiency
°F	degrees Fahrenheit
F/M	Food-to-Microorganism ratio
ft	foot or feet
ft/s	feet per second
ft^2	square feet
ft^3	cubic feet
f^3/min	cubic feet per minute
ft^3/s	cubic feet per second
g	gram(s) or gravity (Note context)
gal	gallon(s)
gpcpd	gallons per capita per day
gpd	gallons per day

gph	gallons per hour
gpm	gallons per minute
gr-eq	gram equivalent weight
hp	horsepower
hr	hour(s)
HTH	High Test Hypochlorite
in.	inch(es)
kg	kilogram
kW	Kilowatt(s)
lb	pounds
L	Liter or Length (Note context)
m	meter
m^3	cubic meters
M	Mole(s)
mA	milliamp(s)
MCRT	Mean Cell Residence Time
MCL	Maximum Contaminant Level
ME	Motor Efficiency
meq	milliequivalent
mg	milligrams
mgd	million gallons per day
mg/L	milligrams per liter
mhp	motor horsepower
mil	million
mil gal	million gallons
min	minute
mL	milliliter
MLSS	Mixed Liquor Suspended Solids
MLVSS	Mixed Liquor Volatile Suspended Solids

MR	Mix Ratio		temp.	temperature
N	Normality		TDH	Total Dynamic Head
ntu	nephelometric turbidity units		TDS	Total Dissolved Solids
oz	ounce(s)		TS	Total Solids
pH	Hydrogen ion concentration		TSS	Total Suspended Solids
PE	Pump Efficiency		TH	Total Head
%	percent		V	Volume or Velocity or Volt(s) Note context
ppm	parts per million		VR	Volatilization Rate
psi	pressure per square inch absolute		VS	Volatile Solids
psig	pressure per square inch gauge		VSC	Volatile Solids Concentration
Q	Flow		VSR	Volatile Solids Reduction
r	radius		W	Width
RAS	Return Activated Sludge		w	watt
RBC	Rotating Biological Contactor		WAS	Waste Activated Sludge
s	second(s)		whp	water horsepower
sec	second(s)		wt	weight
sed	sedimentation		WVS	Waste Volatile Solids
soln.	Solution		yr	year
sp gr	specific gravity		yd^3	cubic yards
SRT	Solids Retention Time			
SS	Suspended Solids			
SSC	Sludge Solids Concentration			

BIBLIOGRAPHY

American Water Works Association. *AWWA Wastewater Operator Field Guide.* Denver, Colo.: American Water Works Association, 2006.

American Water Works Association. *Basic Science Concepts and Applications.* 3rd ed. Denver, Colo.: American Water Works Association, 2003.

Boikess, Robert S. *How to Solve General Chemistry Problems.* 8th ed. Englewood Cliffs, N.J.: Prentice Hall Inc., 2008.

Forster, Christopher. *Wastewater Treatment and Technology.* London: Thomas Telford Limited, 2003.

Frey, Paul R. *Chemistry Problems and How to Solve Them.* 8th ed. Barnes and Noble College Outline Series. New York: Barnes and Noble Inc., 1985.

Giorgi, John. *Math for Water Treatment Operators: Practice Problems to Prepare for Water Treatment Operator Certification Exams.* Denver, Colo.: American Water Works Association, 2007.

Giorgi, John. *Math for Distribution System Operators: Practice Problems to Prepare for Distribution System Operator Certification Exams.* Denver, Colo.: American Water Works Association, 2007.

Idaho Department of Environmental Quality. *Wastewater Land Application Operators Study and Reference Manual.* Boise, Idaho: Idaho Department of Environmental Quality, 2005.

Lin, Shun Dar. *Water and Wastewater Calculations Manual.* 2nd ed. New York: McGraw-Hill, 2007.

Kerri, Kenneth D. *Operation of Wastewater Treatment Plants, Volume 1: A Field Study Training Program.* 6th ed. Sacramento, Calif.: California State University, Sacramento School of Engineering, 2004.

Kerri, Kenneth D. *Operation of Wastewater Treatment Plants, Volume 2: A Field Study Training Program.* 6th ed. Sacramento, Calif.: California State University, Sacramento School of Engineering, 2003.

Price, Joanne Kirkpatrick. *Applied Math for Wastewater Plant Operators*. Boca Raton, Fla.: CRC Press, 1998.

Price, Joanne Kirkpatrick. *Basic Math Concepts for Water and Wastewater Plant Operators*. Lancaster, Pa: Technomic Publishing Co., 1991.

Skoog, Douglas A., Donald M. West, F. James Holler, and Stanley R. Crouch. *Fundamentals of Analytical Chemistry*. 8th ed. Pacific Grove, Calif.: Brooks Cole, 2003.

Spellman, Frank R. *Mathematics Manual for Water and Wastewater Plant Operators*. New York: CRC Press, 2004.

ADDITIONAL RESOURCES

The following books are all available from the AWWA Bookstore: www.awwa.org, or contact AWWA Customer Service, 1.800.926.7337 or custsvc@awwa.org.

MATH FOR WASTEWATER TREATMENT OPERATORS GRADES 1 & 2

An introduction to math concepts and problems for beginning level and operators.
Published by AWWA, 2009, Softbound, 360 pp.
ISBN 1583215875   Catalog No. 20662

AWWA WASTEWATER OPERATOR FIELD GUIDE

This time-saving book is packed with all the day-to-day information on math, conversion factors, chemistry, safety, collection, pumps, common abbreviations and acronyms, and more.
Published by AWWA, 2006, Softbound, 443 pp.
ISBN 1583213864   Catalog No. 20600

BASIC CHEMISTRY FOR WATER & WASTEWATER OPERATORS

A basic chemistry primer tailored for operators of drinking water or wastewater systems.
Published by AWWA, 2005, Softbound, 196 pp.
ISBN 1583211489   Catalog No. 20494

WASTEWATER MICROBIOLOGY: A HANDBOOK FOR OPERATORS

Learn about wastewater treatment systems, general microscopy, bacteria, protozoa, metazoans, filamentous bacteria, and microbiology and process control.
Published by AWWA, 2005, Softbound, 182 pp.
ISBN 1583213430   Catalog No. 20563

BASIC SCIENCE APPLICATIONS FOR WASTEWATER

Designed for individual or classroom study, this text covers math, hydraulics, chemistry, and electricity as they apply to wastewater treatment. Student Workbook also available, catalog no. 20543
Published by AWWA, 2005, Softbound, 520 pp.
ISBN 1583212906   Catalog No. 20544

WATER & WASTEWATER TREATMENT: A GUIDE FOR THE NON-ENGINEERING PROFESSIONAL

Processes involved in drinking water and wastewater treatment are described in jargon-free language.
Published by Technomic Publishing, 2001, Hardback, 316 pp.
ISBN 1587160498   Catalog No. 20482